World Food Problem: A Selective Bibliography of Reviews

World Food Problem:

A Selective Bibliography of Reviews

Author

Miloslav Rechcigl, Jr. B.S., M.N.S., Ph.D.

Nutrition Advisor and Acting Director
Office of Research
Agency for International Development
U.S. Department of State
Washington, D.C.

CRC Press
Taylor & Francis Group
Boca Raton London New York

CRC Press is an imprint of the
Taylor & Francis Group, an **informa** business

First published 1975 by CRC Press
Taylor & Francis Group
6000 Broken Sound Parkway NW, Suite 300
Boca Raton, FL 33487-2742

Reissued 2018 by CRC Press

Library of Congress Cataloging in Publication Data

Rechcigl, Miloslav.
World food problem.

Includes index.
1. Food supply – Bibliography. 2. Nutrition – Bibliography. 3. Agriculture – Bibliography. I. Title.
Z7164.F7R42 016.3381'9 74-30748
ISBN 0-8493-5098-0

Publisher's Note
The publisher has gone to great lengths to ensure the quality of this reprint but points out that some imperfections in the original copies may be apparent.

Disclaimer
The publisher has made every effort to trace copyright holders and welcomes correspondence from those they have been unable to contact.

ISBN 13: 978-1-315-89866-7 (hbk)
ISBN 13: 978-1-351-07776-7 (ebk)

Visit the Taylor & Francis Web site at http://www.taylorandfrancis.com and the
CRC Press Web site at http://www.crcpress.com

FOREWORD

The population of developing countries will double by about the year 2000. This means that food production must be doubled over the next quarter of century just to keep the teeming masses of people at current levels of nutrition. If the quality and quantity of the food is to perceptibly improve, production has to much more than double. Most of that increase must come from developing countries themselves, where food production and nutritional levels are low. Furthermore, because frontiers of virgin land yet to be put into cultivation are rapidly shrinking, most of the required production increases must be derived from improvements of land already under cultivation. In short, to make two blades of grass grow where one is growing now will take a massive transfer and adaptation of technology.

This bibliography is a most valuable contribution to the War-on-Hunger tool kit, in that it will serve as catalyst for bridging the communications gap between those who know and those who need to know about the extent of the world food problem as well as the available technology on matters related to production, processing and marketing of food.

Frank Meissner
Executive Secretary
Work Group on Strategy and
Action Program in Agriculture
Inter-American Development Bank

Washington, D.C.
March, 1975

the world food problem as well as those that offer alternative strategies and specific technological measures for alleviating the problem.

It was our intention initially to include this bibliography in the earlier CRC text, *MAN, FOOD AND NUTRITION. STRATEGIES AND TECHNOLOGICAL MEASURES FOR ALLEVIATING THE WORLD FOOD PROBLEM,*[2] edited by the present author, however, the manuscript has grown to such proportions that it was decided to publish it separately. Close to 5,000 publications are listed in this book. There has been virtually a "literature explosion" in this area which has not been systematically canvassed prior to this undertaking. The present bibliography which is comprised of monographs, review articles, and relevant chapters from books, including government publications and publications of the international organizations, should fill an important gap which has so long existed in this field. As such, it should serve, together with the earlier CRC title, *MAN, FOOD, and NUTRITION,* as an important reference guide to all serious researchers, students, practitioners, as well as policy makers, who are concerned with the food problem issue.

The organization of entries in our book, as shown in the Table of Contents, is based on subject categorization, according to the major emphasis of individual entries. Under each heading, the listings are arranged alphabetically by first author or corporate name, and in those cases where the author is not given, then by the title. The author index and subject index will facilitate cross references of specific citations. All index references are to page numbers and not to entry numbers. For the convenience of the reader, the book has been appended with the list of most frequently used acronyms and abbreviations.

The author wishes to acknowledge with deep appreciation and gratitude the encouraging and gracious guidance and assistance of the Editorial and Production staffs of the CRC Press, particularly Mrs. Susan Nigoff and Miss Joyce Gavelek. He is also indebted to Miss Karen Rechcigl who rendered indispensable help during the compilation of the subject index.

Miloslav Rechcigl, Jr.

Washington, D.C.
February 1975

REFERENCES

1. United Nations Economic and Social Council, Report of the World Food Conference, Rome, 5–16 Nov. 1974 (E5587. 22 Nov. 1974).
2. **Rechcigl, M., Jr., Ed.,** *Man, Food and Nutrition. Strategies and Technological Measures for Alleviating the World Food Problem,* CRC Press, Cleveland, 1973.
3. **Rechcigl, M., Jr., Ed.,** *Food, Nutrition and Health. A Multidisciplinary Treatise Addressed to the Major Nutrition Problems from a World Wide Perspective,* S. Karger, Basel, 1973.
4. **Bengoa, J. M. and Donoso, G.,** Prevalence of protein-calorie malnutrition, 1963 to 1973, *PAG Bull.,* 4(1), 24–35, 1974.
5. **Puffer, R. R. and Serrano, C. V.,** Patterns of Mortality in Childhood. Report of the Inter-American Investigation of Mortality in Childhood, Pan American Health Organization, Washington, D.C., 1973.
6. **Meadows, H. D., Meadows, D. L., Randers, J., and Behrens, W. W., III,** *The Limits to Growth. A Report for the Club of Rome's Project on the Predicament of Mankind,* Universe Books, New York 1972.
7. **Malthus, T. R.,** *Essay on the Principles of Population,* Oxford University Press, London, 1798; reprinted by Macmillan, 1926.
8. U.S. Senate Select Committee on Nutrition and Human Needs, National Nutrition Policy Study. Report and Recommendation VI, U.S. Gov. Print. Off., Washington, D.C., 1974.

PREFACE

Following the World Food Conference which was held under the United Nations auspices in Rome from 5 to 16 November, 1974, there is little doubt that the world food problem ranks high among the major problems that confront mankind today.[1]

It is estimated that over half of today's world population is afflicted by hunger and malnutrition.[2–4] The effects of malnutrition on the health of the preschool child are particularly alarming. Two thirds to four fifths of all deaths that occur in developing countries are within the under-five age group, a group which constitutes 20% of the population. Some 30% of the children in the developing countries fail to reach the age of five.

In a recent investigation of the underlying causes of mortality in childhood in the Americas, carried out by the Pan American Health Organization with the assistance of the Agency for International Development, malnutrition was found to be directly or indirectly responsible for 50% of the deaths of children under five years of age.[5]

However, not all malnourished children die. Millions of those who live have been malnourished and do suffer serious deprivation of the opportunity to realize their full potential. The FAO and WHO estimate the toll to be as many as two thirds of all surviving children in the developing countries.

Overpopulation is certainly one of the major factors responsible for the current situation. More than one and a half billion people have been added to the world population since 1900, and it is estimated that by the year 2000, the globe will carry in excess of six billion inhabitants.[2]

There is some fear that food production will not be able to keep pace with population growth.[6] This has resulted in the reemergence of the Malthusian thought.[7] This view was strengthened, particularly during 1972, when as a result of widespread bad weather, the world food production had fallen for the first time since the Second World War. The consequent increase in import demand, especially for staple cereals, had seriously depleted the stocks of the main exporting countries. World prices for most agricultural commodities had risen steeply, and food prices had been a major factor in a rapid inflation of retail prices that caused severe hardship for the poorest people, and even brought difficulties in the affluent countries. Despite the improved weather conditions that prevailed in the following years, the difficulties in the agricultural sector were compounded by additional problems, including inflation, worsening unemployment, monetary problems, the energy crisis, fears of world-wide recession, and a general atmosphere of uncertainty about the future.[1]

There is a genuine need for concern. However, there is no need to despair. Our earth resources and the technological know-how, if properly used, should not only avert any future catastrophe, but could actually lead to the improvement of the nutritional status and the well-being of the many malnourished peoples throughout the world.[2] What is needed is a coordinated effort involving nutritionists, family planners, agronomists, plant breeders, animal husbandrymen, food technologists, economists, and other social scientists, as well as national planners who shape the economic and health policies of their nations. In seeking specific solutions to existing problems, not infrequently it might be a question of simply applying what we already know. In many other situations, however, solutions will have to be sought through experimentation and new imaginative research efforts. In this connection, it should be noted that the recently issued report of the Panel on Nutrition and the International Situation[8] recommended that there should be a significant increase in the support of agricultural research devoted to the problem of increasing food production in the developing countries.

The aim of this publication is to provide the interested reader with an authoritative and comprehensive up-to-date bibliography on all important facets of the world food problem, encompassing such questions as the availability of natural resources, the present and future sources of energy, environmental quality, population growth, world malnutrition, the state of food production, food consumption patterns, future food needs, toxicological aspects of food, agricultural and industrial aspects of food production, and family planning. It is the first compilation of its kind in that it covers the subject from a multidisciplinary point of view, including publications that deal with the description and analysis of

To Eva, Jack and Karen, with my love

THE AUTHOR

Miloslav Rechcigl, Jr. is Nutrition Advisor and Acting Director of the Office of Research in the Agency for International Development, U.S. Department of State.

He has a B.S. in Biochemistry (1954), a Master of Nutritional Science degree (1955), and a Ph.D. in nutrition, biochemistry, and physiology (1958), all from Cornell University. He was formerly a Research Biochemist in the National Cancer Institute, National Institutes of Health and subsequently served as Special Assistant for Nutrition and Health in the Health Services and Mental Health Administration, U.S. Department of Health, Education, and Welfare.

Dr. Rechcigl is a member of some 30 scientific and professional societies, including being a Fellow of the American Association for the Advancement of Science, Fellow of the Washington Academy of Sciences, Fellow of the American Institute of Chemists, and Fellow of the International College of Applied Nutrition. He holds membership in the Cosmos Club, the Honorary Society of Phi Kappa Pi, and the Society of Sigma Xi, and is recipient of numerous honors, including an honorary membership certificate from the International Social Science Honor Society Delta Tau Kappa. In 1969, he was a delegate to the White House Conference on Food, Nutrition, and Health and in the last two years served as President of the District of Columbia Institute of Chemists and a Councilor of the American Institute of Chemists.

His bibliography extends over 100 publications, including contributions to books, articles in periodicals, and monographs in the fields of nutrition, biochemistry, physiology, pathology, enzymology, and molecular biology. Most recently he authored and edited *Man, Food, and Nutrition: Strategies and Technological Measures for Alleviating the World Food Problem* (CRC Press, 1973), *Food, Nutrition and Health: A Multidisciplinary Treatise Addressed to the Major Nutrition Problems from a World Wide Perspective* (Karger, 1973), following his earlier pioneering treatise on *Enzyme Synthesis and Degradation in Mammalian Systems* (Karger, 1971), and that on *Microbodies and Related Particles. Morphology, Biochemistry and Physiology* (Academic Press, 1969).

TABLE OF CONTENTS

I. GENERAL REFERENCES

A. Bibliographies

1. **Baranson, J.,** *Technology for Underdeveloped Areas. An Annotated Bibliography,* Pergamon Press, Oxford, 1967, 81 pp.
2. Food and Agriculture Organization, Bibliography on the Analysis and Projection of Demand and Production, 1963, Commodity Ref. Ser. No. 2, FAO, Rome, 1963, 279 pp.
3. **Henderson, E.,** Food Aid: A Selective Annotated Bibliography on Food Utilization for Economic Development, FAO, Rome, 1964, 203 pp.
4. **Katz, S. M. and McGowan, F.,** A Selected List of U.S. Readings on Development, Agency for International Development, Washington, D.C., 1963, 363 pp.
5. **Lundeen, G. and Lundeen, B.,** *World Food Problems Bibliography,* Food Science Research Center, Fresno, Calif., 1969, 46 pp.
6. **Mallowe, C.,** *A Bibliography on World Hunger,* St. Joseph's College, Academy of Food Marketing, Philadelphia, 1967, 59 pp.
7. **Rechcigl, M., Jr.,** Reviews relating to food, nutrition and health. A selected bibliography, in *Food Nutrition and Health,* Vol. 16, Rechcigl, M., Jr., Ed., *World Rev. Nutr. Diet.,* S. Karger, Basel, 1973, 398-445.
8. **ReQua, E. G. and Statham, J.,** *The Developing Nations: A Guide to Information Sources Concerning Their Economic, Political, Technical and Social Problems,* Gale Research Co., Detroit, 1965, 339 pp.
9. **Winton, H. N. M.,** *Man and the Environment. A Bibliography of Selected Publications of the United Nations System 1946–1971,* UNIPUB/R.R. Bowker, New York, 1972, 305 pp. (1219 ref.)

B. Surveys and Treatises

1. The General World Food Problem

1. **Abercrombie, K. C.,** Population growth and agricultural development, *FAO Mon. Bull. Agric. Econ. Stat.,* 18(4), 1–9, 1969.
2. **Abercrombie, K. C.,** Changing views on the man-food relationships, *Ceres,* 4(2), 22–26, 1971.
3. American Society of Agronomy, *World Population and Food Supplies, 1980,* Publ. 6, ASA, Madison, WI, 1965.
4. **Basil, P. C.,** *India's Food Resources and Population. A Historical and Analytical Study,* Vora & Co., Bombay, 1958, 25 pp.
5. **Bennett, I. L., Jr.,** Food and population, an overview, *Agric. Sci. Rev.,* 6(1), 10–17, 1968.
6. **Bennett, M. K.,** Population and food supply. The current scare, *Sci. Mon.,* 68, 17–26, 1949.
7. **Blodgett, J. H.,** Relations of Population and Food Products in the United States, 1850–1900, U.S. Dep. Agric. Div. Stat. Bull. 24, 1903, 86 pp.
8. **Borgstrom, G.,** *Too Many: A Study of Earth's Biological Limitations,* Macmillan, New York, 1969.
9. **Borgstrom, G.,** *The Hungry Planet,* Macmillan, New York, 1965.
10. **Boyd Orr, J.,** The food problem, *Sci. Am.,* 183, 111–115, 1950.
11. **Brown, H.,** *The Challenge of Man's Future,* Viking, New York, 1954.
12. **Brown, L. R.,** Man, Land, Food, U.S. Gov. Print. Off., Washington, D.C., 1965.
13. **Brown, L. R. and Finsterbusch, G. W.,** *Man and His Environment: Food,* Harper & Row, New York, 1972, 208 pp.
14. **Brown, L. R.,** *World Without Borders,* Random House, New York, 1972, 395 pp.
15. **Cépède, M., Houtart, F., and Grond, L.,** *Population and Food,* Sheed & Ward, New York, 1964, 461 pp.
16. **Clark, C.,** *Starvation or Plenty?* Seeker and Warburg, London, 1970.
17. **Clark, C.,** Food production and population, in *Proc. 3rd Western Hemisphere Nutrition Congress,* Futura Publishing, Mount Kisco, N.Y., 1972, 223–225.
18. **Clark, F. L. G. and Pirie, N. W., Eds.,** *Four Thousand Million Mouths. Scientific Humanism and the Shadow of World Hunger,* Books for Libraries Press, Freeport, N.Y., 1951.
19. **Coale, A. J. and Hoover, E. M.,** *Population Growth and Economic Development in Low-Income Countries,* Princeton University Press, Princeton, N.J., 1958.
20. **Cook, R. C.,** Population and food supply, in *The Population Crisis and the Uses of World Resources,* Mudd, S., Ed., Junk, The Hague, 1964, 451–477.
21. **Cuthbertson, D. P.,** Food and the future, *J. Franklin Inst.,* 261, 121–131, 1956. (21 ref.)
22. **Cuthbertson, D. P.,** The world food situation as related to knowledge of science and its application, in *Progress in Nutrition and Allied Sciences,* Cuthbertson, D. P., Ed., Oliver & Boyd, Edinburgh, 1963, 421–444.
23. **Desrosier, N. W.,** *Attack on Starvation,* Avi Publishing, Westport, Conn., 1961, 312 pp.
24. **DeTurk, E. E., Ed.,** Freedom from want. A survey of the possibilities of meeting the world's food needs. A symposium, *Chron. Bot.,* 11(4), 207–284, 1948. (7 papers)
25. **Drogat, N.,** *The Challenge of Hunger,* Newman Publishing, Westminster, MD, 1962, 158 pp.
26. Editorial: World food and population, *Outlook Agric.,* 3(1), 2–8, 1960.
27. **Elbing, F. J. and Heath, G. W., Eds.,** *The Future of Man. A Symposium,* Academic Press, New York, 1972, 212 pp.

28. **Farber, S. M., Wilson, N. L., and Wilson, R. H. L., Eds.,** *Food and Civilization,* Charles C Thomas, Springfield, IL., 1966.
29. Feeding the world, *Sci. J.,* 4 (May), 1968, Special issue.
30. Food and Agriculture Organization, Aspects of Economic Development. The Background to Freedom from Hunger Campaign, Basic Study No. 8, FAO, Rome, 1962, 84 pp.
31. Food and Agriculture Organization, Millions Still Go Hungry, FAO, Rome, 1962, 102 pp.
32. Food and Agriculture Organization, Man and Hunger, World Food Problems No. 2, FAO, Rome, 1962, 77 pp.
33. Food and Agriculture Organization, Six Billion to Feed, World Food Problems No. 4, FAO, Rome, 1963, 41 pp.
34. Food supplies and population growth, *J. R. Stat. Soc., Ser. A (Gen.),* 125, 373–450, 1962. (6 papers with discussion)
35. **Goldblith, S. A.,** The world food crisis, *Technol. Rev.,* 70, 20–29, 1968.
36. **Halacy, D. S., Jr.,** *The Geometry of Hunger,* Harper & Row, New York, 1972, 280 pp.
37. **Hauser, P. M.,** *Toward the Year 2018,* Cowles Books, New York, 1968.
38. **Holdren, J. P. and Ehrlich, P. R.,** Human population and the global environment, *Am. Sci.,* 62, 282–292, 1974.
39. **Leisner, R. S. and Kormondy, E. J.,** *Population and Food,* Brown, Dubuque, IA, 1971, 83 pp. (14 papers)
40. **Leonard, W. H.,** World population in relation to potential food supply, *Sci. Mon.,* 85, 113–125, 1957.
41. McMillan bicentenary discussion on fertility, mortality and world food supplies, *J. R. Stat. Soc., Ser. A (Gen.),* 129, 210–248, 1966. (3 papers with discussion)
42. **McMillen, W.,** *Land of Plenty: The American Farm Story,* Holt, Rinehart & Winston, New York, 1961, 138 pp.
43. **Malthus, T. R.,** *Essay on the Principles of Population,* Oxford University Press, London, 1798; reprinted by Macmillan, 1926.
44. **Meadows, D. H., Meadows, D. L., Randers, J., and Behrens, W. W., III,** *The Limits to Growth. A Report for the Club of Rome's Project on the Predicament of Mankind,* Universe Books, New York, 1972, 206 pp.
45. **Mudd, S., Ed.,** *The Population Crisis and the Use of World Resources,* Junk, The Hague, 1964, 563 pp.
46. **Myers, M. B. and Abbott, J. W.,** *Resource Guide on World Hunger,* Church World Service, Div. of Overseas Ministries, National Council of Churches, New York, 1968, 207 pp.
47. Meeting the challenges of the coming decades, *Proc. 1st Western Hemisphere Nutrition Congress,* American Medical Association, Chicago, 1966, 225–244. (5 papers)
48. **Nicol, H.,** *The Limits of Man,* Constable, London, 1967, 283 pp.
49. Organization for Economic Cooperation and Development, The Food Problem of Developing Countries, OECD, Paris, 1967.
50. **Oser, J.,** *Must Men Starve? The Malthusian Controversy,* Abelard-Schuman, New York, 1957, 331 pp.
51. **Pawley, W. H.,** In the year 2070, *Ceres,* 4(4), 22–27, 1971.
52. **Pawley, W. H.,** World food and population problems, *FAO Nutr. Newsletter,* 8(2), 1–7, 1970.
53. **Pearson, F. A. and Harper, F. A.,** *The World's Hunger,* Cornell University Press, Ithaca, N.Y., 1945.
54. **Rechcigl, M., Jr., Ed.,** *Man, Food, and Nutrition. Strategies and Technological Measures for Alleviating the World Food Problem,* CRC Press, Cleveland, Ohio, 1973, 344 pp.
55. **Revelle, R.,** Population and food supplies. The edge of the knife, *Proc. Natl. Acad. Sci. U.S.A.,* 56, 328–351, 1966.
56. **Revelle, R., Khosla, A., and Vinovskis, M.,** *The Survival Equation. Man, Resources and His Development,* Houghton Mifflin, Boston, 1971, 512 pp.
57. **Robinson, H. F.,** Dimensions of the world food crisis, *Bioscience,* 19(1), 24–29, 1969.
58. Royal Statistical Society, *Food Supplies and Population Growth,* Oliver & Boyd, Edinburgh, 1963.
59. **Russell, E. J. and Wright, N. C., Eds.,** *Hunger: Can it Be Averted?* British Association for the Advancement of Science, London, 1961, 80 pp.
60. **Russell, J.,** *World Population and World Food Supply,* Macmillan, New York, 1954.
61. **Sen, B. R.,** The Basic Freedom, Freedom from Hunger, FAO, Rome, 1960.
62. **Sen, B. R.,** Food, Population and Development, FAO, Rome, 1965.
63. Symposium on World Needs and Food Resources, Proc. Fifth Int. Congr. Nutr., Washington, D.C., 1960, *Fed. Proc.,* 20(1), III, Suppl. 7, 365–406, 1961.
64. **Thompson, L. M.,** Iowa Agriculture. World Food Needs and Educational Response, CAED Rep. 156, Ames, IA, 1962.
65. **Toynbee, A.,** Population and Food Supply, McDougall Memorial Lecture 1959, FAO, Rome, 1959, 20 pp.
66. **Underwood, E. J.,** Man, land and food, *Aust. J. Sci.,* 29(II), 395–401, 1967.
67. United Nations, Agricultural production, population growth, and food consumption per capita, *Econ. Bull. Afr.,* 9(2), 1–9, 1969.
68. United Nations, Agricultural development and population growth, *Econ. Bull. Afr.,* 9(2), 41–43, 1969.
69. U.S. Department of Agriculture, The World Food Budget 1962 to 1966, USDA, ERS, Foreign Agric. Econ. Rep. 4, 1961; revised, 1962.
70. U.S. Department of Agriculture, World Food Forum, Proc. Commemorating Centennial U.S. Dep. Agric., 1862–1962, Washington, D.C., May 15–17, 1962, U.S. Gov. Print. Off., Washington, D.C., 1963.
71. U.S. Department of Agriculture, The World Food Budget 1970, U.S. Dep. Agric., Foreign Agric. Econ. Rep. 19, 1965.

72. U.S. Department of the Interior, The Population Challenge, Washington, D.C., 1966.
73. War on Hunger, *Intercom*, 8(6), 24–72, 1966.
74. World Food – Population Levels, Report to the President by the Secretary of Agriculture and the Secretary of State, Washington, D.C., April 9, 1970.
75. **Wyotinsky, W. S. and Wyotinsky, E. S.,** *World Population and Production: Trends and Outlook,* Twentieth Century Fund, New York, 1953, 1268 pp.

2. The Protein Problem

1. **Abbott, J. C.,** Protein supplies and prospects: The problem, in *World Protein Resources,* Adv. Chem. Ser. No. 57, American Chemical Society, Washington, D.C., 1966, 1–15. (14 ref.)
2. **Altschul, A. M.,** *Proteins, Their Chemistry and Politics,* Basic Books, New York, 1965, 337 pp.
3. **Altschul, A. M.,** Food: Proteins for humans, *Chem. Eng. News,* 47(11), 68–81, 1969.
4. **Altschul, A. M.,** Protein food technologies and the politics of food: An overview, in *New Protein Foods,* Vol. IA, Altschul, A. M., Ed., Academic Press, New York, 1974, 1–38.
5. **Autret, M.,** World protein supplies and needs, in *Proteins as Human Food,* Lawrie, R. A., Ed., Avi Publishing, Westport, Conn., 1970, 3–19.
6. **Aylward, F.,** An assessment of the world's protein needs and supplies, in *The Better Use of the World's Fauna for Food,* Symp. Inst. Biol. No. 11, Institute of Biology, London, 1963, 1–15.
7. **Bacobo, L.,** The protein problem in the Asia and the Far East Region, *Ind. J. Nutr. Diet.,* 7, 307–316, 1970.
8. **Bean, L. H.,** Closing the World's Nutritional Gap with Animal or Vegetable Protein? FAO Bull., 6, 1966.
9. **Flodin, N. W.,** Amino acids and proteins. Their place in human nutrition problems, *J. Agric. Food Chem.,* 1, 222–235, 1953. (138 ref.)
10. Food and Agriculture Organization, Protein: At the Heart of the World Food Problem, World Food Problems No. 5, FAO, Rome, 1964, 62 pp.
11. Food and Agriculture Organization, Lives in Peril: Protein and the Child, World Food Problems No. 12, FAO, Rome, 1970, 52 pp.
12. **Jansen, G. R. and Howe, E. E.,** World food problems in protein nutrition, *Am. J. Clin. Nutr.,* 15, 262–274, 1964.
13. **Jones, F. G. W., Ed.,** *The Biological Efficiency of Protein Production,* Cambridge University Press, London, 1973.
14. **Lawrie, R. A., Ed.,** *Proteins as Human Food,* Avi Publishing, Westport, Conn., 1970, 525 pp.
15. **Milner, M.,** Introduction to the protein problem, *Econ. Bot.,* 22, 3–7, 1968.
16. National Research Council, *Progress in Meeting Protein Needs of Infants and Pre-School Children,* Proc. Int. Conf. Publ. 843, National Academy of Sciences, Washington, D.C., 1960.
17. **Parpia, H. A. B.,** Bridging the protein-calorie gap, in Improving Protein by Nuclear Techniques, IAEA, Vienna, 1970, 3–12.
18. **Porter, J. W. G. and Rolls, B. A., Eds.,** *Protein in Human Nutrition,* Academic Press, London, 1973, 560 pp. (37 papers)
19. Protein Advisory Group of the United Nations System, *PAG Bull.,* Vol.1–, 1973–.
20. Protein Advisory Group, Statement on the Nature and Magnitude of the Protein Problem, PAG Statement No. 3, Oct. 27, 1969.
21. Protein Advisory Group, The World Protein Problem. Research and Development Needs, PAG Statement No. 12, April 10, 1972.
22. **Sebrell, W. H., Jr.,** The prospect of meeting protein needs, *Fed. Proc.,* 20(1), III, Suppl. 7, 393–397, 1961.
23. **Simm, S. R., Cesario, F., Kline, O. L., and Yates, D. M.,** A System Approach to the World Protein Problem. Report prepared by the Battelle Memorial Institute for the U.N. Protein Advisory Group, May 1970.
24. **Subramaniam, C.,** A strategy for fighting protein hunger, *Ind. J. Nutr. Diet.,* 8, 268–280, 1971.
25. Symposium. Protein for human food, *Proc. N.Z. Soc. Anim. Prod.,* 32, 137–234, 1972.
26. Symposium. Selection of alternative strategies for meeting protein needs, in *Proc. 3rd Western Hemisphere Nutrition Congress,* Futura Publishing, Mont Kisco, N.Y., 1972, 27–53. (5 papers)
27. United Nations Economic and Social Council, *Increasing the Production and Use of Edible Proteins. Feeding the Expanding World Population: Recommendations for International Action to Avert the Impending Protein Crisis,* Report of the Advisory Committee on the Application of Science and Technology to Development, May 25, 1967, United Nations, New York, 1968, 110 pp.
28. **Woodham, A. A.,** The world protein shortage: Prevention and cure, *World Rev. Nutr. Diet.,* 13, 2–42, 1971. (102 ref.)
29. *World Protein Resources, Advances in Chemistry Ser. No. 57,* American Chemical Society, Washington, D.C., 1966, 285 pp.

II. THE STATE OF WORLD FOOD AND NUTRITION

A. The State of World Nutrition

1. The World as a Whole

1. **Bailey, K. V.,** World hunger and malnutrition, *Philipp. J. Nutr.*, 25(2), 161–170, 1972.
2. **Bengoa, J. M.,** The state of world nutrition, in *Man, Food, and Nutrition. Strategies and Technological Measures for Alleviating the World Food Problem*, Rechcigl, M., Jr., Ed., CRC Press, Cleveland, 1973, 1–13. (58 ref.)
3. **Bengoa, J. M.,** Hunger and malnutrition in the world-today, *Cajanus*, 7(4), 127–133, 1974.
4. **Bengoa, J. M. and Donoso, G.,** Prevalence of protein-calories malnutrition, 1963 to 1973, *PAG Bull.*, 4(1), 24–35, 1974. (81 ref.)
5. **Cépède, M. and Lengellè, M.,** *Economie Alimentaire du Globe*, Libraire de Médicis, Paris, 1953.
6. **Clements, F. W.,** The geography of hunger, *Food Nutr. Notes Rev.*, 24, 65–79, 1967.
7. **Darby, W. J.,** Research developments in international nutrition, *Am. J. Public Health*, 53, 1789–1794, 1963. (49 ref.)
8. **Gounelle de Pontanei, H.,** Major human nutrition problems today, *Fed. Proc.*, 20(1), III, Suppl. 7, 389–392, 1961.
9. **Hollingsworth, D. and Russell, M., Eds.,** *Nutritional Problems in a Changing World*, John Wiley & Sons, New York, 1973.
10. **Jelliffe, D. B.,** Infant Nutrition in the Subtropics and Tropics, 2nd ed., WHO Monogr. No. 29, 1968.
11. **Levine, S. Z. and Kowlessar, O. D.,** World nutrition problems, *Annu. Rev. Med.*, 13, 41–60, 1962.
12. **Livet, R.,** *Géographie de l'alimentation*, Les Éditions Ouvrières, Paris, 1969, 317 pp.
13. *Nutrition under Various Geographic and Climatic Conditions*, Vol. 3, Proc. 7th International Congress of Nutrition, Hamburg, 1966, Pergamon Press, Oxford, 1967, 371 pp.
14. **Platt, B. S.,** The clinical nutrition problems, *Proc. Nutr. Soc.*, 5, 2–15, 1946.
15. *Problems of World Nutrition*, Vol. 4, Proc. 7th International Congress of Nutrition, Hamburg, 1966, Pergamon Press, Oxford, 1967, 1091 pp.
16. **Rechcigl, M., Jr., Ed.,** *Food, Nutrition, and Health. A Multidisciplinary Treatise Addressed to the Major Nutrition Problems from a World Wide Perspective*, S. Karger, Basel, 1973, 511 pp.
17. **Robson, J. R. K.,** Global nutritional problems, in *Malnutrition: Its Causation and Control*, Vol. 1, Gordon & Breach, New York, 1972, 19–79. (83 ref.)
18. **Schaefer, A. E.,** Nutritional deficiencies in developing countries, *J. Am. Diet. Assoc.*, 42, 295–298, 1963.
19. **Schaefer, A. E. and Berry, F. B.,** U.S. interest in world nutrition, *Public Health Rep.*, 75, 677–686, 1960. (22 ref.)
20. **Scrimshaw, N. S.,** Progress in solving world nutrition problems, *J. Am. Diet. Assoc.*, 35, 441–448, 1959. (28 ref.)
21. **Scrimshaw, N. S.,** World-wide importance of protein malnutrition and progress toward its prevention, *J. Public Health*, 53, 1781–1788, 1963.
22. **Scrimshaw, N. S. and Béhar, M.,** Malnutrition in underdeveloped countries, *N. Engl. J. Med.*, 272, 137–144, 193–198, 1965. (49 ref.)
23. **Streeter, D. W.,** A geography lesson in food, *Nutr. Food Sci.*, No. 22, 1971.
24. **Swaminathan, N.,** Nutrition and the world food problem, *Borden's Rev. Nutr. Res.*, 28, 1–31, 1967.
25. Symposium. Nutritional problems in developing countries, *Proc. 8th International Congress of Nutrition*, Prague, 1969, Excerpta Medica, Amsterdam, 1970, 575–610. (10 papers)
26. Symposium. Nutritional problems in the tropics, *Proc. Nutr. Soc.*, 29, 191–236, 1970. (7 papers)
27. World nutrition surveys, *Public Health Rep.*, 75(8), 677–757, 1960. (9 papers)

2. Europe

a. Europe in General

1. **May, J.,** *The Ecology of Malnutrition in Five Countries of Eastern and Central Europe*, *Stud. Med. Geogr.*, Vol. 4, Hafner Publishing, New York, 1964, 292 pp.
2. **May, J.,** *The Ecology of Malnutrition in Central and Southeastern Europe*, *Stud. Med. Geogr.*, Vol. 6, Hafner Publishing, New York, 1966, 290 pp.

b. Individual Countries

(1) Czechoslovakia

1. **Hejda, S.,** Epidemiological investigations of the nutritional status of the population, *Acta Inst. Aliment. Hum. Prague*, 4, 17, 1969.
2. **Hrubý, J.,** *Příspěvek ke koncepci vývoje výživy v ČSSR*, Studie ČSAV č. 7, Academia, Prague, 1971.
3. **Kajaba, J.,** Development on the nutrition of the population of Slovakia, *Rev. Czech. Med.*, 15, 185–197, 1969.
4. **Křikava, L.,** The trend of the nutritional status of our population, *Acta Inst. Aliment. Hum. Prague*, 2, 13, 1958.
5. **Křikava, L., Mašek, J., and Malek, P.,** The Czech experience in nutrition improvement, in *Nutrition, National Development, and Planning*, Berg, A., Scrimshaw, N. S. and Call, D. L., Eds., M.I.T. Press, Cambridge, Mass., 1973, 329–334.

6. **Mašek, J.**, Activities of the Institute of Human Nutrition, 1962–1968, *Acta Inst. Aliment. Hum. Prague*, 4, 7, 1969.
7. **Mašek, J.**, Nutritional problems of working people in Czechoslovakia, in *Nutrition and Physical Activity*, Blix, G., Ed., Almqvist & Wiksells, Stockholm, 1967, 78–83.
8. **Ošancová, K. and Hejda, S.**, Nutrition and its reflection on the health status of the population, *Rev. Czech. Med.*, 14, 101–117, 1968. (29 ref.)

(2) France

1. Thi-nguyen-Huu, Liautard, Y., and Richard, D., *Conditions de Vie et Consommation Alimentaire des Francais, Anneé 1967*, Paris, 1969, 155 pp.

(3) Germany

1. **Gräfe, H. K. and Schmidt, H. E.**, *Ernährungsatlas der Deutschen Demokratischen Republik*, Akademie Verlag, Berlin, 1967, 84 pp.
2. *Moderne Ernährung im Berufsleben*, Umschau Verlag, Frankfurt a.M., 1969, 96 pp.

(4) Great Britain

1. Great Britain, Ministry of Health, A Pilot Survey of the Nutrition of Young Children in 1963, HMSO, London, 1968, 79 pp.
2. **Lambert, R.**, *Nutrition in Britain 1950–60. A Critical Discussion of the Standards and Findings of the National Food Survey*, Codicote Press, Hertfordshire, 1964, 46 pp.
3. Office of Health Economics, London, Malnutrition in the 1960s? London, 1967, 32 pp.
4. Symposium. Food habits and nutritional status of minority groups in the United Kingdom, *Proc. Nutr. Soc.*, 26, 191–229, 1967. (6 papers)

(5) Ireland

1. **Edwards, R. D. and Williams, T. D.**, *The Great Famine*, Brown & Nolan, Dublin, 1956.
2. **Woodham-Smith, C.**, *The Great Hunger: Ireland 1845–9*, Hamilton, London, 1962.

(6) Italy

1. **Galeotti, G.**, *Condizioni ed evaluzione dei consumi alimentari in Italia*, Unione nazionale consumatori, Rome, 1969, 161 pp.
2. **Proja, M.**, Nutrition problems and activities in Italy, *Nutr. Rev.*, 20, 129–131, 1962.

(7) Netherlands

1. **Hartog, C. den**, Nutrition in the Netherlands, *World Rev. Nutr. Diet.*, 6, 90–123, 1966. (46 ref.)

(8) Spain

1. Interdepartmental Committee on Nutrition for National Defense, Spain: Nutrition Survey of the Armed Forces, National Institutes of Health, Bethesda, MD, 1958, 106 pp.
2. **Varela, G.**, Nutritive state of the population in Spain, *World Rev. Nutr. Diet.*, 13, 86–104, 1971. (24 ref.)

(9) Switzerland

1. **Verzar, F. and Gsell, D.**, The nutritional status of mountain populations in Switzerland, *World Rev. Nutr. Diet.*, 4, 35–51, 1964. (16 ref.)

(10) U.S.S.R.

1. **Fitzpatrick, W. H.**, Nutrition Research in the U.S.S.R., 1961–1970, U.S. Gov. Print. Off., Washington, D.C., 1972, 143 pp.

(11) Yugoslavia

1. **Ferber, E.**, Human nutrition research in Yugoslavia, *Am. J. Clin. Nutr.*, 22, 1531–1535, 1969. (50 ref.)
2. **Ferber, E.**, Nutritional conditions in Yugoslavia and their reflection on health, *World Rev. Nutr. Diet.*, 18, 263–274, 1973. (36 ref.)

3. Near East

a. Near East in General

1. **Adolph, W. H.**, Nutrition in the Near East, *J. Am. Diet. Assoc.*, 30, 753–756, 1954.
2. **McLaren, D. S. and Pellet, P. L.**, Nutrition in the Middle East, *World Rev. Nutr. Diet.*, 12, 43–127, 1970. (95 ref.)
3. **Patwandhan, V. N. and Darby, W. J.**, *The State of Nutrition in the Arab Middle East*, Vanderbilt University, Nashville, Tenn., 1972.
4. **Pharaon, H. M.**, The influence of ecological factors on malnutrition in the Middle East, *Bibl. Nutr. Dieta.*, No. 14, S. Karger, Basel, 1970, 73–82.

b. Individual Countries

(1) Egypt

1. **Abdou, I.,** Nutrition problems in the Egyptian Region, *J. Egypt. Public Heath Assoc.*, 35, 137–144, 1960.
2. **Abdou, I.,** A study of the pattern and trends of the Egyptian diet and their relation to health, *J. Egypt. Public Health Assoc.*, 33, 21–32, 1958.

(2) Iran

1. **Browe, J. H. et al.,** A nutrition survey of the armed forces of Iran, *Am. J. Clin. Nutr.*, 9, 470–514, 1961.
2. **Hedayat, H.,** The present situation in Iran, in *CENTO Conference on Combating Malnutrition in Preschool Children,* Istamabad, Pakistan, March 18–22, 1968, CENTO, Ankara, 1968, 75–80.
3. **Hedayat, H.,** Nutritional work in Iran, in *Progress in Human Nutrition,* Vol. 1, Margen, S., Ed., Avi Publishing, Westport, Conn., 1971, 37–43.
4. Interdepartmental Committee on Nutrition for National Defense, Iran: Nutrition Survey of the Armed Forces, National Institutes of Health, Bethesda, MD, 1956, 39 pp.
5. **Sadre, M. and Donoso, G.,** The changing pattern of malnutrition in Iran, in *Progress in Human Nutrition,* Vol. 1, Margen, S., Ed., Avi Publishing, Westport, Conn., 1971, 29–36.

(3) Iraq

1. **Jalili, M. A., Georges, F., and Fadhil, A. S.,** The state of nutrition in Iraq, *J. Fac. Med. Baghdad, Irag,* 73–141, 1950.

(4) Israel

1. **Bavly, S.,** Nutritional patterns among seven rural communities in Israel, *World Rev. Nutr. Diet.*, 9, 32–55, 1968. (46 ref.)

(5) Jordan

1. Interdepartmental Committee on Nutrition for National Defense, Hashemite Kingdom of Jordan: Nutrition Survey, National Institutes of Health, Bethesda, MD, 1963, 327 pp.
2. Interdepartmental Committee on Nutrition for National Defense, The Hashemite Kingdom of Jordan: Nutrition Survey of Infants and Preschool Children in Jordan, National Institutes of Health, Bethesda, MD, 1964, 180 pp.

(6) Lebanon

1. Interdepartmental Committee on Nutrition for National Defense, Lebanon: Nutrition Survey, National Institutes of Health, Bethesda, MD, 1962, 205 pp.
2. **Sabry, J. H. and Sabry, Z. I.,** Nutrition in Lebanon, *Can. Nutr. Notes*, 19, 121–124, 1963.

(7) Turkey

1. **Baysal, A.,** Present situation in Turkey regarding malnutrition in infants and preschool children-types of problems, magnitude of problem, location, in CENTO Conference on Combating Malnutrition in Preschool Children, Istamabad, Pakistan, March 18–22, 1968, CENTO, Ankara, 1968.
2. **Cura, S.,** The social aspects of child nutrition in Turkey, *Turk. J. Pediatr.*, 3, 145–152, 1961.
3. Interdepartmental Committee on Nutrition for National Defense, Turkey: Nutrition Survey of the Armed Forces, National Institutes of Health, Bethesda, MD, 1958, 87 pp.
4. **Thomson, J. C.,** Reports on a Nutrition Survey in Turkey, WHO, 1956.

4. Far East

a. Far East In General

1. **Bell F. J.,** Relating village-level food production development to the problem of malnutrition in "underprivileged areas" in Southeast Asia, *Bibl. Nutr. Dieta,* No. 14, S. Karger, Basel, 1970, 83–96.
2. Interdepartmental Committee on Nutrition for National Defense, Far East Symposium on Nutrition. Transactions, lst, Saigon 1962, ICNND, Washington, D.C., 1963, 301 pp.; 2nd, Taipeh, Formosa 1964, ICNND, Bethesda, MD, 1965, 332 pp.; 3rd, Manila 1967, ICNND, Bethesda, MD, 1968, 317 pp.
3. **May J.,** *The Ecology of Malnutrition in the Far and Near East,* Vol. 3, *Stud. Med. Geogr.*, Hafner Publishing, New York, 1961.
4. **Niedringhaus, T. E.,** The Food Geography of Mainland Southeast Asia, U.S. Army Natick Laboratories, Natick, Mass., 1968, 86 pp.

b. Individual Countries

(1) Bangladesh

1. Nutrition Section, Office of International Research, Pakistan: Nutrition Survey of East Pakistan, National Institutes of Health, Bethesda, MD, 1966, 426 pp.

(2) Burma

1. Interdepartmental Committee on Nutrition for National Defense, Union of Burma: Nutrition Survey, National Institutes of Health, Bethesda, MD, 1963, 287 pp.

(3) China

1. **May, J. M.,** Food technology and nutrition in Communist China, in *Food Technology the World Over,* Vol. 2 Peterson, M. S. and Tressler, D. K., Eds., Avi Publishing, Westport, Conn., 1965, 245–274.
2. **White, R. O.,** *Rural Nutrition in China,* Oxford University Press, London, 1972, 54 pp.

(4) India

1. **Bhatia, B. M.,** *Famines in India,* Asia Publishing, London, 1967.
2. **Gangulu, N.,** *Bibliography of Nutrition in India,* Oxford University Press, New York, 1940, 79 pp.
3. **Gopalan, C.,** Nutritional problems of India, in *Proc. 8th International Congress of Nutrition,* Prague, 1969, Excerpta Medica, Amsterdam, 1970, 579–581.
4. **Pandit, C. G. and Someswara Rao, K.,** *Nutrition in India (1946–58),* Indian Council of Medical Research, New Delhi, 1960.
5. **Passmore, R.,** Famine in India. An historical survey, *Lancet,* II(7), 303–307, 1951.
6. **Patwandhan, V. N.,** *Nutrition in India,* Indian Journal of Medical Science Publishing, Bombay, 1961.
7. **Porter, A.,** *The Diseases of the Madras Famine of 1877–79,* Madras, 1889.
8. **Verghese, B. G.,** Beyond the Famine, Bihar Relief Committee, New Delhi, 1967.

(5) Indonesia

1. **Postmus, S. and Rijst, P. J., van der,** *Nutrition Bibliography of Indonesia,* University of Hawaii Press, Honolulu, 1955, 145 pp.

(6) Japan

1. **Arimoto, K. and Sakural, Y.,** Food and nutrition in Japan, in *Food Technology the World Over,* Vol. 2, Peterson, M. S. and Tressler, D. K., Eds., Avi Publishing, Westport, Conn., 1965, 359–394.
2. **Innami, S. and Mickelsen, O.,** Nutritional status – Japan, *Nutr. Rev.,* 27, 275–278, 1969. (10 ref.)
3. **Insull, W., Jr., Oiso, T., and Tschuchiya, K.,** Diet and nutritional status of Japanese, *Am. J. Clin. Nutr.,* 21, 753–777, 1968.

(7) Korea

1. Interdepartmental Committee on Nutrition for National Defense, Korea: Nutrition Survey of the Armed Forces, National Institutes of Health, Bethesda, MD, 1957, 66 pp.
2. **Williams, R. R., Combs, G. F., McGanity, W. J., and Kertesz, Z. I.,** A nutrition survey of the Republic of Korea, *J. Nutr.* (Suppl. 1), 68, 1–80, 1959.

(8) Malaya

1. Interdepartmental Committee on Nutrition for National Defense, Malaya: Nutrition Survey, National Institutes of Health, Bethesda, MD, 1964, 355 pp.
2. **Leong, P. C.,** *Nutrition Bibliography of Malaya,* University of Hawaii Press, Honolulu, 1952, 23 pp.

(9) Nepal

1. **Brown, M. L., Worth, R. M., and Shah, N. K.,** Health survey of Nepal: Diet and nutritional status of the Nepalese people, *Am. J. Clin. Nutr.,* 21, 875–881, 1968.

(10) Pakistan

1. **French, C. E. et al.,** A nutrition survey of the armed forces of Pakistan, *J. Nutr.* (Suppl. 2), 68, 1–69, 1959.
2. Interdepartmental Committee on Nutrition for National Defense, Pakistan: Nutrition Survey of the Armed Forces, National Institutes of Health, Bethesda, MD, 1956, 42 pp.
3. **Rahman, M. H.,** Present situation of nutritional status of infants and preschool children in Pakistan, in CENTO Conference on Combating Malnutrition in Preschool Children, Istamabad, Pakistan, March 18–22, 1968, CENTO, Ankara, 1968, 61–74.

(11) Philippines

1. **Darby, W. J. et al.,** A nutrition survey of the armed forces of the Philippines, *Am. J. Clin. Nutr.,* 7, 657–710, 1959.
2. Interdepartmental Committee on Nutrition for National Defense, Philippines: Nutrition Survey of the Armed Forces, National Institutes of Health, Bethesda, MD, 1957, 126 pp.

3. National Institute of Science and Technology, Food and Nutrition Research Center, Nutrition Survey of Ilocos-Mountain Province Region, Manila, 1963, 88 pp; Nutrition Survey of Metropolitan Manila. Manila, 1962, 79 pp.; Nutrition Survey of the Cagayan Valley-Batanas Region. Manila, 1963, 89 pp.; Nutrition Survey of the Southern Tagalog Region. Manila, 1965, 88 pp; Nutrition Survey of Western Visayas Region, 1968, 104 pp; Report of the 6th Post-Nutrition Survey Conference, Cebu City, 1968, 68 pp.

(12) Taiwan

1. **Hsu, S. C.,** Nutrition in Taiwan, the Republic of China, progress, problems, and planning, *World Rev. Nutr. Diet.*, 17, 39–52, 1973. (6 ref.)
2. Interdepartmental Committee on Nutrition for National Defense, Taiwan: Nutrition Survey of the Armed Forces, National Institutes of Health, Bethesda, MD, 1961, 80 pp.

(13) Thailand

1. Interdepartmental Committee on Nutrition for National Defense, Thailand: Nutrition Survey, National Institutes of Health, Bethesda, MD, 1962, 285 pp.

(14) Vietnam

1. Interdepartmental Committee on Nutrition for National Defense, Vietnam: Nutrition Survey, National Institutes of Health, Bethesda, MD, 1960, 257 pp.

5. Africa

a. Africa In General

1. **Amann, V. F., Belshaw, D. G. R., and Stanfield, J. P., Eds.,** *Nutrition and Food in African Economy,* Makerere University, Kampala, Uganda, 1972, 581 pp. (2 vols.)
2. **Aylward, F.,** Some African problems, *Proc. Nutr. Soc.*, 20, 112–117, 1961.
3. **Brock, J. F. and Autret, M.,** *Kwashiorkor in Africa,* WHO Monogr. Ser. 8, WHO, Geneva, 1952, 78 pp.
4. **Edozien, J. C.,** Malnutrition in Africa – Need and basis for action, in *Bibl. Nutr. Diet,* No. 14, S. Karger, Basel, 1970, 64–72. (13 ref.)
5. Food and Agriculture Organization, Food Composition Tables for Use in Africa, FAO, Rome, 1968.
6. **Kraut, H. and Cremer, H. D., Eds.,** *Investigations into Health and Nutrition in East Africa,* Humanities Press, New York, 1969, 342 pp.
7. **Latham, M.,** Human Nutrition in Tropical Africa, with Special Reference to Community Health Problems in East Africa, FAO, Rome, 1965, 268 pp.
8. **Leung, W. T. W. and Butrum, R. D.,** A Selected Bibliography on African Foods and Nutriton, Center for Disease Control, Washington, D.C., 1970, 382 pp.
9. Malnutrition in Africa, *Nutr. Rev.*, 21, 307–309, 1963.
10. **May, J. M.,** *The Ecology of Malnutrition in Northern Africa, Stud. Med. Geogr.*, Vol. 7 Hafner Publishing, New York, 1967, 275 pp.
11. **May, J.,** *The Ecology of Malnutrition in Middle Africa, Stud. Med. Geogr.*, Vol. 5, Hafner Publishing, New York, 1965, 255 pp.
12. **May, J.,** *The Ecology of Malnutrition in the French Speaking Countries of West Africa and Madagascar, Stud. Med. Geogr.*, Vol. 8, Hafner Publishing, New York, 1968.
13. **May, J. and McLellan, D. L.,** *The Ecology of Malnutrition in Eastern Africa and Four Countries of Western Africa, Stud. Med. Geogr.*, Vol. 9, Hafner Publishing, New York, 1970, 675 pp.
14. **May, J. M. and McLellan, D. L.,** *The Ecology of Malnutrition in Seven Countries of Southern Africa and in Portuguese Guinea, Stud. Med. Geogr.*, Vol. 10, Hafner Publishing, New York, 1971, 432 pp.
15. *Nutrition in Eastern Africa*, Longmans of Kenya, Arusha, 1966, 95 pp.
16. Proc. West African Conference on Nutrition and Child Feeding, Dakar, Senegal, March 25–29, 1968, 289 pp.
17. Proc. East African Conference on Nutrition and Child Feeding, Kenya, 1969, 311 pp.
18. Proc. Third African Conference on Nutrition and Child Feeding, Tunis, Tunisia, May 25–29, 1970, Center for Disease Control, U.S. Public Health Service, Atlanta, Georgia, 1970, 400 pp.
19. **Vahlquist, B., Ed.,** *Nutrition. A Priority in African Development,* Almqvist & Wiksell, Stockholm, 1972, 228 pp.

b. Individual Countries

(1) Algeria

1. **Gast, M.,** *Alimentation des populations de l'Ahaggar. Etude ethnographique,* Arts et metiers graphiques, Algeria, 1968, 456 pp.

(2) Angola

1. **Strangway, A. K.,** Malnutrition in Angola, *J. Am. Diet. Assoc.*, 39, 585–589, 1961.

(3) Botswana

1. **Burgess, H. J. L.**, Nutrition Survey in Botswana, AFR/NUT/52, WHO Regional Off. for Africa, Brazzaville, Congo, 1972.

(4) Congo

1. **De Maeyer, E. M.**, Nutrition in the Congo, *Nutr. Rev.*, 20, 225–227, 1962.

(5) Ethiopia

1. Interdepartmental Committee on Nutrition for National Defense, Ethiopia: Nutrition Survey, National Institutes of Health, Bethesda, MD, 1959, 210 pp.
2. **Woodruff, C. W. and Hoerman, K.**, Nutrition in infants and preschool children in Ethiopia, *Public Health Rep.*, 75, 724–730, 1960.

(6) Ghana

1. **Aylward, F.**, Report to the Government of Ghana on Foods and Nutrition, UNDP Rep. TA 1449, FAO, Rome, 1961, 76 pp.

(7) Kenya

1. **Bohdal, M., Gibbs, N. E., and Simmons, W. K.**, Nutrition Survey and Campaign Against Malnutrition in Kenya 1964–1968, WHO/FAO/UNICEF, Nairobi, 1968.

(8) Libya

1. **Ferro-Luzzi, G.**, Report to the Government of Libya on Nutrition, FAO Rep. 920, FAO, Rome, 1958.
2. Interdepartmental Committee on Nutrition for National Defense, Libya: Nutrition Survey of the Armed Forces and Civilians, National Institutes of Health, Bethesda, MD, 1957, 76 pp.
3. **Yang, Y. H.**, Food and Nutrition Policy. Report to the Government of Libya, FAO, Rome, 1963.

(9) Malawi

1. **Burgess, H. J. L., Burgess, A., and Wheeler, E. F.**, Results and appraisal of a nutritional survey in Malawi, *Trop. Geogr. Med.*, in press.

(10) Nigeria

1. **Aall, C.**, Relief, nutrition and health in the Nigerian/Biafran war, *J. Trop. Pediatr.*, 16, 70–90, 1970.
2. **Mayer, J.**, Famine in Biafra, *Postgrad. Med.*, 45(4), 236–240, 1969.
3. Nutrition Section, Office of International Research, Republic of Nigeria: Preliminary Nutrition Survey, National Institutes of Health, Bethesda, MD, 1966, 287 pp.
4. **Oyenuga, V. A.**, The present and future state of food and nutrition in Nigeria, in *Proc. 8th International Congress of Nutrition*, Prague, 1969, Excerpta Medica, Amsterdam, 1970, 601–608.

(11) Sierra Leone

1. **Blankhart, D. M.**, *Nutrition Survey in Sierra Leone*, AFR/NUT/17, WHO, Geneva, 1964.
2. **Thomas, H. M.**, Some aspects of food and nutrition in Sierra Leone, *World Rev. Nutr. Diet.*, 14, 48–58, 1972. (22 ref.)

(12) South Africa

1. Evidence for the existence of dietary deficiency diseases and particularly protein deficiency in South Africa, in Food Enrichment in South Africa, National Nutrition Research Institute, South African Council for Scientific and Industrial Research, Pretoria, 1959, 6–52.
2. **Walker, A. R. P.**, Nutritional, biochemical, and other studies on South African populations, *S. Afr. Med. J.*, 40(34), 814–852, 1966.

(13) Tanzania

1. **Latham, M. C.**, Nutritional studies in Tanzania (Tanganyika), *World Rev. Nutr. Diet.*, 7, 31–71, 1966. (57 ref.)

(14) Uganda

1. **Burgess, H. J. L., Burgess, A. P., and Dean, R. F. A.**, Symposium on nutritional work in Uganda carried out for the Scientific Committee on Human Nutrition of the Ministry of Health, *East Afr. Med. J.*, 39, 357–493, 1962.
2. London School of Hygiene and Tropical Medicine, Applied Nutrition Unit, Nutrition in Uganda, London, 1957.
3. **Stanfield, J. P., Ed.**, Recent approaches to malnutrition in Uganda, *J. Trop. Pediatr. Environ. Child Health*, 17(1), Monogr. 13, 1971, 87 pp.

6. Oceania

a. General

1. **Pascual, C. R.,** Nutritional status in the Pacific, *Philipp. J. Nutr.,* 24, 149–159, 1971.

b. Australia

1. **Hutchinson, R. C.,** *Food for the People of Australia,* Angus and Robertson, Sydney, 1958, 261 pp.

c. New Zealand

1. **McGeorge, M.,** Infant feeding in New Zealand in 1960–1961, *Q. Rev. Pediatr.,* 17, 49–58, 1962.

d. Pacific Islands

1. **Fanning, R. J.,** *Pacific Islands Nutrition Bibliography,* University of Hawaii Press, Honolulu, 1951, 70 pp.
2. **McKee, H. S.,** Some Food Problems in the Pacific Islands, South Pacific Commission Tech. Paper No. 106, May 1957.
3. **Malcolm, S.,** Nutrition Investigation in New Caledonia, South Pacific Commission Tech. Paper No. 50, October 1953.
4. **Malcolm, S.,** Diet and Nutrition in American Samoa, South Pacific Commission Tech. Paper No. 63, August 1954.
5. **Malcolm, S.,** Diet and Nutrition in the Trust Territory of the Pacific Islands, South Pacific Commission Tech. Paper No. 85, April 1955.
6. **Murai, M.,** Nutrition Study in Micronesia, Atoll Res. Bull. No. 27, Jan. 31, 1954, Pacific Scientific Board, NAS-NRC, Washington, D.C., 1954.
7. **Oomen, H. A. P. C. and Malcolm, S. H.,** Nutrition and the Papuan Child, South Pacific Commission, Nouméa, New Caledonia, April 1958, South Pacific Commission Tech. Paper No. 118.
8. **Parkinson, S.,** Problems of Applied Nutrition in the Pacific Islands, South Pacific Commission Tech. Information Circ. No. 52, 1962, 6 pp.
9. Report of the New Guinea Nutrition Survey Expedition 1947, A. H. Pettifer, Sydney, Australia, 1947, 308 pp.
10. Report on Nutrition Investigation by the South Pacific Commission in 1950, South Pacific Commission Tech. Paper No. 18, November 1951.

7. North America

a. North America In General

1. **Fraser, D.,** Nutritional problems in North America, *Proc. Western Hemisphere Nutrition Congress,* American Medical Association, Chicago, 1966, 57–59.

b. Individual Countries

(1) Canada

1. **Campbell, J. A.,** The national nutrition survey of Canada, *Can. Nutr. Notes,* 28, 25–30, 1970.
2. Canada Dep. of National Health and Welfare, A Report on Nutrition and Production and Distribution of Food, Ottawa, 1946, 205 pp.
3. Canada Dep. of National Health and Welfare, Canadian Food and Nutrition Statistics, Ottawa, 1959, 29 pp.
4. **Sinclair, D.,** Canadian food and nutrition statistics 1935 to 1965, *Can. Nutr. Notes,* 25, 109–116, 1969.
5. Toronto Nutrition Committee, Food Customs of New Canadians, Toronto, 1967, 60 pp.

(2) Mexico

1. **García, S.,** Malabsorption and malnutrition in Mexico, *Am. J. Clin. Nutr.,* 21, 1066–1076, 1968.
2. **May, J. M. and McLellan, D. L.,** *The Ecology of Malnutrition in Mexico and Central America, Stud. Med. Geogr.,* Vol. 11, Hafner Publishing, New York, 1972, 395 pp.
3. **Miranda, F. P.,** La alimentación en México, Instituto Nacional de Nutriologia, 1947, 47 pp.

(3) U. S.

1. American Medical Association Council on foods and nutrition, Malnutrition and hunger in the United States, *JAMA,* 213, 272–275, 1970.
2. Citizens' Board of Inquiry into Hunger and Malnutrition in the United States, *Hunger U.S.A. A Report,* Beacon Press, Boston, 1968, 96 pp.
3. **Davis, T. R. A., Gershoff, S. N., and Gamble, D. F.,** Review of studies of vitamin and mineral nutrition in the United States (1950–1968), *J. Nutr. Educ.,* Suppl. 1, 1(2), 41–57, 1969.
4. **Emerson, G. A.,** Nutritional status, USA, *J. Nutr.,* Suppl. 1, 91, 51–54, 1967. (23 ref.)
5. **Eppright, E. S., Fox, H. M., Fryer, B. A., Lamkin, G. H., Vivian, V. M., and Fuller, E. S.,** Nutrition of infants and preschool children in the North Central Region of the United States of America, *World Rev. Nutr. Diet.,* 14, 269–332, 1972.
6. **Gershoff, S. N.,** Childhood malnutrition in the United States, in *Progress in Human Nutrition,* Vol. 1, Margen, S., Ed., Avi, Publishing, Westport, Conn., 1971, 68–74.

7. **Goldsmith, G. A.,** Clinical nutrition problems in the United States today, *Nutr. Rev.*, 23, 1–3, 1965.
8. Interdepartmental Committee on Nutrition for National Defense, Alaska: An Appraisal of the Health and Nutritional Status of the Eskimo, National Institutes of Health, Bethesda, MD, 1959, 165 pp.
9. Interdepartmental Committee on Nutrition for National Defense, Blackfeet Indian Reservation: Nutrition Survey, National Institutes of Health, Bethesda, MD, 1964, 121 pp.
10. Interdepartmental Committee on Nutrition for National Defense, Fort Belknap Indian Reservation: Nutrition Survey, National Institutes of Health, Bethesda, MD, 1964, 107 pp.
11. **Joffe, N. F. and Walker, T. T.,** Some Food Patterns of Negroes in the U.S.A. and their Relationship to Wartime Problems of Food and Nutrition, National Research Council, Committee on Food Habits, Washington, D.C., 1942, 50 pp.
12. **Kelsay, J. L.,** A compendium of nutritional status studies and dietary evaluation studies conducted in the United States, 1957–1967, *J. Nutr.*, Suppl. 1, 99(1), II, 123–166, 1969.
13. **Kotz, N.,** *Let Them Eat Promises. The Politics of Hunger in America,* Prentice-Hall, Englewood Cliffs, N.J., 1969, 272 pp.
14. **Mann, G. V., Scott, E. M., Hursh, L. M., Heller, C. A., Youmans, J. B., Consolazio, C. F., Bridgforth, E. B., Russell, A. L., and Silverman, M.,** The health and nutritional status of Alaskan Eskimos, *Am. J. Clin. Nutr.*, 11, 31–76, 1962.
15. **Mayer, J.,** The nutritional status of American Negroes, *Nutr. Rev.*, 23, 161–164, 1965.
16. **Mayer, J., Ed.,** *U.S. Nutrition Policies in the Seventies,* W. H. Freeman Publishing, San Francisco, 1973, 250 pp.
17. **Moore, W. M., Silverberg, M. M., and Read, M. S., Eds.,** Nutrition, Growth and Development of North American Indian Children, U.S. Gov. Print. Off., Washington, D.C., 1972, 246 pp.
18. **Morgan, A. F.,** Nutritional status U.S.A., *Calif. Agric. Exp. Stn. Bull.*, 769, 1959.
19. Problems of Assessment and Alleviation of Malnutrition in the United States, Proc. of Workshop, Vanderbilt University, Nashville, Tenn., Jan. 13–14, 1970, 186 pp.
20. **Rechcigl, M., Jr.,** Nutrition of health – A national challenge, *J. Appl. Nutr.*, 22, 9–16, 1970.
21. **Rechcigl, M., Jr.,** Malnutrition and hunger, USA – Problems and solutions, in Proc. of Workshop Conference on Nutrition, South Carolina Regional Medical Program, Columbia, South Carolina, May 1, 1970.
22. **Schaefer, A. E.,** The national survey, *J. Am. Diet. Assoc.*, 54, 371–375, 1969.
23. **Schaefer, A. E.,** Findings and implications of the National Nutrition Survey, in *Dimensions of Nutrition,* Dupont, J., Ed., Colorado Associated University Press, Boulder, 1970, 243–261.
24. U.S. Senate Select Committee on Nutrition and Human Needs, The Food Gap: Poverty and Malnutrition in the United States, 91st Congress, 1st Session, August 1969.
25. White House Conference on Aging. Recommendations of the Nutrition Section, Washington, D.C., 1971.
26. White House Conference on Food, Nutrition and Health. Final Report, U.S. Gov. Print. Off. Washington, D.C., 1970; Follow-up Report, 1971.

8. Latin America

a. General

1. **Esquef, L. de,** Food and nutrition in the overall development of Latin America, *FAO Nutr. Newsletter,* (4), 9–15, 1971.
2. Food and Agriculture Organization, Report, Nutrition Conference, Montevideo, Uruguay, 1948, FAO Montevideo, 1950. FAO Nutr. Meet. Rep. Ser. No. 1, 286 pp.
3. Food and Agriculture Organization, Report, 2nd Conference on Nutrition Problems in Latin America, Rio de Janeiro, 1950, FAO, Washington, D.C., 1950. FAO Nutr. Meet. Rep. Ser. No. 4, 34 pp.
4. Food and Agriculture Organization, Report, 3rd Conference on Nutrition Problems in Latin America, Caracas, 1953, Oficina Sanitaria Panamericana, 1954. FAO Nutr. Meet. Rep. Ser. No. 8, 54 pp.
5. Food and Agriculture Organization, Report, 4th Conference on Nutrition Problems in Latin America, Guatemala City, 1957, FAO, Rome, 1959. FAO Nutr. Meet. Rep. Ser. No. 18, 82 pp.
6. **Horwitz, A.,** The physician's view of nutritional needs in the Western Hemisphere, *Proc. Western Hemisphere Nutrition Congress,* 1965, American Medical Association, Chicago, 1966, 3–6.
7. **Kevany, J.,** Nutritional problems in Latin America, *Proc. Western Hemisphere Nutrition Congress,* 1965, American Medical Association, Chicago, 1966, 50–57.
8. **Monckeberg, F.,** Factors conditioning malnutrition in Latin America with special reference to Chile. Advice for a volunteers action, *Bibl. Nutr. Dieta,* S. Karger, Basel, 1970, 23–33.
9. **Puffer, R. R. and Serrano, C. V.,** The role of nutritional deficiency in mortality. Findings of the Inter-American investigation of mortality in childhood, *Bol. Ofic. Sanit. Panam.*, English ed., 7(1), 1–25, 1973.
10. **Schaefer, A. E.,** Magnitude of the problem of malnutrition in the preschool child in Latin America, Proc. 4th Inter-American Conference on Malnutrition, Panama, June 1966.

b. South America

(1) General

1. **Castro, J. de.,** El probleme de la alimentación en América del Sur, Editorial Sudamericana, Buenos Aires, 1950, 65 pp.

(2) Individual Countries

(a) Argentina

1. La alimentacion de la familia en Buenos Aires, in Boletin del Instituto Nacional de la Nutrition, año 1–12 (no. 1–135), Instituto Nacional de la Nutrición, Buenos Aires, 1938–1949.

(b) Bolivia

1. Interdepartmental Committee on Nutrition for National Defense, Bolivia: Nutrition Survey, National Institutes of Health, Bethesda, MD, 1964, 281 pp.

(c) Brazil

1. Nutrition Section, Office of International Research, Northeast Brazil: Nutrition Survey, National Institutes of Health, Bethesda, MD, 1965, 294 pp.
2. **Waterlow, J. C. and Vergara, A.,** Protein Malnutrition in Brazil, FAO Int. Stud. 14, FAO, Rome, 1956, 40 pp.

(d) Chile

1. Interdepartmental Committee on Nutrition for National Defense, Chile: Nutrition Survey, National Institutes of Health, Bethesda, MD, 1961, 357 pp.

(e) Colombia

1. Interdepartmental Committee on Nutrition for National Defense, Colombia: Nutrition Survey, National Institutes of Health, Bethesda, MD, 1961, 263 pp.

(f) Ecuador

1. Interdepartmental Committee on Nutrition for National Defense, Ecuador: Nutrition Survey, National Institutes of Health, Bethesda, MD, 1960, 230 pp.

(g) Guyana

1. Pan American Health Organization, The National Food and Nutrition Survey of Guyana, PAHO, Washington, D.C., in press.

(h) Paraguay

1. National Center for Chronic Disease Control, Nutrition Program, Republic of Paraguay: Nutrition Survey of the Armed Forces (May–August 1965), Bethesda, MD, 1967, 99 pp.
2. National Center for Chronic Disease Control, Nutrition Program, Republic of Paraguay: Nutrition Survey (May–August 1965), Bethesda, MD, 1967, 439 pp.

(i) Peru

1. Interdepartmental Committee on Nutrition for National Defense, Peru: Nutrition Survey of the Armed Forces, National Institutes of Health, Bethesda, MD, 1960, 181 pp.

(j) Surinam

1. Studies on the physiology of nutrition in Surinam, *Nutr. Rev.*, 21, 131–135, 1963.

(k) Uruguay

1. Interdepartmental Committee on Nutrition for National Defense, Uruguay: Nutrition Survey, National Institutes of Health, Bethesda, MD, 1963, 294 pp.

(l) Venezuela

1. Interdepartmental Committee on Nutrition for National Defense, Venezuela: Nutrition Survey, National Institutes of Health, Bethesda, MD, 1964, 407 pp.

c. Central America

(1) Central America In General

1. **May, J. M. and McLellan, D. L.,** *The Ecology of Malnutrition in Mexico and Central America, Stud. Med. Geogr.,* Vol. 11, Hafner Publishing, New York, 1972, 395 pp.

(2) Individual Countries

(a) British Honduras

1. **Nocks, B. N.,** Nutrition study in British Honduras, *Am. J. Clin. Nutr.*, 20, 661–671, 1967.

(b) Costa Rica

1. Institute of Nutrition of Central America and Panama, Evaluación nutricional de la población de Centro América y Panamá: Costa Rica, INCAP, Guatemala, 1969.

(c) El Salvador

1. Institute of Nutrition of Central America and Panama, Evaluación nutricional de la población de Centro América y Panamá: El Salvador, INCAP, Guatemala, 1969.

(d) Guatemala

1. Institute of Nutrition of Central America and Panama, Evaluación nutricional de la población de Centro América y Panamá: Guatemala, INCAP, Guatemala, 1969.
2. **Salomon, J. B., Mata, L. J., and Gordon, J. E.,** Malnutrition and the common communicable diseases of childhood in rural Guatemala, *Am. J. Public Health,* 58, 505–516, 1968.

(e) Honduras

1. Institute of Nutrition of Central America and Panama, Evaluación nutricional de la población de Centro América y Panamá: Honduras, INCAP, Guatemala, 1969.

(f) Nicaragua

1. Institute of Nutrition of Central America and Panama, Evaluación nutricional de la población de Centro América y Panamá: Nicaragua, INCAP, Guatemala, 1969.

(g) Panama

1. Institute of Nutrition of Central America and Panama, Evaluación nutricional de la población de Centro América y Panamá: Panama, INCAP, Guatemala City, 1969.

d. The Caribbean

(1) The Caribbean In General

1. **Aykroyd, W. R.,** Nutrition in the Caribbean, *J. Hyg. (Camb.),* 63, 137–153, 1965.
2. **Cook, R.,** Nutrition and mortality under five years in the Caribbean area, in *Proc. 8th International Congress of Nutrition,* Prague, 1969, Excerpta Medica, Amsterdam, 1970, 387–390.
3. Interdepartmental Committee on Nutrition for National Defense, West Indies: Trinidad and Tobago; St. Lucia; St. Christopher; Nevis and Anguilla; Nutrition Survey, National Institutes of Health, Bethesda, MD, 1962, 187 pp.
4. **May, J. M. and McLellan, D. L.,** *The Ecology of Malnutrition in the Caribbean, Stud. Med. Geogr.,* Vol. 12, Hafner Publishing, New York, 1973, 490 pp.
5. **Platt, B. S.,** Nutrition in the British Indies, HMSO, London, 1946.
6. Underwood, E. J., Report on Food and Nutrition in the Caribbean, FAO, Rome, 1966, 22 pp.

(2) Individual Countries

(a) Barbados

1. Caribbean Food and Nutrition Institute, Barbados Nutrition Survey. Preliminary Report, CFNI, Kingston, 1969.
2. Pan American Health Organization, The National Food and Nutrition Survey of Barbados, Sci. Publ. 237, PAHO, Washington, D.C., 1972, 139 pp.

(b) Dominican Republic

1. **Sebrell, W. H., King, K. W., Webb, R. E., and Daza, C. H.,** Nutritional status of middle and low income groups in the Dominican Republic, *Archivos Latinoam. Nutr.,* 22 (Numero especial), 1–190, 1972.

(c) Haiti

1. **Beghin, I., Fougere, W., and King, K. W.,** *L'alimentation et la nutrition en Häiti,* Presses Universitaires de France, Paris, 1970, 248 pp.

(d) Jamaica

1. Caribbean Food and Nutrition Institute, Preliminary Results of the Jamaica Nutrition Survey, March 1970, CFNI, Kingston, 1970.
2. **Fox, H. C., Campbell, V. S., and Morris, J. C.,** The dietary and nutritional status of Jamaican infants and toddlers, *Inf. Bull. Sci. Res. Counc. Jam.,* 8, 33–51, 1968.
3. **Gurney, J. M., Fox, H., and Neill, J.,** A rapid survey to assess the nutrition of Jamaican infants and young children in 1970, *Trans. R. Soc. Trop. Med. Hyg.,* 66(4), 653–662, 1972.
4. **Standard, K. L.,** Nutrition Survey of Jamaica, Nutrition Committee of Jamaica Government, 1958.

(e) Puerto Rico

1. **Asenjo, C. F.,** Publications Concerning Nutrition in Puerto Rico Up to 1967, San Juan, 1968, 40 pp.
2. **Fernández, N. A., Burgos, J. C., Asenjo, C. F., and Rosa, I. R.,** Nutritional status of the Puerto Rican population – Master sample survey, *Bol. Assoc. Med. P. R.,* 63(4), Suppl. 1–46, 1971.
3. **Zayas, Ester Seijo de,** Cultural Nutrition Patterns of Puerto Ricans, 1962, 5 pp.

(f) St. Lucia

1. **Lus, R. E. M.,** Malnutrition. The pattern and prevention in St. Lucia, *West Indies Med. J.,* 13, 97–102, 1964.

(g) St. Vincent

1. Caribbean Food and Nutrition Institute, A Rapid P.C.M. Survey in St. Vincent, Preliminary Report, Kingston, CFNI, 1968.

(h) Trinidad and Tobago

1. Caribbean Food and Nutrition Institute, Report on National Household Food Consumption Survey in Trinidad and Tobago, 1970, CFNI, St. Augustine, Trinidad, 1971.

B. World Food Consumption

1. General

1. **Bennett, M. K.,** International contrasts in food consumption, *Geogr. Rev.,* 31, 365–376, 1941.
2. **Bennett, M. K.,** *The World's Food,* Harper & Row, New York, 1954.
3. **Brothwell, D. and Brothwell, P.,** *Food in Antiquity,* Praeger, New York, 1969, 284 pp.
4. **Brown, L. R.,** Food Consumption and Expenditures: India, Japan, United States, USDA, ERS, Washington, D. C., 1962. 15 pp.
5. **Cochrane, W.,** *The Economics of Consumption; Economics of Decision Making in the Household,* McGraw-Hill, New York, 1956.
6. Food and Agriculture Organization, Factors influencing the trend in food consumption, in The State of Food and Agriculture, FAO, Rome, 1957, 70–110.
7. Food and Agriculture Organization, Review of Food Consumption Surveys, FAO, Rome, 1958; Suppl., 1964.
8. Food and Agriculture Organization, Bibliography of Food Consumption Surveys, 1964, FAO, Rome, 1963, 180 pp.; 1965 Suppl., FAO, Rome, 1965, 23 pp.
9. Food and Agriculture Organization, Bibliography of Food Consumption Surveys, FAO, Rome, 1964, 97 pp.
10. Food patterns in different parts of the world, *Proc. 3rd International Congress of Dietetics,* London, July 10-14, 1961, Newman Books, London, 1961, 13–45.
11. **François, P.,**Nutrimetric study on dietary disparities, *FAO Nutr. Newsletter,* 7(3), 10–33, 1969.
12. **François, P. J.,** Food consumption surveys. Study on a general formula for the estimation of per caput, households and group consumption, *FAO Nutr. Newsletter,* 8(4), 10–26, 1970.
13. **Fujimuto, T.,** Food consumption in Western countries, *J. Sci. Labor* (Tokyo), 47, 712–754, 1971.
14. **Goreux, L.,** Income and food consumption, *FAO Mon. Bull. Agric. Econ. Stat.,* 9(10), 1–13, 1960.
15. **Hand, D. B., Schaefer, A. E., and Wilson, C. S.,** A comparative study of food consumption patterns in Latin America, Middle Eastern and Far Eastern countries, *Proc. 1st International Congress Food Science and Technology,* London, 1962, Vol. 5, Gordon & Breach, New York, 1967, 251–270.
16. **Houthakker, H. S.,** The present state of consumption theory, *Econometrica,* 29, 704–740, 1961.
17. **Jennings, R. D.,** Consumption of Feed by Livestock, 1909-47, U.S.Dep.Agric. Circ. 836, 1949, 105 pp.
18. **Jurèen, L.,** Long-term trends in food consumption: A multicountry study, *Econometrica,* 24(1), 1–21, 1956.
19. **Lörstad, M. H.,** Food consumption surveys – Cost aspects on the precision of the estimates, *FAO Nutr. Newsletter,* 8(2), 21–36, 1970.
20. **Mathias, M.,** The impact of industrialization and urbanization on food consumption patterns in developing countries, in *Proc. 1st Asian Congress of Nutrition,* Tulpule, P. G. and Jaya Rao, K. S., Eds., Nutrition Society of India, Hyderabad, 1972, pp. 742–750.
21. Organization for Economic Cooperation and Development, Food Consumption Statistics 1954–1966, OECD, Paris, 1968, 690 pp.
22. Organization for Economic Cooperation and Development, Food Consumption Statistics 1960–1968, OECD, Paris, 1970, 596 pp.
23. Organization for Economic Cooperation and Development, Food Consumption in the OEEC Countries. I. Food Balance Sheets, OECD, Paris, 1960.
24. **Paris, Q.,** An Appraisal of "Income" Elasticities for Total Food Consumption in Developing Countries, OECD, Paris, 1970, 52 pp.
25. **Pekkarinen, M.,** World food consumption patterns, in *Man, Food, and Nutrition. Strategies and Technological Measures for Alleviating the World Food Problem,* Rechcigl, M., Jr., Ed., CRC Press, Cleveland, 1973, 15–33. (13 ref.)

26. **Pekkarinen, M.**, Methodology in the collection of food consumption data, *World Rev. Nutr. Diet.*, 12, 145–171, 1970.
27. **Reh, E.**, Manual on Household Food Consumption Survey, FAO Nutritional Studies No. 18, FAO, Rome, 1962, 96 pp.
28. Symposium on food consumption patterns in Asian Countries, *Proc. 1st Asian Congress of Nutrition*, Tulpule, P. G. and Jaya Rao, K. S., Eds., Nutrition Society of India, Hyderabad, 1972, 213–284. (6 papers)
29. Symposium. The changing patterns of consumption of food, *Proc. 1st International Congress of Science and Technology*, Vol. 5; London, Sept. 18–21, 1962, Gordon & Breach, New York, 1967, 159–272. (8 papers with discussion)
30. **Szarf, A.**, Food consumption since the war, *FAO Mon. Bull. Agric. Econ. Stat.*, 5(10), 1–10, 1956.
31. **Woodbury, R. M.**, Food Consumption and Dietary Surveys in the Americas: Results – Methods, International Labour Office, Montreal, 1942, 64 pp.

2. Individual Countries or Regions

a. United Kingdom

1. **Drummond, J. C. and Wilbraham, A.**, *The Englishman's Food: A History of Five Centuries of English Diet*, J. Cape, London, 1939, 574 pp.
2. Food habits in Britain, *Proc. Nutr. Soc.*, 20, 25–51, 1961. (5 papers)
3. **Greaves, J. P. and Hollingsworth, F.**, Trends in food consumption in the United Kingdom, *World Rev. Nutr. Diet.*, 6, 34–89, 1966. (88 ref.)
4. **Hollingsworth, D.**, *Drummond's and Wilbraham's The Englishman's Food*, 2nd ed., Cape, London, 1958.
5. Ministry of Agriculture, Fisheries and Food, National Food Survey Committee, Domestic Food Consumption and Expenditure 1953, HMSO, London, 1955.
6. Ministry of Agriculture, Fisheries and Food, National Food Survey Committee, Household Food Consumption and Expenditure 1957, HMSO, London, 1959.
7. Ministry of Agriculture, Fisheries and Food, National Food Survey Committee, Household Food Consumption and Expenditure 1965, HMSO, London, 1967.

b. United States

1. **Bennett, M. K. and Pierce, R. H.**, Change in the American national diet, 1879–1959, *Food Res. Inst. Stud. Agric. Econ. Trade Dev. (Stanford)*, 2(2), 95–119, 1961.
2. **Burk, M. C.**, *Influences of Economics and Social Factors on U. S. Food Consumption*, Burgess Publishing, Minneapolis, 1961, 135 pp.
3. **Burk, M. C.**, Measures and Procedures for Analysis of U. S. Food Consumption, Agric. Handb. 206, USDA, ERS, Washington, D. C., 1961, 118 pp.
4. **Clark, F.**, Dietary levels of families in the United States, *J. Am. Diet. Assoc.*, 34(4), 378–382, 1958.
5. **Clark, F., Murray, J., Weiss, G. S., and Grossman, E.**, Food Consumption of Urban Families in the United States, Agric. Inf. Bull. 132, U.S.Dep.Agric., Washington, D.C., October 1954, 203 pp.
6. **Cummings, R. O.**, *The American and His Food*, University of Chicago Press, Chicago, 1941.
7. **Sherck, C. K.**, Changes in food consumption patterns, *Food Technol.*, 25, 914–916, 1971.
8. U.S. Department of Agriculture, Family Food Consumption in the United States, Spring 1942, Misc. Publ. No. 550, U.S.Dep.Agric., Washington, D. C., 1944.
9. U.S. Department of Agriculture, Agricultural Marketing Service, Consumption of Food in the United States, 1909-52, Agric. Handb. 62, U.S.Dep.Agric., Washington D.C., 1953, 249 pp.
10. U.S. Department of Agriculture, Dietary Levels of Households in the United States, Household Food Consumption Survey 1955, Rep. 6, U.S.Dep.Agric., Washington, D.C., 1957.
11. U.S. Department of Agriculture, Consumption and Utilization of Agricultural Products in Major Statistical Series of the U.S. Department of Agriculture, How they are constructed and used, U.S.Dep.Agric., Agric. Handb. 118, 5, 1957, 91 pp.
12. U.S. Department of Agriculture, Food Consumption and Utilization Section, U.S. Food Consumption Sources of Data and Trends, 1909–63, U.S.Dep.Agric. Stat. Bull. No. 364, USDA, ERS, Washington, D.C., 1965, 194 pp.
13. U.S. Department of Agriculture, Dietary Levels of Households in the United States, ARS Rep. 6, U.S.Dep.Agric., Washington, D.C., 1965.
14. U.S. Department of Agriculture, Food Consumption of Households in the United States, Spring 1965-66, ARS Rep. 1, U.S.Dep.Agric., Washington, D.C., 1968.
15. U.S. Department of Agriculture, *Food Consumption, Prices, Expenditures*, USDA, ERS, Agric. Econ. Rep. 138, Washington, D.C., 1968.

c. India

1. **Aykroyd, W. R.**, *Notes on the Results of Diet Surveys in India, Burma and Ceylon*, Indian Research Fund Association, Special Rep. 16, Cawnpote, 1947.

2. **Gopalan, C., Balasubramanian, S. C., Rama Sastri, B. V. and Visweswara Rao, K.,** *Diet Atlas of India*, National Institute of Nutrition, Hyderabad, 1971, 133 pp.
3. **Mitra, K.,** A Supplement to the Results of Diet Surveys in India 1935-48, Special Rep. Ser. 25, Indian Council of Medical Research, New Delhi, 1953.
4. Results of Diet Surveys in India 1935-48, Special Rep. Ser. 20, Indian Council of Medical Research, New Delhi, 1951.
5. **Sukhatme, P. V.,** The food and nutrition in India. I, *Indian J. Agric. Econ.*, 17(2), 1–28, 1962.
6. **Visweswara Rao, K.,** Patterns and trends in food consumption in India, *J. Nutr. Diet.*, 4, 79–87, 1967.

d. Korea

1. **Chai Re Suk,** Food Consumption of Korean Individuals, Ministry of Health and Social Affairs, National Chemical Laboratories, Korea, 1956.

e. Lebanon

1. **Cowan, J. W.,** Dietary survey in rural Lebanon. II. *J. Am. Diet. Assoc.*, 47, 466–469, 1966.
2. **Cowan, J. W., Chopra, S., and Houry, G.,** Dietary survey in rural Lebanon, *J. Am. Diet. Assoc.*, 45, 130-133, 1964.

f. Africa

1. **Aylward, F.,** Food habits in western tropical Africa, *Chem. Ind.*, 1624–1627, 1966.
2. **Johnston, B. F.,** *The Staple Food Economics of Western Tropical Africa*, Stanford University Press, Stanford, Calif., 1958, 305 pp.

C. The State of World Food Production

1. **Abel, M. E. and Rojko, A. S.,** The World Food Situation, FAE Rep. No. 35, U.S. Dep. Agric., Econ. Res. Service, Washington, D.C., 1967.
2. **Anderson, J. R.,** *A Geography of Agriculture*, W. C. Brown, Dubuque, IA, 1970.
3. **Anderson, J. R.,** The state of world food production, in *Man, Food, and Nutrition. Strategies and Technological Measures for Alleviating the World Food Problem*, Rechcigl, M., Jr., Ed., CRC Press, Cleveland, 1973, 35 –54.
4. **Black, J. D.,** Food Resources for Expanding World Production, General Electric, Santa Barbara, Calif., 1959, 70 pp.
5. **Blakeslee, L. L., Heady, E. O., and Framingham. C. F.,** *World Food Production, Demand, and Trade*, Iowa State University Press, Ames, 1973, 417 pp.
6. **Brown, L. R.,** Increasing World Food Output, Problems and Prospects, Foreign Agric. Econ. Rep. No. 25, U.S. Dep. Agric., Econ. Res. Service, Washington, D.C., 1965, 140 pp.
7. **Carter, L. J.,** World food supply: Problems and prospects, *Science*, 155, 56 – 58, 1967.
8. **Cesal, L., Blakeslee, L. L., and Heady, E. O.,** The world food situation, *World Rev. Nutr. Diet.*, 7, 1–23, 1967.
9. **Clark, C.,** Future sources of food supply: Economic problems, *J. R. Stat. Soc. Ser. A* (Gen.), 125, 418–444, 1962.
10. **De Vries, E.,** Food production trends in selected countries, *Proc. World Food Forum*, U.S. Dep. Agric. May 15, 1962, U.S. Gov. Print. Off. Washington, D.C., 1963, 79 – 86.
11. Food and Agriculture Organization, Index numbers of food production by countries, *Mon. Bull. Agric. Econ. Stat.*, 11(4), 4, 1962.
12. Food and Agriculture Organization, Third World Food Survey, Freedom from Hunger Campaign Basic Study No. 11, FAO, Rome, 1963.
13. Food and Agriculture Organization, Agricultural Commodities. Projections for 1975 and 1985, FAO, Rome, 1967. (2 vols.)
14. Food and Agriculture Organization, Agricultural Commodity Projections, 1970 – 80, FAO, Rome, 1971.
15. Food and Agriculture Organization, Food Balance Sheets Average 1964 – 66, FAO, Rome, 1971, 766 pp.
16. Food and Agriculture Organization, State of Food and Agriculture 1972, FAO, Rome, 1972, 189 pp.
17. Food and Agriculture Organization, FAO Commodity Review and Outlook 1972 – 1973, FAO, Rome, 1973, 261 pp.
18. **Glen, R.,** The world food situation – today, in *The Food Resources of Mankind*, Agri-World Press, Montreal, 1968, 15-31.
19. **Guidry, N. P.,** *A Graphic Summary of World Agriculture*, USDA, ERS, Misc. Publ. 705, 1964, 64 pp.
20. **Heady, E. O.,** *Economics of Agricultural Production and Resource Use*, Prentice-Hall, Englewood Cliffs, NJ, 1952.
21. **Hutchinson, J.,** *Farming and Food Supply: The Interdependence of Countryside and Town*, Cambridge University Press, New York, 1972, 146 pp.
22. **Johnson, D. G.,** Prospects of food supply in highly developed regions, in *Food. One Tool in International Economic Development*, Iowa State University Press, Ames 1962, 244 – 266.
23. **Johnson, D. G. and Gustafson, R. L.,** *Grain Yields and the American Food Supply*, University of Chicago Press, Chicago, 1962.

24. **Kellogg, C. E.,** Potentials for food production, in Farmer's World. The Yearbook of Agriculture 1964, U.S. Gov. Print. Off. Washington, D.C., 1964, 57 – 69.
25. **Kellogg, C. E.,** World food prospects and potentials: A long-run look, in *Alternatives for Balancing World Food Production and Needs,* Iowa State University Press, Ames, 1967, 98 – 111.
26. **Koffsky, N.,** The food potential of developing nations, *J. Farm Econ.*, 49, 1108 – 1113, 1967.
27. **Lucas, J. W.,** Food production in the year 2000 A.D., *Plant Foods Hum. Nutr.*, 1, 265 – 267, 1969.
28. **Malin, K. M.,** Food resources of the earth, U.N. Population Conference, Belgrade, Sept. 1965.
29. **Mehren, G. L.,** Anticipating the year 2000, *J. Am. Diet. Assoc.*, 52, 467 – 474, 1968.
30. **Moseman, A. H.,** World food prospects and potentials: A short-run look, in *Alternatives for Balancing World Food Production and Needs,* Iowa State University Press, Ames, 1967, 86 – 97.
31. National Research Council, *Prospects of the World Food Supply. A Symposium,* National Academy of Sciences, Washington, D.C., 1966, 84 pp.
32. Organization for Economic Cooperation and Development, Agricultural Projections for 1975 and 1985. Europe, North America, Japan, Oceania: Country Studies, OECD, Paris, 1968.
33. Organization for Economic Cooperation and Development, Agricultural Projections for 1975 and 1985. Europe, North America, Japan, Oceania: Production and Consumption of Major Foodstuffs, OECD, Paris, 1968, 128 pp.
34. **Paarlberg, D.,** Prospects for improving world food supply and nutrition, in *Food, Nutrition and Health,* Rechcigl, M., Jr., Ed., *World Rev. Nutr. Diet.*, Vol. 16, S. Karger, Basel, 1973, pp. 253-285.
35. **Penman, H. L.,** The earth's potential, *Sci. J.*, 4(5), 43-47, 1968.
36. **Pirie, N. W.,** Future sources of food supply: Scientific problems, *J. R. Stat. Soc. Ser. A* (Gen.), 125, 399 – 417, 1962.
37. **Sanders, J. H. and Hoyt, R. C.,** The world food problem: Four recent empirical studies, *Am. J. Agric. Econ.*, 52, 132-135, 1970.
38. **Schmitt, W. R.,** The planetary food-potential, *Ann. N.Y. Acad. Sci.*, 18(17), 645 – 718, 1965.
39. **Schulte, W., Becker, K., Naiken, L., and Chinappi-Favilli, A.,** Food balance sheets and world food supplies, *FAO Nutr. Newsletter,* 11(2), 15 – 25, 1973.
40. **Schultz, T. W.,** Some reflections on agricultural production, output and supply, *J. Farm Econ.*, 38(3), 748 – 762, 1956.
41. **Shaw, B. T.,** Prospective world production and distribution of food, *Fed. Proc.* 20, (1), III, Suppl. 7, 373 – 377, 1961.
42. **Sprague, G. F.,** Agricultural production in the developing countries, *Science,* 157, 774 – 778, 1967.
43. **Sukhatme, P. V.,** The world's food supplies, *J. R. Stat. Soc., Ser. A.* (Gen.), 129, 222 – 241, 1966.
44. Symposium: The part the International Biological Programme will play in increasing world food supplies, *Proc. Nutr. Soc.,* 26, 125 – 140, 1967.
45. **Tauber, A.,** Social and economic aspects of world food production, *Impact Sci. Soc.* (Eng. Ed.), 12 (1), 39 – 59, 1962.
46. U.S. Department of Agriculture, Agricultural Production and Food Consumption in Western Europe, Agric. Monogr. No. 10, U.S. Dep. Agric., Washington, D.C., 1951, 66 pp.
47. U.S. Department of Agriculture, Changes in Agriculture in Developing Nations, 1948 to 1963, FAER 27, U.S.Dep. Agric., Econ. Res. Service, Washington, D.C., 1965.
48. U.S. Department of Agriculture, Food Balances in Foreign Countries, U.S. Dep. Agric., Foreign Agric. Service, Washington, D.C., 1960 – 61.
49. U.S. Department of Agriculture, Food Balances of the Western Hemisphere, 1959 – 61, U.S. Dep. Agric., Econ. Res. Service, ERS Foreign 86, Washington, D.C., 1964.
50. U.S. Department of Agriculture, Food Balances for 16 Countries of Western Europe, 1959 – 61, U.S. Dep. Agric., Econ. Res. Service, ERS Foreign 87, Washington, D.C., 1964.
51. U.S. Department of Agriculture, Food Balances for 12 Countries in the Far East and Oceania, 1959-61, U.S. Dep. Agric., Econ. Res. Service, ERS Foreign 88, Washington, D.C., 1964.
52. U.S. Department of Agriculture, Food Balances for 30 Countries in Africa and West Asia, 1959 – 61, U.S. Dep. Agric., Econ. Res. Service, ERS Foreign 119, Washington, D.C., 1965.
53. U.S. Department of Agriculture, Food Balances for 8 Eastern European Countries, 1959 – 61, U.S. Dep. Agric., Econ. Res. Service, ERS Foreign 124, Washington, D.C., 1965.
54. U.S. Department of Agriculture, The Agricultural Data Book for the Far East and Oceania, U.S.Dep.Agric., ERS Foreign 189, Washington, D.C., 1967.
55. U.S. Department of Agriculture, The 1968 Agricultural Data Book for the Far East and Oceania, U.S.Dep.Agric., ERS Foreign 219, Washington, D.C., 1968.
56. U.S. Department of Agriculture, The World Food Budget 1970, U.S.Dep.Agric., Econ. Res. Service, FAER 19, Washington, D.C., 1964.
57. U.S. Department of Agriculture, The World Food Deficit. A First Approximation, U.S.Dep.Agric. Foreign Agric. Service, Washington, D.C., 1961.
58. **Visher, S. S.,** Comparative agricultural potentials of the world's regions, *Econ. Geogr.*, 31, 82–86, 1955.

59. **von Blanckenburg, P. and Cremer, H. D.,** *Handbuch der Landwirtschaft und Ernährung in den Entwicklungsländern. Band 2. Pflanzliche and tierische Produktion in den Tropen und Subtropen,* Ulmer, Stuttgart, 1971.
60. **West Q. M.,** The world food supply: Progress and promise, in *Food–One Tool in International Economic Development,* Iowa State University Press, Ames, 1962, 97–111.
61. **West, Q. M.,** Outlook for calorie production, *Econ. Bot.,* 22, 8–15, 1968.
62. **Wright, N. C.,** The current food supply situation and present trends, in *Hunger: Can it Be Averted?* Russell, E. J. and Wright, N. C., Eds., British Association for the Advancement of Science, London, 1961, 1–14.

D. Future World Food Needs

1. Agricultural Research Institute, *World Food Needs and Production. Present and Future,* Proc. 14th Annu. Meeting of the Agric. Res. Inst., October 18–19, 1965, National Academy of Sciences, Washington, D.C., 1965.
2. **Allen, G. R.,** The world's food shortage: Nutritional requirements and the demands for food, in *Food – One Tool in International Economic Development,* Iowa State University Press, Ames, 1962, 32–59.
3. **Brown, L. R.,** Man, Land and Food: Looking Ahead at World Food Needs, FAE Rep. No. 11, U.S.Dep.Agric., Econ. Res. Service, Washington, D.C., 1963, 161 pp.
4. **Brown, L. R.,** Population growth, food needs and environmental stress, in *Proc. 3rd Western Hemisphere Nutrition Congress,* Futura Publishing, Mount Kisco, N.Y., 1972, 235–242.
5. **Clark, C. and Turner, J. B.,** World population growth and future food trends, in *Man, Food, and Nutrition. Strategies and Technological Measures for Alleviating the World Food Problem,* Rechcigl, M., Jr., Ed., CRC Press, Cleveland, 1973, 55–77. (55 ref).
6. **Cochrane, W. W.,** The world food budget: A forward look to 2000 and beyond, Proc. World Food Forum, U.S.Dep.Agric., U.S. Gov. Print. Off., Washington, D.C., 1963, 86–95.
7. Food and Fiber for the Future, Report of the National Advisory Commission on Food and Fiber, U.S. Gov. Print. Off., Washington, D.C., 1967, 361 pp.
8. **Hutchinson, J., Ed.,** *Population and Food Supply. Essays on Human Needs and Agricultural Prospects,* Cambridge University Press, Cambridge, Mass., 1969, 144 pp.
9. **McMeekan, C. P.,** The world situation tomorrow, in *The Food Resources of Mankind,* Agri-World Press, Montreal, 1968, 109–121.
10. **Phipard, E. and Shepherd, G. S.,** Nutritional needs by world regions, in *Food – One Tool in International Economic Development,* Iowa State University Press, Ames, 1962, 60–71.
11. President's Science Advisory Committee, Population and nutritional demands, in The World Food Problem, Vol. 2, Report of the Panel on The World Supply, U.S. Gov. Print. Off., Washington, D.C., 1967, 1–135.
12. **Schaefer, A. E.,** Nutritional aspects of the world food needs, in The Land-Grant University and World Food Needs, University of Illinois College of Agriculture, Special Publ. No. 13, Urbana, 1968, 13–26.
13. **Sukhatme, P. V.,** The world's hunger and future needs in food supplies, *J. R. Stat. Soc. Ser. A* (Gen.), 124, 463–525, 1961.
14. Symposium on World Food Needs and Food Resources, Proc. Fifth International Congress Nutrition, Washington, D.C., Sept. 1–7, 1960, *Fed. Proc.,* 20(1), III, Suppl. 7, 365–406, 1961.
15. Symposium. Food requirements as related to agricultural production in the future, in *Proc. 7th International Congress Nutrition,* Vol. 4, Pergamon Press, Oxford, 1967, 857–935. (13 papers)
16. **West, Q. M.,** World Food Needs, USDA, ERS, Washington, D.C., Feb. 1966.
17. **Yudkin, J.,** The nutritional requirement, *Sci. J.,* 4(5), 48–52, 1968.

III. THE WORLD FOOD PROBLEM: POSSIBLE SOLUTIONS

A. The Role of Agriculture

1. **Adolph, W. H.**, Agricultural science and improved human nutrition, *Borden's Rev. Nutr. Res.*, 18, 15–27, 1957.
2. Symposium. Agriculture's contribution to the solution of world food supply, *Fed. Proc.*, 22, 134–151, 1963.
3. **Arnon, I.**, *Organization and Administration of Agricultural Research*, Elsevier, Amsterdam, 1968, 342 pp.
4. **Barndt, K.**, *The Reconstruction of World Agriculture*, Norton, New York, 1945, 416 pp.
5. **Bentley, C. F.**, Food for all: Can agriculture provide? in *The Food Resources of Mankind*, Agri-World Press, Montreal, 1968, 133–163.
6. **Borton, R. E., Ed.**, Selected Readings To Accompany "Getting Agriculture Moving," Agricultural Development Council, New York, 1966. (2 vols.)
7. **Breireim, K.**, The world's agriculture and food supplies, *Tidsskr. Landoknomi*, 4, 188–216, 1961.
8. **Brown, L. R.**, Increasing World Food Output. Problems and Prospects, U.S. Gov. Print. Off., Washington, D.C., 1965, 140 pp.
9. **Brown, L. R.**, The world outlook for conventional agriculture, *Science*, 158, 604–611, 1967.
10. **Brown, L. R.**, New directions seen in world agriculture, *Foreign Agric.*, 6(13), 8–10, 1968.
11. **Brown, L. R.**, Breakthrough against hunger, in *Science Year*, Field Educational Enterprises, Chicago, 1969, 115–125.
12. **Brown, L. R.**, *Seeds of Change: The Green Revolution and Development in the 1970's*, Praeger, New York, 1970.
13. **Bunting, A. H. and Harrison, A.**, Improving traditional agriculture, *Sci. J.*, 4(5), 61–65, 1968.
14. **Chang, C. W.**, Increased Food Production through Education, Research and Extension, Freedom from Hunger Campaign Basic Study No. 9, FAO, Rome, 1962, 78 pp.
15. **Dalrymple, D. G.**, Development and Spread of High-Yielding Varieties of Wheat and Rice in the Less Developed Nations, FAER Agric. Econ. Rep. 95, U.S. Dep. Agric., Washington, D.C., 1974.
16. **Dalrymple, D. G.**, The Green Revolution: Past and Prospects, Economic Research Service, U.S. Dep. Agric., Agency Int. Devel. Washington, D.C., July 1974.
17. **Eicher, C. K. and Witt, L. W.**, *Agriculture in Economic Development*, McGraw-Hill, New York, 1964.
18. **Ewell, P. R.**, *Exploring Agribusiness*, Interstate, Danville, Ill., 1967.
19. **Ewell, R.**, Agriculture's crucial role in the next decade, *Agric. Chem.*, 20(July), 32, 35, 38, 86, 1965.
20. Food and Agriculture Organization, Possibilities for Increasing World Food Production, FFHC Basic Stud. No. 11, FAO, Rome, 1963.
21. Food and Agriculture Organization, A Strategy for Plenty: The Indicative World Plan for Agricultural Development, FAO, Rome, 1970.
22. Food and Agriculture Organization, Toward a Strategy for Agricultural Development, FFHC Basic Stud. No. 21, FAO, Rome 1969, 66 pp.
23. Food and Agriculture Organization, Agricultural Development. A Review of FAO's Field Activities, FFHC Basic Stud. No. 23, FAO, Rome 1970, 194 pp.
24. **Freeman, O. L.**, Agriculture in the year 2000 AD., *Vital Speeches*, 33, 290–294, 1967.
25. **Freeman, O. L.**, *World Without Hunger*, Praeger, New York, 1968, 190 pp.
26. **Hardin, C. M., Ed.**, *Overcoming World Hunger*, Prentice-Hall, Englewood Cliffs, N.J., 1969, 177 pp. (4 papers)
27. **Hayami, Y. and Ruttan, V. W.**, *Agricultural Development: An International Perspective*, Johns Hopkins Press, Baltimore, 1971, 367 pp.
28. **Heady, E. O.**, *Agricultural Policy Under Economic Development*, Iowa State University Press, Ames, 1962, 682 pp.
29. **Heady, E. O. and Mayer, L. V.**, Food needs and U.S. agriculture in 1980, National Advisory Committee on Food and Fiber Tech. Paper 1, 1967.
30. **Hendricks, S. B.**, Food from the land, in *Resources and Man. A Study and Recommendations by the Committee on Resources and Man*, W. H. Freeman Publishing, San Francisco, 1969, 65–85.
31. **Johnston, B. F. and Mellor, J. W.**, The role of agriculture in economic development, *Am. Econ. Rev.*, 51, 566–593, 1961.
32. **King, C. G.**, Agriculture and health, *Food Technol.*, 18, 1162–1163, 1964.
33. **Kuznets, S. S.**, Economic growth and the contribution of agriculture: Notes on measurement, *Int. J. Agric. Aff.*, 3(2), 56–75, 1961.
34. The Land-Grant University and World Food Needs, Proc. Symposium, University of Illinois College of Agriculture, Special Publ. No. 13, Urbana, 1968, 162 pp.
35. **Lucas, J. W. and Wokes, F.**, Agricultural sciences and the world food problem, *Plant Foods Hum. Nutr.*, 1, 93–97, 1969.
36. **Mellor, J. W.**, The process of agricultural development in low income countries, *J. Farm Econ.*, 44, 700–716, 1962.
37. **Millikan, M. F. and Hapgood, N. D.**, *No Easy Harvest. The Dilemma of Agriculture in Underdeveloped Countries*, Little, Brown & Co., Boston, 1967, 178 pp.
38. **Moseman, A. H., Ed.**, Agricultural Sciences for the Developing Countries, AAAS, Washington, D.C., 1964, 221 pp.
39. **Mosher, A. T.**, *Getting Agriculture Moving*, Praeger, New York, 1966.

40. **Myrdal, G.,** United Nations agriculture and the world economic revolution, *J. Farm Econ.*, 47, 889–899, 1966; 48, 471–472, 1966.
41. **Nicholls, W. H.,** The importance of an agriculture surplus in underdeveloping countries, J. S. McLean Memorial Lecture, Nashville, Tenn., 1962, 55 pp.
42. **Palmer, I.,** Food and New Agricultural Technology, U.N. Research Institute for Social Development, Geneva, 1972, 85 pp.
43. **Palmer, I.,** Science and Agriculture, U.N., Research Institute for Social Development, Geneva, 1972, 100 pp.
44. **Pawley, W. H.,** *Possibilities of Increasing World Food Production*, FAO, Rome, 1967.
45. **Phillips, J.,** *The Development of Agriculture and Forestry in the Tropics*, Praeger, New York, 1962, 205 pp.
46. **Pirie, N. W.,** Implementing the possibilities, *Sci. J.*, 4(May), 102–106, 1966.
47. **Schertz, L. P.,** The success of agriculture in meeting world food needs, *Ecol. Food Nutr.*, 1, 207–212, 1972.
48. **Schultz, T. W.,** *Economic Growth and Agriculture*, McGraw-Hill, New York, 1968.
49. **Schultz, T. W.,** *The Economic Organization of Agriculture*, McGraw-Hill, New York, 1953, 374 pp.
50. **Schultz, T. W.,** *Transforming Traditional Agriculture*, Yale University Press, New Haven, Conn., 1964.
51. **Stakman, E. C., Bradfield, R., and Mangelsdorf, P. C.,** *Campaigns Against Hunger*, Harvard University Press, Cambridge, Mass., 1967.
52. Symposium. Agricultural problems in developing countries, *Proc. Nutr. Soc.*, 21, 84–120, 1962. (5 papers)
53. Symposium. The part the International Biological Programme will play in increasing world food supplies, *Proc. Nutr. Soc.*, 26, 125–140, 1967. (4 papers)
54. Symposium on agricultural development in relation to national nutritional needs, *Proc. 1st Asian Congress of Nutrition*, Tulpule, P. G. and Jaya Rao, K. S., Eds., Nutrition Society of India, Hyderabad, 1972, 17–99. (5 papers)
55. **Thompson, L. M.,** Impact of world food needs on American agriculture, *J. Soil Water Conserv.*, 23, 3–9, 1968.
56. **Thorbecke, E., Ed.,** *The Role of Agriculture in Economic Development*, National Bureau of Economic Research, New York, 1970, 480 pp.
57. **Thurston, H. D.,** Tropical agriculture. A key to the world food crisis, *BioScience*, 19, 29–34, 1969.
58. U.S. Department of Agriculture, Power to Produce. The Yearbook of Agriculture 1960, U.S. Gov. Print. Off., Washington, D.C., 1960.
59. U.S. Department of Agriculture, After a Hundred Years. The Yearbook of Agriculture 1962, U.S. Gov. Print. Off., Washington, D.C., 1962.
60. U.S. Department of Agriculture, Farmers World. The Yearbook of Agriculture 1964, U.S. Gov. Print. Off., Washington, D.C., 1964.
61. U.S. Department of Agriculture, Changes in Agriculture in 26 Developing Nations: 1948 to 1963, FAER No. 27, USDA, ERS, 1965, 134 pp.
62. U.S. Department of Agriculture, Science for Better Living. The Yearbook of Agriculture 1968, U.S. Gov. Print. Off., Washington, D.C., 1968.
63. **Webster, C. C. and Wilson, P. N.,** *Agriculture in the Tropics*, Longmans, London, 1969, 488 pp.
64. **Wharton, C. R., Ed.,** *Subsistence Agriculture and Economic Development*, Aldine Publishing, Chicago, 1969, 481 pp.
65. **Wharton, C. R., Jr.,** The Green Revolution – Cornucopia or Pandora's box, *Foreign Affairs*, 47, 464–476, 1969.
66. **Whitney, R. S., Ed.,** Challenge to Agronomy for the Future, Special Publ. 10, American Society of Agronomy, Madison, Wis., 1967.

B. The Role of Food Science and Technology

1. **Altschul, A.,** Combating malnutrition: New strategies through food science, The 2nd Joint Meeting of the American Institute of Chemical Engineers and Institutio de Ingenieros Quimicos de Puerto Rico, Tampa, Florida, May 21, 1968.
2. **Aylward, F.,** Food conservation and utilization in developing countries, in *Progress in Human Nutrition*, Vol. 1, Margen, S., Ed., Publishing, Westport, Conn., 1971, 53-67.
3. **Aylward, F.,** Synthesis, preparation and preservation of food, in Chemistry and Industry in the 1990's, Society of Chemical Industry, London, 1970.
4. **Berg, A.,** Industry's struggle with world malnutrition, *Harvard Business Review*, January-February 1972, 130–141.
5. **Hand, D. B.,** Food technology and the world food problem, *Food Technol.*, 20, 1031–1034, 1966.
6. **Heinz, H. J., II,** A "five-point program" for industry's role in improving global nutrition, *Fed. Proc.*, 20, (1), III, Suppl. 7, 411–414, 1961.
7. **Jackson, J. M.** A challenge to food technologists – Food for developing countries, *Food Technol.*, 16 (7) 14–18, 1962.
8. **Kosikowski, F.,** Establishing principles and training for international food development, *J. Milk Food Technol.*, 29, 305–308, 1966.
9. **Kosikowski, F.,** College course in international food development, *J. Dairy Sci.*, 51, 242–245, 1968.
10. **Leniger, H. A.,** Role of food science in the development of developing countries, in Proc. SOS/70, 3rd International Congress Food Sci. Technol., Institute of Food Technologists, Chicago, Ill., 1971, 505–517.

11. **Mautner, M.**, Food industry in developing countries, in Proc. SOS/70, 3rd International Congress Food Sci. Technol., Institute of Food Technologists, Chicago, 1971, 538–543.
12. **Milner, M.**, Food technology and the world food needs, *Food Technol.*, 17, 846–849, 1963.
13. **Mrak, E. M.**, Food, in "Contributions of the Biological Sciences to Human Welfare," *Fed. Proc.*, 31, (6), II, TF81–TF90, 1972.
14. **Mrak, E. M.**, Some of the developments in food production and their impact on nutrition, *J. Nutr.*, 91, 55–61, 1967.
15. National Research Council, *Food Science in Developing Countries: A Selection of Unsolved Problems*, National Academy of Sciences, Washington, D.C., 1974, 79 pp.
16. **Peterson, M. S. and Tressler, D. K.**, *Food Technology World Over*, Avi Publishing, Westport, Conn., 1963. (2 Vol.)
17. **Rafson, H. F.**, The IFT in international technical assistance, *Food Technol.*, 17, 972–976, 1963.
18. **Scrimshaw, N. S.**, Worldwide opportunities for food scientists and technologists, *Food Technol.*, 17, 850–853, 1963.
19. **Stewart, G. F.**, Stimulating international developments in foods, *Food Technol.*, 22, 856–861, 1968.
20. **Subrahmanyan, V.**, The place of food technology in combating hunger and malnutrition, *Food Technol.*, 16, 23–27, 1962.
21. **Subrahmanyan, V.**, Food technology in developing countries, *Food Technol.*, 20, 1570–1574, 1966.
22. The place of food science and technology in the campaign against malnutrition, *Proc. Nutr. Soc.*, 20, 91–137, 1961. (12 papers)

C. The Role of Nutrition

1. **Bengoa, J. M.**, The prevention of malnutrition in young children, *Proc. 1st Western Hemisphere Nutrition Congress*, Am. Med. Assoc., Chicago, Ill., 1966, 36–43.
2. **Bengoa, J. M.**, Nutrition rehabilitation programmes, *J. Trop. Pediatr.*, 10, 63–64, 1964.
3. **Blix, G., Hofvander, Y., and Vahlquist, B., Eds.**, *Famine. A Symposium Dealing with Nutrition and Relief Operations in Times of Disaster*, 9th Symposium of the Swedish Nutrition Foundation, Almqvist & Wiksells, Uppsala, 1971, 200 pp.
4. **Elder, A. L.**, The application of nutritional research to practical problems of nutrition, in *Proc. 6th International Congress of Nutrition*, Livingstone, Edinburgh, 1964, 460–464.
5. **Goldsmith, G. A.**, Malnutrition and hunger – Whose responsibility is it? *Proc. 3rd Western Hemisphere Nutrition Congress*, Futura Publishing, Mount Kisco, 1972, 4–9.
6. **Gordon, J. E. and Scrimshaw, N. S.**, Field evaluation of nutrition intervention programs, *World Rev. Nutr. Diet.*, 17, 1–38, 1973.
7. Improvement of the nutritional status of infants and children (8 papers), *Proc. 3rd Western Hemisphere Nutrition Congress*, Futura Publishing, Mount Kisco, 1972, 83–112.
8. **Jelliffe, D. B.**, The assessment of the Nutritional Status of the Community, WHO, Geneva, 1966.
9. **Martin, E. A.**, *Nutrition in Action*, 2nd ed., Holt, Rinehart & Winston, New York, 1965, 298 pp.
10. **Masefield, G. B.**, *Famine: Its Prevention and Relief*, Oxford University Press, London, 1963.
11. **Masefield, G. B.**, Food and Nutrition Procedures in Times of Disaster, FAO Nutr. Stud. 21, FAO, Rome, 1967.
12. National Research Council, *Conference on Prevention of Malnutrition of the Preschool Child*, National Academy of Sciences, Washington, D.C., 1964.
13. **Ritchie, J. A. S.**, Learning Better Nutrition, FAO, Rome, 1967.
14. **Stare, F. J.**, Nutritional improvement and world health potential, *J. Am. Diet. Assoc.*, 57, 107–110, 1970.
15. **Trémolières, J.**, Nutrition and underdevelopment, in *Advances in Human Nutrition*, Vol. 1, Margen, S., Ed., Avi Publishing, Westport, Conn., 1971, 1-28.
16. **Trémolières, J.**, Nutrition and public health, *World Rev. Nutr. Diet.*, 18, 275–319, 1973. (32 references)
17. World Health Organization, Prevention and Treatment of Severe Malnutrition in Times of Disaster, WHO Tech. Rep. Ser. 45, WHO, Geneva, 1951.

IV. NATURAL RESOURCES AND MANAGEMENT

A. Bibliographies

1. **Ahmad, J.,** *Natural Resources in Low Income Countries: An Analytical Survey of Socio-Economic Research,* University of Pittsburgh Press, Pittsburgh, 1960, 121 pp.
2. **Anglemyer, M.,** Natural Resources: A Selection of Bibliographies, 2nd ed., U.S. Engineer Agency for Resources Inventories, Washington, D.C., 1970, 145 pp.
3. **Carvajal, J. and Munzer, M. E.,** *Conservation Education. A Selected Bibliography,* Interstate, Danville, Ill., 1968, 98 pp.
4. **Durrenberger, R. W.,** *Environment and Man. A Bibliography,* National Press Books, Palo Alto, Calif., 1970, 118 pp.
5. **Hoadley, I. B.,** *Natural Resources Bibliography,* Ohio State Universities Libraries, Office of Educational Services, Columbus, 1970, 242 pp.
6. **McDonald, R.,** *Guide to Literature on Environmental Sciences,* Washington, D.C., American Society for Engineering Education, 1970, 10 pp.
7. **Outub, M. Y.,** *Environmental Science Bibliography,* Mid-Continent Scientific, Chicago, 1971, 103 pp.
8. **Paulsen, D. F.,** *Natural Resources in the Governmental Process. A Bibliography Selected and Annotated,* University of Arizona Press, Tucson, 1970, 99 pp.
9. U.S. Department of Agriculture, Abstracts of Recent Published Material on Soil and Water Conservation, U.S.Dep. Agric., Washington, D.C., 1949.
10. **Watkins, J. B.,** *Ecology and Environmental Quality. A Selected Bibliography and Annotated Bibliography for Biologists and Earth Scientists,* Syracuse University Libraries, Syracuse, N.Y., 1971, 127 pp.

B. General Surveys and Treatises

1. **Addison, H.,** *Land, Water and Food,* Chapman & Hall, London, 1961.
2. **Ahmad, J.,** *Natural Resources in Low Income Countries: An Analytical Survey of Socio-Economic Research,* University of Pittsburgh Press, Pittsburgh, Pa., 1960, 118 pp.
3. American Society of Agronomy, Our Heritage of Land and Water Resources, Am. Soc. Agron. Special Pub. 7, Madison, Wis., 1966, 44 pp.
4. **Amoss, H. L. and McNickle, R. K., Eds.,** *Land and Water: Planning for Economic Growth,* University of Colorado Press, Boulder, 1961, 219 pp.
5. **Barnett, H. J. and Morse, C.,** *Scarcity and Growth. The Economics of Natural Resource Availability,* Johns Hopkins Press, Baltimore, 1963, 320 pp.
6. **Ciriacy-Uantrup, S. V. and Parsons, J. J., Eds.,** *Natural Resources: Quality and Quantity,* University of California Press, Berkeley, 1967, 217 pp. (14 papers)
7. **Clawson, M., Ed.,** *Natural Resources and International Development,* Johns Hopkins Press, Baltimore, 1964, 475 pp.
8. **Cloud, P.,** Our Disappearing Earth Resources, in *Science Year,* Field Enterprises, Chicago, 1969, 166–181.
9. **Christensen, R. P.,** Efficient Use of Food Resources in the United States, U.S.Dep. Agric. Tech. Bull, 963, 1948, 98 pp.
10. **Darling, F. F. and Milton, J. P.,** *Future Environments of North America,* Natural History Press, Garden City, N.Y., 1966, 767 pp.
11. **Deju, R. A., Baez, A. P., Bhappu, R., and Evans, G. J.,** *The Environment and Its Resources,* Gordon & Breach, New York, 1972, 340 pp.
12. Economic Research Service, Major Uses of Land and Water in the United States with Special Reference to Agriculture, Summary for 1964, Agric. Econ. Rep. 149, U.S. Gov. Print. Off., Washington, D.C., 1968.
13. **Ehrlich, P. R. and Ehrlich, A. H.,** *Population, Resources, Environment. Issues in Human Ecology,* 2nd ed., W. H. Freeman Publishing, San Francisco, 1972, 509 pp.
14. **Firey, W. I.,** *Man, Mind, and Land. A Theory of Resources Use,* Free Press, Chicago, 1960, 256 pp.
15. **Fisher, J. L. and Potter, N.,** World Prospects for Natural Resources, Resources for the Future, Washington, D.C., 1964, 73 pp.
16. **Frank, B.,** *Water, Land and People,* Knopf, New York, 1950.
17. **Freeman, S. D.,** *Energy, The New Era,* Walker, New York, 1974, 386 pp.
18. **Hare, F. K.,** Land, water and climate as parameters of food production, in *The Food Resources of Mankind,* Agri-World Press, Montreal, 1968, 33–51.
19. **Hatt, P. K., Ed.,** *World Population and Future Resources,* Proc. 2nd Centennial Academic Conference of Northwestern University, Evanston, Ill., 1951, American Book Co., New York, 1952, 262 pp.
20. Indian Agricultural Research Institute, Recent Research on the Improvement of Protein and Nutritive Properties of Food and Feed Plants, New Delhi, 1971, 156 pp.
21. **Jarrett, H., Ed.,** *Comparisons in Resource Management,* Johns Hopkins Press, Baltimore, 1961. 271 pp.
22. **Lansberg, H. H.,** *Natural Resources for U.S. Growth. A Look Ahead to the Year 2000,* Johns Hopkins Press, Baltimore, 1964, 261 pp.

23. **Landsberg, H. H., Fischman, L. L., and Fischer, J. L.,** *Resources in America's Future: Patterns of Requirements and Availabilities, 1960–2000,* Johns Hopkins Press, Baltimore, 1963, 1040 pp.
24. **Mouzon, O. T.,** *International Resources and National Policy,* Harper & Row, New York, 1959, 752 pp.
25. **Mudd, S., Ed.,** *The Population Crisis and the Use of World Resources,* Junk, The Hague, 1964, 562 pp.
26. **Murdoch, W. W., Ed.,** *Environment, Resources, Pollution and Society,* Sinauer Associates, Stamford, Conn., 1971, 440 pp.
27. National Research Council, *Resources and Man. A Study and Recommendations by the Committee on Resources and Man,* W. H. Freeman Publishing, San Francisco, 1969, 259 pp.
28. **Over, J. A., Ed.,** *Energy Conservation. Ways and Means,* Future Shape of Technology Foundation, The Hague, 1974, 182 pp.
29. **Owen, O. S.,** *Natural Resource Conservation. An Ecological Approach,* Macmillan, New York, 1971. 593 pp.
30. **Perloff, H. S., Dunn, E. S., Lampard, E. E., and Muth, R. F.,** *Regions, Resources and Economic Growth,* Johns Hopkins Press, Baltimore, 1960, 760 pp.
31. Political and Economic Planning, *World Population and Resources, A Report,* Allen & Unwin, London, 1955, 339 pp.
32. **Pollak, F. S., Ed.,** *Resources Development: Frontiers for Research,* University of Colorado Press, Boulder, 1960, 333 pp.
33. President's Science Advisory Committee, Water and land, in the World Food Problem. Report of the Panel on the World Food Supply, Vol. 2, U.S. Gov. Print. Off., Washington, D.C., 1967, 405-469.
34. President's Materials Policy Commission, Resources for Freedom. A Report to the President, U.S. Gov. Print. Off., Washington, D. C., 1952. (5 vols.)
35. **Sargent, F., II,** Natural resources, in Contributions of the biological sciences to human welfare, *Fed. Proc.,* 31, TF131–TF139, 1972.
36. **Scarseth, G. D.,** *Man and His Earth,* Iowa State University Press, Ames, 1962.
37. **Skinner, B. J.,** *Earth Resources,* Prentice-Hall, Englewood Cliffs, N. J., 1969, 150 pp.
38. **Smith, G. H.,** *Conservation of Natural Resources,* John Wiley & Sons, New York, 1965.
39. **Spengler, J. J., Ed.,** *Natural Resources and Economic Growth,* Johns Hopkins Press, Baltimore, 1961, 306 pp.
40. **Strahler, A. N. and Strahler, A. H.,** *Introduction to Environmental Science,* Hamilton Publishing, Santa Barbara, Calif., 1973, 633 pp.
41. **Stamp. L. D.,** Land use and food production, in *Hunger: Can it be Averted?* Russell, E. J. and Wright, N. C. Eds., British Association for the Advancement of Science, London, 1961, 34–39.
42. **Steinhart, C. and Steinhart, J.,** *Energy. Sources, Use, and Role in Human Affairs,* Duxbury Press, North Scituate, Mass., 1974, 362 pp.
43. **Thorne, W., Ed.,** *Land and Water Use,* Publ. No. 73, AAAS, Washington, D.C., 1963.
44. United Nations, A Review of the Natural Resources of the African Continent, Natural Research No. 1, UNESCO, 1963, 437 pp.
45. United Nations, *Atlas of Physical, Economic and Social Resources of the Lower Mekong Basin,* United Nations, 1968, 257 pp.
46. United Nations, Natural Resources, in Science and Technology for Development, Vol. 2, Report of the U.N. Conference on the Application of Science and Technology for the Benefit of the Less Developed Areas. United Nations, 1963, 243 pp.
47. United Nations, Natural Resources of Developing Countries: Investigation, Development and Rational Utilization, Report of the Advisory Committee on the Application of Science and Technology to Development, United Nations, 1970, 174 pp.
48. U. S. Department of Agriculture, Land and Water Policy Committee, Land and Water Resources. A Policy Guide, U.S. Gov. Print. Off., Washington, D.C., 1962, 73 pp.
49. **Van Royen, W.,** *Atlas of the World's Resources,* Prentice-Hall, Englewood Cliffs, N.J., 1954.
50. **Wadleigh, C. H. and Klingebiel, A. A.,** An inventory of land and water resources, Am. Soc. Agron. Special Publ. No. 7, 1966, 1–23.
51. **Watt, K. E. F.,** *Ecology and Resource Management,* McGraw-Hill, New York, 1968.
52. **Wengert, N. I.,** The Administration of Natural Resources, Asia Publishing, New York, 1962, 71 pp.

C. Food Resources

1. **Addison, H.,** *Land, Water and Food,* Chapman & Hall, London, 1961.
2. **Black, J. D.,** Food Resources for an Expanding World Production, General Electric, Santa Barbara, Calif., 1959, 70 pp.
3. **Borgstrom, G.,** *World Food Resources,* Intext Educational Publishers, New York, 1973, 237 pp.
4. **Christensen, R. P.,** Efficient Use of Food Resources in the United States, U.S. Dep. Agric Tech. Bull. 963, 1948, 98 pp.
5. Food for Tomorrow. Resources – Environment – Distribution, Proc. 17th Annual Meeting of the Agricultural Research Institute, Washington, D.C., 1968, 198 pp.

6. **Hendricks, S. B.,** Food from land, in *Resources and Man,* National Research Council Committee of Resources and Man, W. H. Freeman Publishing, San Francisco, 1969, 65–85.
7. **Malin, K. M.,** Food resources of the earth, U.N. Population Conference, Belgrade, Sept. 1965.
8. **Mangelsdorf, P. C.,** Genetic potentials for increasing yields of food crops and animals, *Proc. Natl. Acad. Sci.,* 56, 370–375, 1966.
9. **Penman, H. L.** The earth's potential, *Sci. J.,* 4 (5), 43–47, 1968.
10. **Steppler, H. A., Ed.,** *The Food Resources of Mankind,* Agri-World Press Montreal, 1968, 163 pp. (10 papers with discussion)
11. **Tanner, J. W. and Jones, G. E.,** Production physiology, in Genes to Genus – A Symposium, International Mineral and Chemical Corp., Skokie, Ill., Jan. 21–22, 1965.
12. **Underwood, E. J.,** Man, land and food, *Aust. J. Sci.,* 29 (11), 395–401, 1967.
13. **Wittwer, S. H.,** Approaching maximum capacity in production: Biological limits of productivity, *Proc. 16th Annu. Meeting of the Agricultural Research Institute,* Oct. 16–17, 1967, National Academy of Sciences, Washington, D.C., 1967, 97–124.

D. Biological Resources

1. **Black, J. D.,** *The Management and Conservation of Biological Resources,* Davis, Philadelphia, 1968, 339 pp.
2. **Clements, F. E. and Shelford, V. E.,** *Bio-Ecology,* John Wiley & Sons, New York, 1939, 425 pp.
3. **Ehrenfeld, D. W.,** *Biological Conservation,* Holt, Rinehart & Winston, New York, 1970, 226 pp.
4. **Huxley, J.,** The Conservation of Wild Life and Natural Habitats in Central and East Africa, Report on a Mission Accomplished by UNESCO, July–September 1960, UNESCO, 1961, 113 pp.
5. National Research Council, *Land Use and Wildlife Resources,* National Academy of Sciences, Washington, D.C., 1970, 262 pp.
6. **Riney, T.,** *Conservation and Management of African Wildlife,* African Special Project Stage III Reports, FAO, Rome, 1967, 35 pp; Riney, T. and Hill, P., English-Speaking Country Reports, 145 pp; Riney, T. and Hill, P., Rapports sur les pays francophones, 135 pp.
7. **Riney, T.,** Wildlife conservation and management, *Unasylva,* 24 (4) 15–17, 1970.
8. United Nations, Wildlife and Fish Resources, in Proc. U.N. Scientific Conference on the Conservation and Utilization of Resources, Vol. 8, August 17–September 6, 1949, Lake Success, N.Y., United Nations, 1951, 259 pp.
9. United Nations, Report of the International Technical Conference on the Conservation of the Living Resources of the Sea, April 18–May 10, 1955, United Nations, Rome, 1955, 17 pp.
10. United Nations, Papers at the International Technical Conference on the Conservation of the Living Resources of the Sea, Rome, April 18–May 10, 1955, 371 pp.

E. Energy Resources

1. General

1. **Cambel, A. B.,** Energy for a restless world, in *Science and Technology in the World of the Future,* Bronwell, A. B., Ed., Interscience, New York, 1970, 165–183.
2. **Cheney, E. S.,** U.S. energy resources: Limits and future outlook, *Am. Sci.,* 62, 14–22, 1974.
3. **Cook, C. S.,** Energy: Planning for the future, *Am. Sci.,* 61, 61–65, 1973.
4. **Häfele, W.,** A systems approach to energy, *Am. Sci.,* 62, 438–447, 1974.
5. **Hammond, A. L., Metz, W. D., and Maugh, T. H., II,** *Energy for the Future,* American Association for the Advancement of Science, Washington, D.C., 1973, 184 pp.
6. **Hottel, H. C. and Howard, J. B.,** *New Energy Technology. Some Facts and Assessment,* M.I.T. Press, Boston, Mass., 1971, 364 pp.
7. **Hubbert, M. K.,** *Energy Resources,* National Research Council Publication 1000–D, National Academy of Sciences, Washington, D.C., 1962.
8. **Hubbert, M. K.,** Energy resources for power production, in *Resources and Man,* National Research Council Committee on Resources and Man, W. H. Freeman Publishing, San Francisco, 1969, 157–242.
9. **Hubbert, M. K.,** Energy resources, in *Environment, Resources, Pollution and Society,* Murdock, W. W., Ed., Snauer, Stamford, Conn., 1971, 89–116.
10. **Landsberg, H. M. and Schurr, S.,** *Energy in the United States,* Random House, New York, 1968.
11. **Lincoln, G. A.,** Energy conservation, *Science,* 180, 155–162, 1973.
12. **Odeum, H. F.,** Energetics of world food production, in The World Food Problem, Vol. 3, President's Science Advisory Committee, Report of the Panel on the World Food Supply, U.S. Gov. Print. Off., Washington, D.C., 1967, 55–94.
13. **Pimental, D., Hurd, L. L., Bellotti, A. C., Forster, M. J., Oka, I. N., Sholes, O. D., and Whitmare, R. J.,** Food production and the energy crisis, *Science,* 182, 443–449, 1973.
14. **Roberts, R.,** Energy resources and conversion techniques, *Am. Sci.,* 61, 66–75, 1973.
15. **Rocks, L. and Runyon, R. P.,** *The Energy Crisis,* Crown, New York, 1972, 190 pp.
16. **Schurr, S. H., Ed.,** *Energy, Economic Growth and the Environment,* Johns Hopkins Press, Baltimore, 1972, 232 pp.

17. **Seale, R. L. and Sierka, R. A., Eds.,** *Energy Needs and the Environment*, University of Arizona Press, Tucson, 1973.
18. **Steinhart, J. S. and Steinhart, C. E.,** Energy use in the U.S. food system, *Science*, 184, 307–316, 1974.
19. **Thomas, T. M.,** World energy resources: Survey and review, *Geogr. Rev.*, 63, 246–258, 1973.
20. United Nations, Fuel and Energy Resources, in Proc. U.N. Scientific Conference on the Conservation and Utilization of Resources, Vol. 3, August 17–September 6, 1949, Lake Success, N.Y., United Nations, 333 pp.
21. U.S. Department of Agriculture, Power to Produce. The Yearbook of Agriculture 1960, U.S. Gov. Print. Off. Washington, D.C., 1960, 480 pp.
22. **Weinberg, A. M.,** Some views of the energy crisis, *Am. Sci.*, 61, 59–60, 1973.
23. **Weinberg, A. M. and Hammond, R. P.,** Limits to the use of energy, *Am. Sci.*, 58, 412–418, 1970.

2. Coal and Lignite

1. United Nations, Coal and Iron Ore Resources of Asia and the Far East, ECAFE Mineral Resources Devel. Ser. No. 7, United Nations, 1952, 155 pp.
2. United Nations, Lignite Resources of Asia and the Far East. Their Exploration, Exploitations and Utilization, ECAFE Mineral Resources Devel. Ser. No. 7, United Nations, 1957, 134 pp.

3. Nuclear Energy

1. **Foreman, H., Ed.,** *Nuclear Power and the Public Opinion*, University of Minnesota, Minneapolis, 1970, 273 pp.
2. **Glasstone, S.,** *Controlled Nuclear Fusion*, U.S. AEC, Oak Ridge, Tenn., 1964.
3. **Glasstone, S.,** *Sourcebook of Atomic Energy*, 3rd ed., Van Nostrand, Princeton, N.J., 1967.
4. **Gough, W. C. and Eastlund, B. J.,** The prospects of fusion power, *Sci. Am.*, 224(2), 50–64, 1971.
5. **Hammond, A.,** Breeder reactors: Power for the future, *Science*, 174, 807–810, 1971.
6. International Atomic Energy Agency, Processing of Low Grade Uranium Ores. Proc. of a Panel, Vienna, June 27–July 1, 1966 AEA, 1967, 247 pp.
7. International Atomic Energy Agency, The Recovery of Uranium. Proc. of a Symposium, São Paulo, August 17–21, 1970, IAEA, 1971, 478 pp.
8. International Atomic Energy Agency, Peaceful Uses of Atomic Energy in Africa, Proc. of a Symposium, Kinshasa, July 28–August 1, 1969, IAEA, 1970, 574 pp.
9. International Atomic Energy Agency, Uranium Exploration Geology, Proc. of a Panel, Vienna, April 13–17, 1970, IAEA, 1970, 384 pp.
10. **Lang, A. H.,** Surveying and Evaluating Radioactive Deposits, Rev. Ser. 1, IAEA, 1959.
11. **Metschert, B. C. and Schurr, S. H.,** *Atomic Energy. Applications with Reference to Underdeveloped Countries*, Johns Hopkins Press, Baltimore, 1957, 129 pp.
12. Organization for Economic Cooperation and Development, Uranium: Resources, Production and Demand, OECD, 1970, 54 pp.
13. **Pinkney, E. T., Lurie, W., and Van Zyl, P. C. N.,** *Chemical Processing of Uranium Ores*, Rev. Ser. 23/24, IAEA, 1962, 133 pp.
14. **Post, R. F. and Ribe, F. L.,** Fusion reactors as future energy sources, *Science*, 186, 397–407, 1974.
15. **Praksh, B., Kantan, S. R., and Rao, N. K.,** Metallurgy of Thorium Production, Rev. Ser. No. 22, IAEA, 1962, 56 pp.
16. **Rose, D. J.,** Controlled nuclear fusion: Status and outlook, *Science*, 172, 797–808, 1971.
17. **Self, M.,** Fusion power: Progress and problems, *Science*, 173, 802–803, 1971.
18. United Nations, Nuclear Fuels, in Proc. 3rd International Conference on the Peaceful Uses of Atomic Energy, Vol. 12, Geneva, August 31–September 9, 1964, United Nations, 1965, 496 pp.
19. United Nations, Production Technology of the Materials Used for Nuclear Energy, in Proc. of the International Conference on the Peaceful Uses of Atomic Energy, Vol. 8, Geneva, August 8–20, 1955, United Nations, 1956, 627 pp.
20. United Nations, Survey of Raw Material Resources, in Proc. of the 2nd International Conference on the Peaceful Uses of Atomic Energy, Vol. 2, Geneva, September 1–13, 1958, United Nations, 1958, 843 pp.; Processing of Raw Materials, in Proc. of the 2nd International Conference on the Peaceful Uses of Atomic Energy, Vol. 3, Geneva, September 1–13, 1958, United Nations, 1958, 607 pp.
21. **Weinberg, A. M.,** Social institutions and nuclear energy, *Science*, 177, 27–34, 1972.
22. **Wendt, G.,** *New Prospects of Nuclear Power and Technology*, Van Nostrand, Princeton, N.J., 1957, 348 pp.
23. **Wood, L. and Nuckolls, J.,** Fusion power, *Environment* (St. Louis), 14(4), 29–33, 1972.

4. Petroleum and Natural Gas

1. **Ayres, E.,** The age of fossil fuels, in *Man's Role in Changing the Face of the Earth*, Thomas, W. L., Jr., Ed., University of Chicago Press, Chicago, 1956, 367–381.
2. **Clark, N. J.,** Elements of Petroleum Reservoirs, Soc. of Petroleum Engineers of the American Institute of Mining Engineers, Dallas, Texas, 1960, 243, pp.
3. United Nations, Techniques of Petroleum Development. Proc. of the U.N. Inter-regional Seminar on Techniques of Petroleum Development, New York, January 23–February 21, 1962, United Nations, 1964, 345 pp.
4. United Nations, Utilization of Oil Shale. Progress and Projects, United Nations, 1967, 112 pp.

5. United Nations, Proc. of the Symposium on the Development of Petroleum Resources of Asia and the Far East ECAFE, Mineral Resources Dev. Ser. No. 10, United Nations, 1959, 251 pp.; Annex, 73 pp.
6. United Nations, Proc. of the 2nd Symposium of the Development of Petroleum Resources of Asia and the Far East, ECAFE, Mineral Resources Dev. Ser. No. 18, United Nations, 1963. (2 vols)
7. United Nations, Proc. of the Seminar on the Development and Utilization of Natural Gas Resources with Special Reference to the ECAFE Region, ECAFE, Mineral Resources Dev. Ser. No. 25, United Nations, 1965, 436 pp.
8. United Nations, Proc. of the 3rd Symposium on the Development of Petroleum Resources of Asia and the Far East, Mineral Resources Dev. Ser. No. 26, United Nations, 1967. (3 vols.)
9. **Wang, F. F. H.,** Latest advances in off-shore petroleum technology, *Nat. Resour. Forum,* 1(1), 22-30, 1971.
10. **Zimmerman, E. W.,** *Conservation in the Production of Petroleum,* Yale University, New Haven, 1957, 417 pp.

5. Wind, Solar, Tidal and Geothermal Energy

1. **Armstead, H. C. H.,** Geothermal Energy, UNESCO, New York, 1973, 186 pp.
2. **Barnes, J.,** Multipurpose exploration and development of geothermal resources, *Nat. Resour. Forum,* 1(1), 55–58, 1971.
3. **Barnes, J.,** Geothermal power, *Sci. Am.,* 226(1), 70–77, 1972.
4. **Bernstein, L. B.,** *Tidal Energy for Electric Power Plants,* Israel Program for Scientific Translations, Jerusalem, 1965.
5. **Bethe, H. A.,** Energy production in stars, *Phys. Rev.,* 55, 103, 434–456, 1939.
6. **Bradbury, J. C. C.,** The economics of geothermal power, *Nat. Resour. Forum,* 1(1), 46–53, 1971.
7. **Brinkworth, B. J.,** *Solar Energy for Man,* Halsted Press, New York, 1973, 252 pp.
8. **Daniels, F.,** *Direct Use of the Sun's Energy,* Yale University Press, New Haven, 1964.
9. **Daniels, F.,** Direct use of the sun's energy, *Am. Sci.,* 55, 15–47, 1967.
10. **Daniels, F.,** *Direct Use of the Sun's Energy,* Yale University Press, New Haven, 1967, 374 pp.
11. **Hammond, A. L.,** Geothermal energy: An emerging major resource, *Science,* 177, 978–980, 1972.
12. **Hammond, A. L.,** Solar energy: The largest resource, *Science,* 177, 1088–1090, 1972.
13. **Lee, W. H. K., Ed.,** *Terrestrial Heat Flow,* American Geophysical Union, Washington, D.C., 1965.
14. **Kruger, P. and Otte, C.,** Geothermal Energy Resources, Production, Stimulation, Stanford University Press, Stanford, Calif., 1973, 360 pp.
15. United Nations, Wind and Solar Energy, Proc. of the New Delhi Symposium, Arid Zone Res. No. 7, UNESCO, 1956, 238, pp.
16. United Nations, New Sources of Energy and Economic Development: Solar Energy, Wind Energy, Tidal Energy, Geothermic Energy and Thermal Energy of the Seas, United Nations 1957, 150 pp.
17. United Nations, New Sources of Energy and Energy Development Report on the U.N. Conference on New Sources of Energy: Solar Energy– Wind Power– Geothermal Energy, Rome, Aug. 21–31, 1961, United Nations, 1962, 65 pp.
18. United Nations. Proc. of the U.N. Conference on New Sources of Energy: Solar Energy, Wind Power and Geothermal Energy, Rome, Aug. 21–31, 1961, United Nations, 1964. (7 vols.)
19. **White, D. E.,** Geothermal Energy, U.S. Geological Survey Circ. 519, Washington, D.C., 1965, 17 pp.

F. Land Resources

1. **Austin, M. E.,** Land Resource Regions and Major Land Resource Areas of the United States, U.S. Dep. Agric., Soil Conservation Service, Agric. Handb. 296, 1965, 82 pp.
2. **Barlowe, R.,** *Land Resource Economics,* Prentice-Hall, New York, 1958, 585 pp.
3. **Clark, C.,** *Population, Growth and Land Use,* St. Martin's Press, New York, 1967, 406 pp.
4. **Clawson, M.,** Land resources, in *Environment, Resources, Pollution and Society,* Murdoch, W. W., Ed., Sinauer, Stamford, Conn., 1971, 117–134.
5. **Clawson, M., Held, B., and Stoddard, C. H.,** *Land for the Future,* Johns Hopkins Press, Baltimore, 1960, 672 pp.
6. **Costin, A. G.,** Replaceable and irreplaceable resources and land use, *J. Aust. Inst. Agric. Sci.,* 25, 3–9, 1959.
7. Center for Agricultural and Economic Development, *Dynamics of Land Use – Needed Adjustment,* Iowa State University Press, Ames, 1961.
8. **Fournier, F.,** The Soils of Africa. A review of the natural resources of the African continent, Natural Resources Research, UNESCO, 1, 221–248, 1967.
9. **Graham, E. H.,** *Natural Principles of Land Use,* Oxford University Press, New York, 1944, 274 pp.
10. **Halcrow, H. G., Ed.,** *Modern Land Policy,* University of Illinois Press, Urbana, 1960.
11. **Henderson, J. M.,** The utilization of agricultural land: A theoretical and empirical inquiry, *Rev. Econ. Stat.,* 41, 242–259, 1959.
12. **Hendricks, S. B.,** Food from the land, in *Resources and Man,* NRC Committee on Resources and Man, W. H. Freeman Publishing, San Francisco, 1969, 65–85.
13. **Hendrie, D. S. and Gilchrist, J. A.,** Use and development of land in the United Kingdom, *Nutr. Abstr. Rev.,* 29, 1085–1101, 1959.

14. **Hochensmith, R. D.,** Land resources and potential uses, in Land and Water Use, Thorne, W., Ed., AAAS Pub. No. 73, Washington, D.C., 1963.
15. **Jacks, G. V. and Whyte, R. O.,** *Vanishing Lands,* Doubleday, New York, 1939, 332 pp.
16. **Kellogg, C. E.,** Expanding farm production: Extending the area of soils used, in *Man, Food, and Nutrition. Strategies and Technological Measures for Alleviating the World Food Problem,* Rechcigl, M., Jr., Ed., CRC Press, Cleveland, 1973, 79–89.
17. **Kellogg, C. E. and Orvedal, A. C.,** Potentially arable soils of the world and critical measures for their use, *Adv. Agron.,* 21, 109–170, 1969.
18. *Maximizing Agricultural Productivity through Efficient Land Use and Application of Technology,* Proc. 16th Annu. Meeting Agricultural Research Institute, Oct. 16–17, 1967, National Academy of Sciences, Washington, D.C., 1967, 284 pp.
19. National Plant Food Institute, Our Land and its Care, 4th ed., Washington, D.C., 1967.
20. **Pereira, H. C.,** Land use and misuse, *Proc. 1st Rhod. Sci. Congr.,* Bulawayo, Rhodesia, 1967.
21. **Ripley, P. O.,** Potential agricultural production in Canada, *Nutr. Abstr. Rev.,* 27, 291–300, 1957.
22. **Ruess, L. A., Wooten, H. H., and Marschner, F. J.,** Inventory of Major Land Uses, United States, U.S. Dep. Agric. Misc. Publ. 663, U.S. Gov. Print. Off., Washington, D.C., 1948, 89 pp.
23. **Salter, R. M.,** World soil and fertilizer resources in relation to food needs, *Science,* 105, 533–538, 1947; *Chron. Bot.,* 11(4), 227–235, 1948.
24. **Scott, R. H.,** Land use and farming potential in New Zealand, *Nutr. Abstr. Rev.,* 26, 565–581, 1956.
25. **Stephens, C. C.,** Land use and development in Australia, *Nutr. Abstr. Rev.,* 24, 731–740, 1954.
26. United Nations, Land Resources, in Proc. U. N. Scientific Conference on the Conservation and Utilization of Natural Resources, Vol. 6, Aug. 17–Sept. 6, 1949, Lake Success, N.Y., United Nations, 1950, 629 pp.
27. U.S. Department of Agriculture, Land, The Yearbook of Agriculture 1958, U.S. Gov. Print. Off. Washington, D.C., 1958, 605 pp.
28. U.S. Department of Agriculture, Agricultural Land Resources: Capabilities, Uses, and Conservation Needs, Agric. Inf. Bull. 263, U.S. Dep. Agric., Washington, D.C., 1962.
29. **Van Royen, W.,** *The Agricultural Resources of the World,* Prentice-Hall, New York, 1954.
30. **Wright, A. C. S. and Bennema, J.,** The Soil Resources of Latin America, World Soil Resources Rep. 18, FAO, Rome, 1965.

G. Water Resources

1. American Association for the Advancement of Science, *Water and Agriculture,* American Association for the Advancement of Science, Washington, D.C., 1960.
2. **Ackerman, E. A. and Löf, G. O. G.,** *Technology in American Water Development,* Johns Hopkins Press, Baltimore, 1959, 710 pp.
3. **Behrman, D.,** *The New World of Oceans,* Little Brown & Co., Boston, 1969, 436 pp.
4. **Briggs, P.,** *Water. The Vital Essence,* Harper & Row, New York, 1967, 223 pp.
5. **Carson, R.,** *The Sea Around Us,* Signet Science Library, New York, 1961.
6. **Cousteau, J. Y. and Dumas, F.,** *The Silent World,* Harper & Row, New York, 1953, 266 pp.
7. **Davis, K. S. and Day, J. A.,** *Water. The Mirror of Science,* Doubleday, Garden City, N.Y., 1961, 195 pp.
8. *Design of Water Resources Systems, New Techniques for Relating Economic Objectives, Engineering Analysis, and Government Planning,* Harvard University Press, Cambridge, Mass., 1962, 620 pp.
9. **Ellis, C. B.,** *Fresh Water from the Ocean for Cities, Industry, and Irrigation,* Ronald Press, New York, 1954.
10. **English, T. S., Ed.,** *Ocean Resources and Public Policy,* University of Washington Press, Seattle, 1973, 184 pp.
11. **Ferrando, R.,** The water problem, *World Rev. Nutr. Diet.* 11, 17–45, 1969.
12. **Fox, C. S.,** *Water, A Study of its Properties, its Constitution, its Circulation on the Earth, and its Utilization by Man,* Technical Press, London, 1951.
13. **Fox, J. K.,** Water resources of the world, *Fed. Proc.,* 20(1), III, Suppl. 7, 378–380, 1961.
14. **Furon, R.,** *The Problem of Water,* American Elsevier, New York, 1967, 208 pp.
15. **Hackensmith, R. D., Ed.,** *Water and Agriculture,* American Association for the Advancement of Science, Washington, D.C., 1960, 198 pp.
16. **Hill, M. N., Ed.,** *The Sea,* John Wiley & Sons, New York, 1963. (3 vols.)
17. **Hirshleifer, J., DeHaven, J. C., and Miliman, J.,** *Water Supply: Economic Technology, and Policy,* 2 ed., University of Chicago Press, Chicago, 1969, 386 pp.
18. **Hutchinson, W. L.,** Water for Millions, U.S. Dep. Agric., Forest Service, San Franscisco, 1956.
19. International Bank for Reconstruction and Development, Water for Economic Development, International Conference on Water for Peace, Washington, D.C., 1967.
20. **Kalinin, G. P. and Bykov, V. D.,** The world's water resources, present and future, in Impact of Science on Society, UNESCO, 19, 135–150, 1969.
21. **King, T.,** *Water: Miracle of Nature,* Macmillan, New York, 1963, 238 pp.
22. **Kuenen, P. H.,** *Realms of Water,* John Wiley & Sons, New York, 1956.
23. **Lear, J.,** The crisis in water. What brought it on? *Saturday Review,* October 23, 1965, 24–28, 78–80.

24. **Leopold, L. B., Davis, K. S., and Editors of Time-Life Books,** *Water,* Time-Life Books, New York, 1966, 200 pp.
25. **Leopold, L. B. and Langbuin, W. B.,** A Primer on Water, U.S. Gov. Print. Off., Washington, D.C., 1960.
26. **Nace, R. L.,** Man in quest of water, *UNESCO Courier* 23(6), 4–13, 1970.
27. **Nace, R. L.,** Water and Man: A World View, UNESCO, 1969, 46 pp.
28. **Nadeau, R. A.,** *The Water Seekers,* Doubleday, New York, 1950.
29. **Pereira, H. C. et al.,** Water resources problems: present and future requirements for life, in Use and Conservation of the Biosphere. Natural Resources Research X, UNESCO, Paris, 1970.
30. **Pereira, H. C.,** Land Use of Water Resources, in *Temperate and Tropical Climates,* Cambridge University Press, New York, 1973, 246 pp.
31. **Sears, M., Ed.,** Oceanography, AAAS, Washington, D.C., 1961, 654 pp.
32. **Sverdrup, H. U., Johnson, M. W., and Fleming, R. H.,** *The Oceans: Their Physics, Chemistry and Biology,* Prentice-Hall, New York, 1942.
33. **Tweedie, A. D.,** *Water and the World,* Nelson Publishing, Melbourne, Australia, 1966, 317 pp.
34. United Nations, Water Resources, in Proc. U. N. Scientific Conference on the Conservation and Utilization of Resources, Vol. 4, August 17 – September 6, 1949, Lake Success, N.Y., United Nations, 1950–51, 466 pp.
35. U.S. Department of Agriculture, Water. The Yearbook of Agriculture 1955, U.S. Gov. Print. Off., Washington, D.C., 1955, 751 pp.
36. **Vallentine, H. R.,** *Water in the Service of Man,* Penguin Books, Ringwood, Victoria, Australia, 1967.
37. **van Hylckama, T. E. A.,** The water balance of the earth, *Publ. Climatol.,* 9, 58–177, 1956.
38. **van Hylckama, T. E. A.,** Water resources, in *Environment, Resources, Pollution and Society,* Murdoch, W. W., Ed., Sinauer, Stamford, Conn., 1971, 135–155.
39. **Vorhis, R. C.,** Bibliography of Publications to Ground Water 1946–1955, U.S. Gov. Print. Off., Washington, D.C., 1957, 203 pp.
40. **Weyl, P. K.,** *Oceanography. An Introduction to the Marine Environment,* John Wiley & Sons, New York, 1970, 535 pp.
41. **Wittfogel, K.,** The hydraulic civilizations, in *Man's Role in Changing the Face of the Earth,* Thomas, W. L., Jr., Ed., University of Chicago Press, Chicago, 1956.
42. **Wornsnop, R. L.,** Water resources and national water needs, Editorial Res. Rep., August 11, 1965, 585.

H. Forests and Timber

1. **Beresford-Peirse, H.,** Forest, Food and People, FFHC Basic Stud. 20, FAO, Rome, 1968, 72 pp.
2. **Bruns, P. E.,** *Applied Forest Management,* Walbridge, Missoula, Mont., 1954, 383 pp.
3. **Duerr, W. A.,** *Fundamentals of Forestry Economics,* McGraw-Hill, New York, 1960, 579 pp.
4. Food and Agriculture Organization, World Forest Inventory, FAO, Rome, 1963.
5. Food and Agriculture Organization, Tropical Silviculture, FAO, Rome, 1957–1958. (3 vols.)
6. **Forbes, R. D. and Meyer, A. B.,** *Forestry Handbook,* Ronald Press, New York, 1955, 1247 pp.
7. **Haden-Guest, S., Ed.,** *World Geography of Forest Resources,* Ronald Press, New York, 1956, 736 pp.
8. **Kittredge, J.,** *Forest Influences,* McGraw-Hill, New York, 1948.
9. **McArdle, R. E.,** Timber Resources for America's Future, U.S. Forest Service, Washington, D.C., 1955.
10. **Meyer, H. A., Recknagel, A. B., Stevenson, D. D., and Bartoo, R. A.,** *Forest Management,* 2nd ed., McGraw-Hill, New York, 1961, 282 pp.
11. **Ovington, J. D.,** *Woodlands,* English Universities Press, London, 1965.
12. **Panshin, A. J.,** *Forest Products: Their Source, Production, and Utilization,* McGraw-Hill, New York, 1950, 549 pp.
13. **Richards, P. W.,** *The Tropical Rain Forest,* Cambridge University Press, Cambridge, Mass., 1952.
14. **Smith, D. M.,** *The Practice of Silviculture,* 7th ed., John Wiley & Sons, New York, 1962.
15. United Nations, Forest Resources, in Proc. U.N. Scientific Conference on the Conservation and Utilization of Natural Resources, Vol. 5, August 17–September 6, 1949, Lake Success, N.Y., United Nations, 1951, 325 pp.
16. U.S. Department of Agriculture, Trees. The Yearbook of Agriculture 1949, U.S. Gov. Print. Off., Washington, D.C., 1949, 944 pp.
17. U.S. Forest Service, Timber Resource Review Sheets, Washington, D.C., 1956.
18. U.S. Forest Service, Timber Trends in the United States, Forest Research Rep. 17, Washington, D.C., 1965.

I. Climate, Weather

1. **Battan, L. J.,** *Harvesting the Clouds: Advances in Weather Modification,* Doubleday, Garden City, New York, 1969.
2. **Battan, L. J.,** *Weather,* Prentice-Hall, Englewood Cliffs, N.J., 1974, 136 pp.
3. **Bean, L. H.,** Crop Yields and Weather, U.S. Dep. Agric. Misc. Publ. 471, U.S. Dep. Agric., Washington, D.C., 1942.
4. **Bedford, F. T.,** *Climates in Miniature,* Philosophical Library, New York, 1955.
5. **Blair, T. A.,** *Climatology. General and Regional,* Prentice-Hall, New York, 1942.
6. **Books, C. E. P.,** *Climate through the Ages,* Benn, London, 1949, 395 pp.
7. **Byers, H. R.,** *General Meteorology,* 3rd ed., McGraw-Hill, New York, 1959, 540 pp.
8. **Chang, J. H.,** *Climate and Agriculture. An Ecological Survey,* Aldine, Chicago, 1968, 304 pp.

9. **Fletcher, J. O.,** Controlling the planet's climate, in *Impact of Science on Society,* Vol. 19(2), UNESCO, Paris, April–June 1969.
10. **Geiger, R.,** *The Climate Near the Ground,* Harvard University Press, Cambridge, Mass., 1957.
11. **Halacy, D. S.,** *The Weather Changes,* Harper & Row, New York, 1968.
12. **Hadlow, L.,** *Climate, Vegetation and Man,* Greenwood Press, New York, 1969, 288 pp.
13. **Huntington, E.,** *Civilization and Climate,* Yale University Press, New Haven, 1924, 453 pp.
14. **Kendrew, W. G.,** *The Climates of the Continents,* Oxford University Press, New York, 1937, 473 pp.
15. **Lee, D. H. K.,** *Climate and Economic Development in the Tropics,* Harper & Row, New York, 1957, 182 pp.
16. **Longstreth, T. M.,** *Understanding the Weather,* Macmillan, New York, 1953, 118 pp.
17. **McDonald, J. E.,** The physics of cloud modification, *Adv. Geophys.,* 5, 223–302, 1958.
19. **Matthews, W. H., Kellogg, W. W., and Robinson, G. D., Eds.,** *Man's Impact on the Climate,* M.I.T. Press, Cambridge, Mass., 1970, 594 pp.
20. **Petterssen, S.,** *Introduction to Meteorology,* 2nd ed., McGraw-Hill, New York, 1958.
21. President's Science Advisory Committee, Tropical soils and climates, in the World Food Problem, Vol. 2, Report of the Panel on the World Food Supply, U.S. Gov. Print. Off., Washington, D.C., 1967, 471–500.
22. **Riehl, H.,** *Tropical Meteorology,* McGraw-Hill, New York, 1954, 392 pp.
23. **Rodenwaldt, E., Ed.,** *World Maps of Climatology,* 2nd ed., Springer-Verlag, New York, 1965.
24. **Sewell, W. R. D., Ed.,** *Human Dimensions of Weather Modifications,* University of Chicago Press, Chicago, 1966.
25. **Smith, L. P.,** Weather and Food, World Meteorological Organization, 1962, 80 pp.
26. **Tannehill, I. R.,** *Hurricanes,* Princeton University Press, Princeton, N.J., 1938, 257 pp.
27. **Tannehill, I. R.,** *Drought: Its Causes and Effects,* Princeton University Press, Princeton, N.J., 1947, 264 pp.
28. **Thornthwaite, C. W.,** An approach toward a rational classification of climate, *Geogr. Rev.,* 38(1), 55–94, 1948.
29. **Trewartha, G. T.,** *Introduction to Climate,* McGraw-Hill, New York, 1954.
30. **Trewartha, G. T.,** *Earth's Problem Climates,* University of Wisconsin Press, Madison, 1961, 334 pp.
31. United Nations, Changes of Climate, Proc. Rome Symposium, UNESCO, 1963, 488 pp.
32. United Nations, Agroclimatological Methods, Proc. Reading Symposium, UNESCO, 1968, 392 pp.
33. U.S. Department of Agriculture, Climate and Man. The Yearbook of Agriculture 1941, U.S. Gov. Print. Off., Washington, D.C., 1941.
34. **Uvarov, B. P.,** Insects and climate, *Trans. Entomol.; Soc. Lond.,* 79, 1–247, 1931.
35. **Weickmann, H. and Smith, W., Eds.,** *Physics of Precipitation,* National Academy of Sciences, Washington, D.C., 1960, 435 pp.
36. **Went, F. W.,** The response of plants to climate, *Science,* 112, 489–494, 1950.
37. **Went, F. W.,** Climate and agriculture, *Sci. Am.,* 196, 83–94, 1957.
38. **Wexler, H.,** Modifying weather on a large scale, *Science,* 128, 1059–1063, 1958.
39. **Williams, C. N. and Joseph, K. T.,** *Climate, Soil and Crop Production in the Humid Tropics,* Oxford University Press, New York, 1971.
40. **Wolfe, J. N.,** The possible role of microclimate, *Ohio J. Sci.,* 51, 134–138, 1951.
41. World Meteorological Organization, *Harvest from Weather,* World Meteorological Organization, 1967, 48 pp.
42. World Meteorological Organization, Meteorology and the Human Environment, World Meteorological Organization, 1971, 40 pp.
43. World Meteorological Organization, Selected Papers on Meteorology as Related to the Human Environment, World Meteorological Organization, 1971, 151 pp.
44. World Meteorological Organization, Weather and Man: The Role of Meteorology in Economic Development, World Meteorological Organization, 1964, 80 pp.

J. Soil and Water Management

1. Soil Science

1. **Ahn, P. M.,** *West African Soils,* Oxford University Press, London, 1970, 332 pp.
2. **Bear, F.,** *Soil in Relation to Crop Growth,* Reinhold, New York, 1965, 320 pp.
3. **Berger, K. C.,** *Introductory Soils,* Macmillan, New York, 1965.
4. **Black, C. A.,** *Soil-Plant Relationships,* John Wiley & Sons, New York, 1968.
5. **Buckman, H. O. and Brady, N. C.,** *The Nature and Properties of Soils,* 7th ed., Macmillan, New York, 1969.
6. **Donahue, R. L.,** *Soils: An Introduction to Soils and Plant Growth,* Prentice-Hall, Englewood Cliffs, N.J., 1965.
7. **Dudal, R.,** Dark Clay Soils of Tropical and Subtropical Regions, FAO Agric. Devel. Papers 83, FAO, Rome, 1965, 161 pp.
8. **Eden, T.,** *Elements of Tropical Soil Science,* Macmillan, London, 1947.
9. **Foth, H. D. and Turk, L. M.,** *Fundamentals of Soil Science,* 5th ed., John Wiley & Sons, New York, 1972, 454 pp.
10. **Hausenbuiller, R. L.,** *Soil Sciences: Principles and Practices,* Brown Reprints, Dubuqe, IA, 1972, 520 pp.
11. **Jacks, G. V.,** *Soil,* Philosophical Library, New York, 1954.
12. **Kellogg, C. E.,** *The Soils That Support Us,* Macmillan, New York, 1941.
13. **Lutz, H. J. and Chandler, R. F.,** *Forest Soils,* John Wiley & Sons, New York, 1946.

14. **Maignien, R.**, Review of Research on Laterites, Nat. Resour. Res. No. 4, UNESCO, 1966, 148 pp.
15. **Millar, C. E., Turk, L. M., and Foth, H. D.**, *Fundamentals of Soil Science*, John Wiley & Sons, New York, 1958, 526 pp.
16. **Mosier, J. G. and Gustafson, A. F.**, *Soil Physics and Management*, Lippincott, Philadelphia, 1917.
17. National Research Council, *Soils of the Humid Tropics*, National Academy of Sciences, Washington, D.C., 1972, 219 pp.
18. **Nye, P. H. and Greenland, D. J.**, *The Soil and Shifting Cultivation*, Commonwealth Bureau of Soils, Harpenden, England, 1960, 156 pp.
19. **Papadakis, J.**, *Soils of the World*, Elsevier, Amsterdam, 1969, 208 pp.
20. **Russell, E. W.**, *Soil Conditions and Plant Growth*, 9th ed., Longmans, New York, 1961.
21. **Thompson, L. M.**, *Soils and Soil Fertility*, McGraw-Hill, New York, 1953.
22. **Troeh, F. R. and Palmer, R. G.**, *Introductory Soil Science*, Iowa State University Press, Ames, 1966.
23. United Nations, Tropical Soils and Vegetation, Proc. Abidjan Symposium, October 20–24, 1959, UNESCO, 1961, 115 pp.
24. U.S. Department of Agriculture, Soils and Men. The Yearbook of Agriculture 1938, U.S. Gov. Print. Off., Washington, D.C., 1938.
25. U.S. Department of Agriculture, Soil. The Yearbook of Agriculture 1957, U.S. Gov. Print. Off., Washington, D.C., 1957, 784 pp.
26. U.S. Department of Agriculture, Soil Survey Staff, Soil Survey Manual, Agric. Handb. No. 18, U.S. Gov. Print. Off., Washington, D.C., 1951, 503 pp.
27. U.S. Department of Agriculture, Soil Survey Staff, Soil Classification – A Comprehensive System. 7th Approximation, Soil Conservation Service, August 1960, 265 pp.; Suppl., 1967.
28. **Vageler, P.**, *An Introduction to Tropical Soils*, Macmillan, London, 1933.
29. **Weir, W. W.**, *Soil Science, Its Principles and Practice*, Lippincott, Chicago, 1936, 615 pp.
30. **Wilde, S. A.**, *Forest Soils. Their Properties and Relation to Silviculture*, Ronald Press, New York, 1958.

2. Soil Erosion and Conservation

1. **Aubert, C., Fournier, F., and Rozanov, V.**, Soils and the maintenance of their fertility as factors affecting the choice of use of land. Use and conservation of the biosphere, Natural Resources Research X, UNESCO, Paris, 1970.
2. **Bennett, H. H.**, *Elements of Soil Conservation*, 2nd ed., McGraw-Hill, New York, 1955, 358 pp.
3. **Blakely, B. D., Coyle, J. J., and Steele, J. H.**, Erosion on Cultivated Land, in Soil. The Yearbook of Agriculture 1957, U.S. Dep. Agric., Washington, D.C., 1957.
4. **Chepil, W. S. and Woodruff, N. P.**, The physics of wind erosion and its control, *Adv. Agron.*, 15, 211–302, 1963.
5. **Cook, R. L.**, *Soil Management for Conservation and Production*, John Wiley & Sons, New York, 1962, 527 pp.
6. **Dale, T. and Carter, V. G.**, *Topsoil and Civilization*, University of Oklahoma Press, Norman, 1955.
7. **Donahue, R. L.**, *Our Soils and Their Conservation*, Interstate, Danville, Ill., 1961, 568 pp.
8. Food and Agriculture Organization, Soil Conservation. An International Study, FAO Agric. Stud. 4, FAO, Rome, 1952, 189 pp.
9. Food and Agriculture Organization, Soil Erosion by Wind and Measures for its Control on Agricultural Lands, FAO Agric. Devel. Papers, No. 71, FAO, Rome, 1960, 88 pp.
10. Food and Agriculture Organization, Soil Erosion by Water: Some Measures for its Control on Cultivated Lands, FAO Agric. Devel. Papers, No. 81, FAO, Rome, 1965, 284 pp.
11. **Kohnke, H. and Bertrand, A. R.**, *Soil Conservation*, McGraw-Hill, New York, 1959, 298 pp.
12. **Parks, W. R.**, *Soil Conservation Districts in Action*, Iowa State University Press, Ames, 1952.
13. **Schwab, G. O., Frevert, R. K., and Barnes, K. K.**, *Soil and Water Conservation Engineering*, 2nd ed., John Wiley & Sons, New York, 1966, 638 pp.
14. **Smith, D. D. and Wischmeier, W. H.**, Rainfall erosion, *Adv. Agron.*, 14, 109–148, 1962.
15. **Stallings, J. H.**, *Soil Conservation*, Prentice-Hall, Englewood Cliffs, N.J., 1957.
16. **Tower, H. E.**, Strip Cropping for Conservation and Production, U.S. Dep. Agric. Farmers' Bull. No. 1381, 1946.
17. U.S. Department of Agriculture, Soil and Water Conservation Needs. A National Inventory, Misc. Publ. 971, Washington, D.C., 1965, 94 pp.
18. **Utz, E. J.**, The coordinated approach to soil erosion control, in Soils and Men. The Yearbook of Agriculture 1938, U.S. Gov. Print. Off., Washington, D.C., 1938.

3. Water Management

1. **Colman, E. A.**, *Vegetation and Watershed Management*, Ronald Press, New York, 1953.
2. **Kramer, P. J.**, *Plant and Soil Water Relations*, McGraw-Hill, New York, 1949.
3. **La Mer, V. K.**, *Retardation of Evaporation by Monolayers*, Academic Press, New York, 1962.
4. **Leversedge, F. M., Ed.**, Priorities in Water Management, Dep. of Geography, University of Victoria, Victoria, British Columbia, 1974, 300 pp.
5. National Research Council, Alternatives in Water Management, Publ. 1408, National Academy of Sciences, Washington, D.C., 1966.

6. **Roe, H. B.,** *Moisture Requirements in Agriculture,* McGraw-Hill, New York, 1950.
7. **Slayter, R. O.,** Absorption of water by plants, *Bot. Rev.,* 26, 331–392, 1960.
8. **Stanhill, G. et al.,** Practical Soil Moisture Problems in Agriculture, World Meteorological Organization, 1968, 69 pp.
9. **Thomas, H. E.,** *The Conservation of Ground Water,* McGraw-Hill, New York, 1951, 327 pp.
10. **White, G.,** *The Strategies of American Water Management,* University of Michigan Press, Ann Arbor, 1969, 155 pp.

4. Salinity

1. **Boyko, H., Ed.,** *Salinity and Aridity – New Approaches to Old Problems,* Monographiae Biologicae, Vol. XVI, Junk, The Hague, 1966, 408 pp.
2. **Boyko, H.,** Salt water agriculture, *Sci. Am.,* 216(3), 89–96, 1967.
3. **Boyko, H., Ed.,** *Saline Irrigation for Agriculture and Forestry, Vol. 4, The UNESCO–WAAS–Italy Symposium, WAAS Ser., World Academy of Art and Sciences,* Junk, The Hague, 1968, 350 pp.
4. **Clawson, M., Landsberg, H. H., and Alexander, L. T.,** Desalted seawater for agriculutre. Is it economic? *Science,* 165, 1141–1148, 1969.
5. **Fuller, W. H.,** Water, Soil and Crop Management Principles for the Control of Salts, Ariz. Agric. Exp. Sta. and Coop. Ext. Serv. Bull. A–43, 1967.
6. International Atomic Energy Agency, Value to Agriculture of High-Quality Water from Nuclear Desalination. Report of a Panel, Vienna, October 30–November 3, 1967, IAEA, 1969, 278 pp.
7. International Atomic Energy Agency, Nuclear Desalination, Proc. of a Symposium, Madrid, November 18–22, 1968, IAEA/Elsevier, 1969, 941 pp.
8. National Research Council, *Desalination Research and the Water Problem,* National Academy of Sciences, Washington, D.C., 1962, 85 pp.
9. UNESCO, Salinity Problems in the Arid Zones, Proc. Teheran Symposium, Arid Zone Res. No. 14, UNESCO, 1961, 395 pp.
10. UNESCO, Utilization of a Saline Water, Reviews of Research, 2nd ed., Arid Zone Programme No. 4, UNESCO, 1956, 102 pp.
11. United Nations, Water Desalination in Developing Countries, United Nations, 1964, 32 pp.
12. United Nations, The Design of Water Supply Systems Based on Desalination, United Nations, 1968, 64 pp.

5. Aridity

1. **Boyko, H.,** Farming the desert, *Sci. J.,* 4 (May), 72–79, 1968.
2. **Dickson, B. T., Ed.,** Guidebook to Research Data for Arid Zone Development, Arid Zone Res. No. 9, UNESCO, 1957, 191 pp.
3. **Hills, E. S., Ed.,** *Arid Lands, A Geographical Appraisal,* Methuen, London, 1966, 461 pp.
4. **Hodge, C., Ed.,** *Aridity and Man,* Publ. 74, AAAS, Washington, D.C., 1963, 584 pp.
5. **Kelso, M. M., Martin, W. E., and Mack, L. E.,** *Water Supplies and Economic Growth in an Arid Environment. An Arizona Case Study,* University of Arizona Press, Tucson, 1973, 328 pp.
6. **McGinnies, W. G. and Goldman, B., Ed.,** *Arid Lands in Perspective,* University of Arizona Press, Tucson, 1969, 421 pp.
7. **Meigs, P.,** Geography of Coastal Deserts, Arid Zone Res. No. 28, UNESCO, 1966, 140 pp.
8. **Stamp, L. D., Ed.,** A History of Land Use in Arid Regions, Arid Zone Res. No. 17, UNESCO, 1965, 388 pp.
9. UNESCO, Problems of the Arid Zone, Proc. Paris Symposium, UNESCO, Paris 1962, 481 pp.
10. **White, G. F., Ed.,** *The Future of Arid Lands,* Publ. 43, AAAS, Washington, D.C., 1956, 453 pp.

6. Irrigation

1. **Booher, L. J.,** Surface Irrigation, FAO, Rome, 1967, 161 pp.
2. **Christiansen, J. E.,** Irrigation in relation to food production, *Agric. Eng.,* June 1953.
3. **Framji, K. K. and Mahajan, I. K.,** *Irrigation and Drainage in the World – A Global Review,* 2nd ed., International Commission of Irrigation and Drainage, New Delhi, 1969, 1344 pp. (2 vols.).
4. **Gulhati, N. D.,** Irrigation in the World, International Commission on Irrigation and Drainage, Delhi, 1955.
5. **Hagan, R. M., Houston, C. E., and Allison, S. V.,** Successful Irrigation: Planning, Development, Management, FAO, Rome, 1968, 53 pp.
6. **Huffman, R. E.,** *Irrigation Development,* Ronald Press, New York, 1953.
7. *Irrigation of Agricultural Lands,* ASA Ser. 11, American Society of Agronomy, Madison, Wis., 1967, 1180 pp.
8. **Israelsen, O. W. and Hansen, V. E.,** *Irrigation Principles and Practices,* 3rd ed., John Wiley & Sons, New York, 1962.
9. **Marsh, A. W.,** Applied Irrigation Research, FAO, Rome, 1967, 54 pp.
10. **Pair, C. H., Ed.,** Sprinkler Irrigation, Sprinkler Irrigation Assoc., Washington, D.C., 1969.
11. **Pillsbury, A. F.,** Sprinkler Irrigation, FAO Agric. Devel. Papers, No. 88, FAO, Rome, 1968, 179 pp.
12. *Sprinkler Irrigation Manual,* Wright Rain, Ringwood, Hampshire, England, 1956.
13. **Thorne, D. W. and Peterson, H. B.,** *Irrigated Soils,* Blakiston, McGraw-Hill, New York, 1954, 392 pp.
14. **van't Wondt, B. D.,** Irrigation development at the farm level, *Econ. Bull. Asia Far East,* 19(3), 32–46, 1968.

15. **Yaron, B., Danfors, E., and Vaadia, Y., Eds.,** *Arid Zone Irrigation*, Springer-Verlag, New York, 1973, 434 pp.
16. **Zimmerman, J. D.,** *Irrigation*, John Wiley & Sons, New York, 1966, 516 pp.

7. Drainage

1. **Jones, L. A.,** Farm Drainage, U.S. Dep. Agric. Farmers' Bull. 2046, 1952.
2. **Jones, L. A.,** Land Drainage – An important tool to increased crop production, *Agric. Eng.*, April 1953, 239–242.
3. **Leopold, L. B. and Maddock, T., Jr.,** *The Flood Control Controversy*, Ronald Press, New York, 1954.
4. **Luthin, J. N., Ed.,** *Drainage of Agricultural Lands*, American Society of Agronomy, Madison, Wis., 1957, 620 pp.
5. **Theobald, G. H.,** Methods and Machines for Tile and Other Tube Drainage, FAO Agric. Dev. Paper No. 78, FAO, Rome 1963, 104 pp.
6. UNESCO, Floods and their Computations, Proc. Leningrad Symposium, August 1967, UNESCO/IASH/WMO, 1969, 985 pp. (2 vols.)
7. U.S. Census Bureau, Drainage, Washington, D.C., 1964.
8. **White, G. F.,** Flood damage prevention policies, *Nat. Resour. Forum*, 1(1), 39–44, 1971.
9. World Meteorological Organization, Assessment of the Magnitude and Frequency of Flood Flows, Transactions of the Inter-regional Seminar, Bangkok, Thailand, April 19–26, 1966, World Meteorological Organization, 1967, 206 pp.
10. World Meteorological Organization, Estimation of Maximum Floods, Report of a Working Group of the Commission for Hydrometeorology, World Meteorological Organization, 1969, 288 pp.

K. Environmental Quality

1. **Appleyard, R. T., Ed.,** *Man and His Environment*, University of Western Australia Press, Needlands, 1970, 198 pp. (11 papers)
2. **Arthur, D. H.,** *Man and His Environment*, American Elsevier, New York, 1969, 218 pp.
3. **Bond, R. G. and Straub, C. P., Eds.,** *CRC Handbook of Environmental Control*, CRC Press, Cleveland, Ohio, 1973. (3 vols.)
4. **Boughey, A. S.,** *Man and the Environment*, Macmillan, New York, 1971, 472 pp.
5. **Brady, N. C., Ed.,** Agriculture and the Quality of Our Environment. A Symposium, Publ. 85, AAAS, Washington, D.C., 1967.
6. **Commoner, B.,** *The Closing Circle. Nature, Man and Technology*, Bantam Books, New York, 1971, 344 pp.
7. **Coulston, F. and Korte, F., Eds.,** *Environmental Quality and Safety. Chemistry, Toxicology and Technology*, Academic Press, New York, 1972.
8. **Dasmann, R. F.,** An Environment Fit for People, Public Affairs Pamphlet 421, 1961.
9. **Dasmann, R. F.,** *Environmental Conservation*, 3rd ed., John Wiley & Sons, New York, 1972.
10. **Detwyler, T. R.,** *Man's Impact on Environment*, McGraw-Hill, New York, 1971, 731 pp.
11. **Ehrlich, P. R., Ehrlich, A. H., and Holdren, J. P.,** *Human Ecology: Problems and Solutions*, W. H. Freeman Publishing, San Francisco, 1973, 304 pp.
12. **Falk, R. A.,** *This Endangered Planet. Prospect and Proposals for Human Survival*, Random House, New York, 1971, 495 pp.
13. **Jackson, W.,** *Man and the Environment*, 2nd ed., Brown, Dubuque, IA, 1973, 324 pp.
14. **Loehr, R.,** Changing practices in agriculture and their effect in the environment, *CRC Crit. Rev. Environ. Contr.*, 1, 69–99, 1970. (67 ref.)
15. **Luck, J. M.,** Man against his environment. The next hundred years, *Science*, 126, 903–908, 1957.
16. **Nicholson, M.,** *The Environmental Revolution*, McGraw-Hill, New York, 1970, 366 pp.
17. **Pavoni, J. L., Hagerty, D. J., and Heer, J. E.,** *Preserving Man's Environment*, Data Courier, Louisville, Ky., 1974, 308 pp.
18. President's Science Advisory Committee, Restoring the Quality of Our Environment. Report of the Environmental Pollution Panel, U.S. Gov. Print. Off., Washington, D.C., 1965, 317 pp.
19. Public Policy Toward Environment 1973: A Review and Appraisal, *Ann. N.Y. Acad. Sci.*, 216, 1–202, 1973.
20. **Rienow, R.,** *Man Against His Environment*, Ballantine Books, New York, 1970, 307 pp.
21. Secretary of Agriculture and Director of the Office of Science and Technology, Control of Agriculture-Related Pollution. A Report to the President, Washington, D.C., 1969.
22. **Southwick, C. H.,** *Ecology and the Quality of Our Environment*, Van Nostrand, Reinhold, New York, 1972, 319 pp.
23. *The Human Environment*, Vol. 1, A Selective, Annotated Bibliography of Reports and Documents on International Environmental Problems, Woodrow Wilson International Center for Scholars, Washington, D.C., 1972, 171 pp.
24. **Thomas, W. A., Goldstein, G., and Wilcox, W. H.,** *Biological Indicators of Environmental Quality: A Bibliography of Abstracts*, Ann Arbor Science Publishers, Ann Arbor, Mich., 1972, 254 pp.
25. U. S. Department of Agriculture, Environmental quality. Survey of attitudes and problems, *Agric. Sci. Rev.*, 6(3), 1–6, 1968. (24 ref.)
26. **Wadleigh, C. H.,** Wastes in Relation to Agriculture and Forestry, U.S. Dep. Agric. Misc. Publ. 1075, 1968.
27. **Wagner, R. H.,** *Environment and Man*, Norton, New York, 1971, 491 pp.
28. **Wilson, T. W., Jr.,** *International Environmental Action: A Global Survey*, Dunellen, New York, 1971, 364 pp.

V. CROP PRODUCTION

A. The Role of Crops in Food Production

1. General

1. **Anderson, E.,** *Plants, Man and Life,* Little, Brown & Co., Boston, 1952.
2. **Baker, H. G.,** *Plants and Civilization,* 2nd ed., Wadsworth, Belmont, Calif., 1970, 194 pp.
3. **Barrau, J., Ed.,** *Plants and Migrations of Pacific Peoples. A Symposium,* Bishop Museum Press, Honolulu, 1963.
4. **Bolhuis, G. G.,** Protein content of tropical food crops, *World Crops,* 14(19), 197–301, 1962.
5. Commonwealth Economic Committee, Grain Crops. A Review of Production, Trade and Consumption and Prices Relating to Wheat, Wheat Flour, Maize, Barley, Oats, Rye and Rice, London, 1963, 193 pp.
6. **De Vries, C. A., Ferwerda, J. D., and Flach, M.,** Choice of food crops in relation to actual and potential production in tropics, *Neth. J. Agric. Sci.,* 15, 241–248, 1967.
7. **Frisch, R. E.,** *Plants that Feed the World,* Van Nostrand, Princeton, N.J., 1966, 104 pp.
8. **Harris, D. R.,** New light on plant domestication and the origins of agriculture, A review, *Geogr. Rev.,* 57, 90–107, 1967.
9. **Harris, R. S. and Munsell, H. E.,** Edible plants of Central America, *J. Home Econ.,* 42, 629–631, 1950.
10. **Harrison, S. G., Masefield, G. B., and Wallis, M.,** *Oxford Book of Food Plants,* Oxford University Press, New York, 1969.
11. **Hill, A. F.,** *Economic Botany,* 2nd ed., McGraw-Hill, New York, 1952, 560 pp.
12. **Hutchinson, J. and Melville, P.,** *The Story of Plants and Their Use to Man,* Gawthorn, London, 1948.
13. **Hyams, E.,** *Plants in the Service of Man,* Dent, London, 1971, 222 pp.
14. **Hylander, C. J. and Stanley, O. B.,** *Plants and Man,* Blakiston, Philadelphia, 1941.
15. **Jalil, M. E. and Tahir, W. M.,** World supplies of plant proteins, in *Proteins in Human Nutrition,* Porter, J. W. G. and Rolls, B. A., Eds., Academic Press, London, 1973, 35–46.
16. **Nicholls, H. A.,** *Food Plants,* Macmillan, London, 1929.
17. **Quisenberry, K. S.,** Crop production potentials in relation to freedom from want, *Chron. Bot.,* 11(4), 237–245, 1948.
18. **Richardson, H. L.,** Increasing world food supplies through greater crop production, *Outlook Agric.,* 3, 9–22, 1960. (28 ref.)
19. **Roberts, L. M.,** World prospects for increasing foods of plant origin, *J. Am. Vet. Med. Assoc.,* 153, 1843–1847, 1968.
20. **Burton, G. W.,** Food resources in the plant kingdom, in *The Food Resources of Mankind,* Agri-World Press, Montreal, 1968, 67–80.
21. **Schery, R. W.,** *Plants for Man,* 2nd ed., Prentice-Hall, Englewood Cliffs, N.J., 1972, 657 pp.
22. **Schuphan, W.,** *Nutritional Values in Crops and Plants,* Faber & Faber, London, 1965, 280 pp.
23. Symposium on the role of plant foods in solving the world food problem, *Plant Foods Hum. Nutr.,* 1, 11–55, 1968.
24. **Wilson, C. M.,** *Grass and People,* University of Florida Press, Gainesville, 1961.

2. Cereals

a. General

1. **Bradley, W. B.,** Cereal grain as food, *Cereal Sci. Today,* 11(6), 240–242, 1966.
2. **Bressani, R.,** Improvement of nutritional status in developing countries by improved food production: Cereals, *Proc. 7th International Congress of Nutrition,* Vol. 4, Hamburg, 1966, Pergamon Press, Oxford, 1967, 54–60.
3. **Coons, C. M.,** Selected References on Cereal Grains in Protein Nutrition: Human and Experimental Studies of Major and Minor Cereals, 1910–1966, USDA, ARS 61–5, 1968, 150 pp.
4. **Geddes, W. F.,** Cereal grains, in *The Chemistry and Technology of Food and Food Products,* 2nd ed., Vol. 2, Jacobs, M. B., Ed., Interscience, New York, 1951, 1022–1133.
5. **Harrell, C. G.,** Cereal grain production and processing, in *Food Processing Operations,* Vol. 3, Joslyn, M. A. and Heid, J. L., Eds., Avi Publishing, Westport, Conn., 1964, 121–169.
6. **Harrel, C. C. and Dirks, B. M.,** Cereals and cereal products, in *Handbook of Food and Agriculture,* Blanck, F. C., Ed., Reinhold, New York, 1955, 411–452.
7. **Hegsted, D. M.,** Nutritional value of cereal proteins in relation to human needs, in *Protein-Enriched Cereal Foods for World Needs,* Milner, M., Ed., American Association of Cereal Chemists, St. Paul, Minn., 1969, 38–48. (38 ref.)
8. **Kent, N. L.,** Structural and nutritional properties of cereal proteins, in *Proteins as Human Food,* Lawrie, R. A., Ed., Avi Publishing, Westport, Conn., 1970, 280–299.
9. **Kent-Jones, D. W. and Amos, A. J.,** *Modern Cereal Chemistry,* 6th ed., Food Trade Press, London, 1968, 730 pp.
10. **Majors, K. R.,** Cereal grains as food and feed, in Crops in Peace and War, The Yearbook of Agriculture 1950–51, U.S. Gov. Print. Off., Washington, D.C., 1951, 331–340.
11. **Matz, S. A., Ed.,** *The Chemistry and Technology of Cereals as Food and Feed,* Avi Publishing, Westport, Conn., 1959.
12. **Matz, S. A., Ed.,** *Cereal Science,* Avi Publishing, Westport, Conn., 1969, 241 pp. (8 papers)

13. **Munck, L.,** Cereals for feed-quality and utilization, *J. Sweed. Seed Assoc.,* 78, 137–201, 1968.
14. **Storck, J. and Teague, W. D.,** *Flour for Man's Bread,* University of Minnesota Press, Minneapolis, 1952.
15. **West, Q. M.,** The quantitative role of cereals as suppliers of dietary protein, in *Protein-Enriched Cereal Foods for World Needs,* American Association of Cereal Chemists, St. Paul, Minn., 1969. (18 ref.)
16. **Dalrymple, D.,** Imports and Plantings of High-Yielding Varieties of Wheat and Rice in the Less Developed Countries, U.S. Dep. Agric., Econ. Dev. Service, Washington, D.C., 1971.

b. Barley

1. **Dickinson, A. D.,** Barley, in *Cereal Science,* Matz, S. A., Ed., Avi Publishing, Westport, Conn., 1969, 97–117.
2. **Pomeranz, Y.,** Food uses of barley, *CRC Crit. Rev. Food Technol.,* 4(3), 377–394, 1974. (67 ref.)

c. Corn or Maize

1. **Cunard, A. C.,** Maize agronomy. Part 1 – Nutrition and silage, *World Crops,* 19(1), 20–27, 1967.
2. **Hruška, J.,** Monografie o kukuřici, Prague, Státní zemědělské nakladatelstvi, 1962, 916 pp.
3. **Inglett, G. E.,** Food uses of corn around the world, in *Corn: Culture, Processing, Products,* Avi Publishing, Westport, Conn., 1970, 138–150.
4. International Maize and Wheat Improvement Center, Mexico, *Bibliography of Corn,* Scarecrow Press, Metuchen, N.J., 1971. (3 vols.)
5. Maize and Maize Diets. A Nutritional Survey, FAO, Rome, 1954, 94 pp.
6. **Matz, S. A.,** Corn, in *Cereal Science,* Matz, S. A., Ed., Avi Publishing, Westport, Conn., 1969, 39–77.
7. **Mertz, E.,** Nutritive value of corn and its products, in *Corn: Culture, Processing, Products,* Avi Publishing, Westport, Conn., 1970, 350–359.
8. Nutritional Value of High-Lysine Corn, Purdue University Agric. Exp. Sta. Res. Progr. Rep. 227, March 1966.

d. Oats

1. **Matz, S. A.,** Oats, in *Cereal Science,* Matz, S. A., Ed., Avi Publishing, Westport, Conn., 1969, 78–96.
2. **Stanton, T. R.,** Oats, in *The Chemistry and Technology of Cereals as Food and Feed,* Matz, S. A., Ed., Avi Publishing, Westport, Conn., 1959.

e. Rice

1. **Athwal, D. S.,** Semidwarf rice and wheat in global food needs, *Q. Rev. Biol.,* 46, 1–34, 1971. (119 ref.)
2. **Dore, W. G.,** Wild Rice, Canada Department of Agriculture, Research Branch, Publ. 1393, Ottawa, 1969.
3. **Hampel, G.,** Reis in der menschlichen Ernährung, *Brot Gebäck,* 24(12), 222–227, 1970. (29 ref.)
4. **Hanks, L. M.,** *Rice and Man. Agricultural Ecology in Southeast Asia,* Aldine-Atherton, Chicago, 1972, 174 pp.
5. **Houston, D. F. and Kohler, G. O.,** *Nutritional Properties of Rice,* National Academy of Sciences, Washington, D. C., 1970, 65 pp.
6. **Huston, D. F., Ed.,** *Rice Chemistry and Technology,* American Association of Cereal Chemists, St. Paul, Minn., 1972, 517 pp. (20 papers)
7. International Rice Research Institute, Supplement to the International Bibliography of Rice Research, 1962, Los Baños, Laguna, Philippines, 1963.
8. **Matz, S. A. and Beachell, H. C.,** Rice, in *Cereal Science,* Matz, S. A., Ed., Avi Publishing, Westport, Conn., 1969, 172–223.
9. **Mickus, R. R.,** Rice (*Oryza sativa*), One of the world's basic foods, *Cereal Sci. Today,* 4(5), 138–149, 1959.
10. Rice and Rice Diets. A Nutritional Survey, FAO, Rome, 1954, 78 pp.
11. **Steeves, T. A.,** Wild rice–Indian food and modern delicacy, *Econ. Bot.,* 6(9), 107–143, 1952.

f. Rye

1. **Schaben, L. J.,** Rye – Source of daily bread, *Foreign Agric.,* 12, 163–168, 1948.
2. **Shands, H. L.,** Rye, in *Cereal Science,* Matz, S. A., Ed., Avi. Publishing, Westport, Conn., 1969, 118–149.

g. Sorghum

1. **Adrian, J. and Jaquot, R.,** *Le sorgho et les mils en alimentation humaine et animale,* Vigot Freres, Paris, 1964, 189 pp.
2. **Deosthale, Y. G.,** Protein quality of grain sorghum and its significance in human nutrition, in *Sorghum in Seventies,* Ganga Prasada Rao, N. and House, L. R., Eds., Oxford and I. B. H. Publishing Co., New Delhi, 1972, 575–581.
3. George Washington University, *Sorghum 1930–1963, Bibliography of World Literature,* George Washington University, Washington, D.C., 1967.
4. **Kramer, N. W. and Matz, S. A.,** Sorghum, in *Cereal Science,* Matz, S. A., Ed., Avi Publishing, Westport, Conn., 1969, 150–171.
5. **Rachie, K. O.,** Sorghum grain: Its worldwide significance and potential, *Cereal Sci. Today,* 14, 271–276, 1969.
6. Rockefeller Foundation, Indian Agricultural Program, *Sorghum. A Bibliography of the World Literature 1964–1969,* Scarecrow Press, Metuchen, N.J., 1973, 393 pp.

7. **Rooney, L. W. and Clark, L. E.,** The chemistry and processing of sorghum grain, *Cereal Sci. Today,* 13, 258–287, 1968.
8. **Voigt, R. L.,** Sorghum, food plant of the ages, *Prog. Agric. Ariz.,* 21, 18–23, 1968.
9. **Wall, J. S. and Ross, W. M., Eds.,** *Sorghum Production and Utilization,* Avi Publishing, Westport, Conn., 1970, 702 pp. (19 papers)
10. **Werler, P. F.,** Manufacture and usage of sorghum flour, Proc. 8th Annu. Symp. Central States Sect. Am. Assoc. Cereal Chemists, St. Louis, Mo., Feb. 17–18 1967.

h. Wheat

1. **Aykroyd, W. R. and Doughty, J.,** Wheat in Human Nutrition, FAO Nutr. Stud. No. 23, FAO, Rome, 1970, 163 pp.
2. **Bailey, C. H.,** *The Constituents of Wheat and Wheat Products,* Reinhold, New York, 1944.
3. **Bennett, M. K.,** World Wheat Survey and Outlook, *Wheat Studies,* Vol. 12, Stanford University Food Research Institute, Stanford, Calif., 1936.
4. **Borlaug, N. E.,** Wheat, rust and people, *Phytopathology,* 55, 1088–1098, 1965.
5. **Farnsworth, H.,** The role of wheat in improving nutritional status and labor productivity in lesser developed countries, *Proc. Int. Wheat Surplus Utilization Conf.,* Brookings, S.D., 1959, 52–65.
6. From Wheat to Flour, Wheat Flour Institute, Chicago, 1965.
7. George Washington University, *Bibliography of Wheat,* Scarecrow Press, Metuchen, N.J., 1971. (3 vols.)
8. **Hegsted, D. M.,** Wheat: Challenge to nutritionists, *Cereal Sci. Today,* 10, 257–259, 1965.
9. **Hegsted, D. M., Trulson, M. F., and Stare, F. J.,** Role of wheat and wheat products in human nutrition, *Physiol. Rev.,* 34, 221–258, 1954. (193 ref.)
10. **Hlynka, I., Ed.,** *Wheat: Chemistry and Technology,* American Association of Cereal Chemists, St. Paul, Minn., 1964, 603 pp.
11. International Wheat Council, Trends in Wheat Consumption, Secretariat Paper No. 4, London, 1964.
12. **Kent, N. L.,** *Technology of Cereals with Special Reference to Wheat,* Pergamon Press, Oxford, 1966, 262 pp.
13. **Mickelsen, O., Bolourchi, S., and Friedmann, C.,** The value of wheat as food for man, *Proc. 4th National Conference on Wheat Utilization,* Boise, ID., 1966.
14. **Moran, T.,** Nutritional significance of recent work on wheat, flour and bread, *Nutr. Abstr. Rev.,* 29, 1–16, 1959.
15. **Pomeranz, Y.,** *Wheat Chemistry and Technology,* 2nd ed., American Association of Cereal Chemists, St. Paul, Minn., 1971, 821 pp.
16. **Reitz, L. P.,** World distribution and importance of wheat, in *Wheat and Wheat Improvement,* Quisenberry, K. S. and Reitz, L. P., Eds., American Society of Agronomy, Madison, Wis., 1967, 1–18.
17. Role of Wheat in the World's Food Supply, Report of Conference, Albany, Calif., 1962, Western Regional Research Laboratory, Albany, Calif., 1962.
18. **Shellenberger, J. A.,** Wheat, in *Cereal Science,* Matz, S. A., Ed., Avi Publishing, Westport, Conn., 1969, 1–38.
19. *Wheat and Wheat Products as Human Food,* Indian Council Med. Res. Special Rep. Ser. 23, 1952.

i. Other Cereals

1. **Anderson, E.,** Millet provides food for millions, *Foreign Agric.,* 12(11), 235–239, 1948.
2. **Anderson, E. and Martin, J. H.,** World production and consumption of millet and sorghum, *Econ. Bot.,* 3, 265–268, 1949.
3. George Washington University, Biological Sciences Communications Project, *The Millets. A Bibliography of the World Literature Covering the Years 1930–1963,* Scarecrow Press, Metuchen, N.J., 1967, 154 pp.
4. George Washington University, *The Millets 1930–1963, Bibliography of World Literature,* George Washington University, Washington, D.C., 1967.
5. **Hulse, J. H. and Laing, E. M.,** *Nutritive Value of Triticale Protein,* International Development Research Centre, Ottawa, 1974, 183 pp.
6. **Matz, S. A.,** Millet, wild rice, adlay, and rice grass, in *Cereal Science,* Matz, S. A., Ed., Avi Publishing, Westport, Conn., 1969, 224–236.
7. **Rachie, K. O.,** *The Millets and Minor Cereals. A Bibliography of the World Literature pre 1930 and 1964–69 and of All Literature of Other Minor Cereals,* Scarecrow Press, Metuchen, N.J., 1974, 201 pp.
8. **Smith, T. E.,** Triticale, grain of the future, *Crops Soils Mag.,* 19, 15–16, 1967.
9. **Stevens, T. A.,** Wild rice – Indian food and a modern delicacy, *Econ. Bot.,* 6, 107–142, 1952.
10. **Tsen, C. C., Ed.,** *Triticale, First Man-Made Cereal. Proceedings of a Symposium,* American Association of Cereal Chemists, St. Paul, Minn., 1974. 292 pp.
11. **von Schaffhausen, R.,** Adlay or Job's tears – A cereal of potentially greater importance, *Econ. Bot.,* 6, 216–227, 1952.

3. Pulses and Oil Seeds

a. General

1. **Aykroyd, W. R. and Doughty, J.,** Legumes in Human Nutrition, FAO Nutr. Stud. No. 19, FAO, Rome, 1964, 138 pp.
2. **Deschamps, I.,** Peas and beans, in *Processed Plant Protein Foodstuffs,* Altschul, A. M., Ed., Academic Press, New York, 1959, 717–737. (57 ref.)
3. **Oyenuga, V. A.,** Improvement of nutritional status in developing countries by improved food production: Legumes, *Proc. 7th International Congress of Nutrition,* Vol. 4, Hamburg, 1966, Pergamon Press, Oxford, 1967, 35–50.
4. **Patwandhan, V. N.,** Pulses and beans in human nutrition, *Am. J. Clin. Nutr.,* 11, 12–30, 1962.
5. Protein Advisory Group of the United Nations System, Upgrading Human Nutrition through the Improvement of Food Legumes, PAG Statement No. 22, March 9, 1973, *PAG Bull.,* 3(2), 1–24, 1973.
6. Symposium on legumes and green leafy vegetables in the nutrition of the infant and young child, *PAG Bull.,* 3(2), 25–53, 1973. (6 papers)

b. Coconut

1. **Abdon, I. C.,** Coconut as a source of protein, *Philipp. J. Nutr.,* 22, 103–113, 1969.
2. **Woodroof, J. G.,** *Coconuts: Production, Processing, Products,* Avi Publishing, Westport, Conn., 1970, 241 pp.

c. Cottonseed

1. **Bailey, A. E., Ed.,** *Cottonseed and Cottonseed Products. Their Chemistry and Chemical Technology,* Interscience, New York, 1948, 936 pp.
2. *Cottonseed and Its By-Products.* Proc. of the Symposium, Hyderabad, Dec. 5–7, 1958, Indian Central Oilseeds Committee, Hyderabad, 1959, 360 pp.
3. **Lambou, M. G., Shaw, R. L., Decossas, K. M., and Vix, H. L. E.,** Cottonseed's role in a hungry world, *Econ. Bot.,* 20, 256–267, 1960. (38 ref.)
4. National Cottonseed Product Association, Cottonseed and its Products, Dallas, 1948.

d. Peanut or Groundnut

1. **Hoffpauir, C. L.,** Peanut composition – relation to processing and utilization, *J. Agric. Food Chem.,* 1, 668–671, 1953.
2. **Milner, M.,** Peanuts as a protein source in international feeding programs, *Food Technol.,* 16, 46–50, 52–53, 1962.
3. **Mottern, H. H.,** Peanuts and human nutrition, in *Peanuts–Culture and Uses. A Symposium,* American Peanut Research and Education Association, Stone Printing Co., Roanoke, Va., 1972, 593–602.
4. **Wilson, C. T.** The present future role of peanuts in meeting the world's need for food, *J. Am. Oil Chem. Soc.,* 49, 343A–344A, 1972.
5. **Woodroof, J. G.,** *Peanuts – Production, Processing, Products,* Avi Publishing, Westport, Conn., 1966, 291 pp.

e. Rapeseed

1. **Appelqvist, L. A. and Ohlson, R.,** *Rapeseed: Cultivation, Composition, Processing and Utilization,* Elsevier, Amsterdam, 1972, 391 pp.

f. Safflower

1. **Kohler, G. O.,** Safflower. A potential source of protein for human food, in *World Protein Resources, Advances in Chemistry Ser.* 57, American Chemical Society, Washington, D.C., 1966, 243–253. (33 ref.)

g. Soybean

1. **Bowdidge, E.,** *The Soya Bean: Its History, Cultivations and Uses,* Oxford University Press, London, 1935.
2. **Caldwell, B. E., Ed.,** *Soybeans: Improvement. Production, and Uses,* American Society of Agronomy, Madison, Wis., 1973, 681 pp.
3. **Chen, P. S. and Chen, H. D.,** *Soybeans for Health, Longevity and Economy,* Chemical Elements, South Lancaster, 1956.
4. **Cowan, J. C.,** Soybeans, in *Encyclopedia of Chemical Technology,* 2nd ed., Vol. 18, John Wiley & Sons, New York, 1969, 599–614.
5. **Cowan, J. C.,** Processing and products, in *Soybeans: Improvement, Production, and Uses,* Caldwell, B. E., Ed., American Society of Agronomy, Madison, Wis., 1973, 619–664.
6. **Drown, M. S.,** Soybeans and Soybean Products as Food, U.S. Dep. Agric. Misc. Publ. 534, 1943.
7. **Judd, R. W.,** The golden bean meets human needs, *Plant Food Rev.,* 12(4), 2–3, 1966.
8. **Leng, E. R.,** Soybeans – Potential for extension to areas of protein shortage, *Econ. Bot.,* 22, 37–41, 1968.
9. **Markley, K. S., Ed.,** *Soybeans and Soybean Products,* Interscience, New York, 1950. (2 vols.)
10. **Morse, W. J. and Cartter, J. L.,** Soybeans for feed, food and industrial products, Farmers' Bull. 2038, U.S. Dep. Agric., Washington, D.C., 1952.

11. **Platt, B. S.**, The soya bean in human nutrition, *Chem. Ind.*, 18, 834–837, 1956.
12. **Smith, A. K. and Circle, S. J., Eds.**, *Soybeans: Chemistry and Technology. Proteins*, Vol. 1. Avi Publishing, Westport, Conn., 1972, 470 pp.
13. **Wolf, W. J. and Cowan, J. C.**, *Soybeans as a Food Source*, CRC Press, Cleveland, Ohio, 1971, 86 pp.
14. **Woodruff, S. and Klaas, H.**, Study of Soybean Varieties with Reference to Their Use as Food, Univ. Ill. Agric. Exp. Sta. Bull. 443, 1938.

4. Roots and Tubers

a. General

1. **Coursey, D. G. and Haynes, P. H.**, Root crops and their potential as food in the tropics, *World Crops*, 22(4) 261–265, 1970. (38 ref.)
2. **Jones, W. O.**, Starchy roots in the dieteries of developing tropical countries, *Proc. Int. Symp. Trop. Root Crops*, University of West Indies, St. Augustine, Trinidad, 2(5), 34–44, 1967.

b. Potato

1. **Salaman, R. N.**, *The History and Social Influence of the Potato*, Cambridge University Press, Cambridge, Mass., 1949.
2. **Smith, O.**, Culinary and nutritive value of potatoes, in *Potatoes: Production, Storing, Processing*, Avi Publishing, Westport, Conn., 1968, 498–557.
3. **Talburt, W. F.**, The nutritive value of potatoes, in *Potato Processing*, Talburt, W. F. and Smith, O., Eds., 2nd ed., Avi Publishing, Westport, Conn., 1967, 218–241.
4. **Wadsworth, G. R. and McKenzie, J. C.**, The potato, with special reference to its use in the United Kingdom, *Nutr. Abstr. Rev.*, 33, 327–344, 1963.

c. Cassava

1. **Holleman, L. W. J. and Aten, A.**, Processing of Cassava and Cassava Products in Rural Industries, FAO Agric. Dev. Paper 54, FAO, Rome, 1966, 115 pp.
2. **Magoon, M. L. and Appan, S. G.**, Cassava, a food for the millions, *Indian Farm.*, 16(1), 12-13, 1966.
3. **Oke, O. L.**, Cassava as food in Nigeria, *World Rev. Nutr. Diet.*, 9, 227–250, 1968. (31 ref.)
4. **Pereira, C. L.**, The cultivation of cassava and its value for human nutrition, *Rev. Agric.* (Mozambique), 3(31), 8–11, 1961.

d. Yam

1. **Coursey, D. G. and Martin, F. W.**, The past and future of the yams as crop plants, *Plant Foods Hum. Nutr.*, 2, 133–138, 1972.
2. **Oke, O. L.**, Yam – A valuable source of food and drugs, *World Rev. Nutr. Diet.*, 15, 156–184, 1972.

5. Vegetables

1. **King, F. W.**, The place of vegetables in meeting the food needs in emerging nations, *Econ. Bot.*, 25, 6–11, 1971.
2. **Lee, F. A.**, Vegetables and mushrooms, in *The Chemistry and Technology of Food and Food Products*, 2nd ed., Vol. 2, Jacobs, M. B., Ed., Interscience, New York, 1951, 1212–1347.
3. **Rajalakshmi, R. and Ramakrishnan, C. V.**, Horticulture in relation to nutritional requirements, *Plant Foods Hum. Nutr.*, 1, 79–92, 1969.
4. **Terra, G. J. A.**, The significance of leafy vegetables, especially of cassava in tropical nutrition, *Trop. Geog. Med.*, 16, 97–108, 1964.

6. Fruits

1. **Cruess, W. V.**, *Commercial Fruit and Vegetable Products*, McGraw-Hill, New York, 1958.
2. **Hulme, A. C., Ed.**, *The Biochemistry of Fruits and Their Products*, Academic Press, New York, 1970, 620 pp.
3. **Kefford, J. F.**, Citrus fruits and processed citrus fruits in human nutrition, *World Rev. Nutr. Diet.*, 6, 197–249, 1966; 18, 61–120, 1973.
4. **Lee, F. A.**, Fruits and nuts, in *The Chemistry and Technology of Food and Food Products*, 2nd ed., Vol. 2, Jacobs, M. B., Ed., Interscience, New York, 1951, 1348–1589.
5. **White, P. L. and Selvey, N., Eds.**, *Nutritional Qualities of Fresh Fruits and Vegetables*, Futura Publishing, Mount Kisco, N. Y., 1974.

B. Technical Measures for Increasing Crop Productivity

1. General

1. **Balls, W. L.**, *The Yields of a Crop*, E. & F. N. Spon, London, 1953, 144 pp.
2. **Bonner, J.**, The upper limit of crop yield, *Science*, 137, 11–16, 1962.
3. **Brickbauer, E. and Mortenson, W. P.**, *Approval Practices in Crop Production*, Interstate, Danville, Ill., 1967.

4. **Bunting, A. H.,** Problems in the improvement of output of arable crops in developing countries, *Proc. Nutr. Soc.*, 21, 91–98, 1962.
5. **Clements, H. F.,** Interaction of factors affecting yield, *Annu. Rev. Plant Physiol.*, 15, 409–442, 1964.
6. **Delroit, R. and Ahlgren, H.,** *Crop Production*, 3rd ed., Prentice-Hall, New York, 1967.
7. **Dungan, G. H. and Ross, W.,** *Growing Field Crops*, McGraw-Hill, New York, 1957, 495 pp.
8. **Dutt, C. P. and Pugh, B. M.,** *Farm Science and Crop Production in India*, 2nd ed., Allahabad, Kitabistan, 1947, 240 pp.
9. **Eastin, J. D. and Munson, R. D., Ed.,** *Moving off the Yield Plateau*, ASA Special Publ. No. 20, American Society of Agronomy, Madison, Wis., 1971, 89 pp. (6 papers)
10. **Helbaek, H.,** Domestication of food plants in the old world, *Science*, 130, 365–372, 1959.
11. **Hughes, H. D. and Henson, E. R.,** *Crop Production*, Macmillan, New York, 1957.
12. **Hughes, H. D. and Metcalfe, D.,** *Crop Production*, 3rd ed., Macmillan, New York, 1972.
13. **Hughes, H. D.,** A half-century of crop production research, *Agron. J.*, 49, 637–642, 1957.
14. **Hutcheson, T. B., Wolfe, T. K., and Kipps, M. S.,** *Production of Field Crops*, 2nd ed., McGraw-Hill, New York, 1936, 445 pp.
15. **Kipps, M. S.,** *Production of Field Crops*, 6th ed., McGraw-Hill, New York, 1970, 790 pp.
16. **Loomis, R. S. and Williams, W. A.,** Maximum crop productivity: An estimate, *Crop Sci.*, 3, 67–72, 1963.
17. **Loomis, R. S., Williams, W. A., and Hall, A. E.,** Agricultural productivity, *Annu. Rev. Plant Physiol.*, 22, 431–468, 1971. (263 ref.)
18. **Martin, J. H. and Leonard, W. H.,** *Field Crop Production*, 2nd ed., Macmillian, New York, 1967.
19. **Martin, J. H. and Leonard, W. H.,** *Principles of Field Crop Production*, Macmillan, New York, 1949, 1176 pp.
20. **Merrill, E. D.,** Domesticated plants in relation to the diffusion of culture, *Bot. Rev.*, 4, 1–20, 1938.
21. *Maximum Crop Yields – The Challenge*, American Society of Agronomy, Madison, Wis., 1967.
22. **Moss, D. N. and Musgrave, R. B.,** Photosynthesis and crop production, *Adv. Agron.*, 23, 317–336, 1971.
23. New Vistas in Crop Yields, Indian Council of Agricultural Research, New Delhi, 1970, 710 pp.
24. **Park, R. D.,** *Crop Husbandry*, Oxford University Press, London, 1961, 237 pp.
25. **Pearson, L. C.,** *Principles of Agronomy*, Reinhold, New York, 1967, 434 pp.
26. **Sprague, G. F.,** Increasing crop yields: Technical measures for increasing productivity, in *Man, Food, and Nutrition. Strategies and Technological Measures for Alleviating the World Food Problem*, Rechcigl, M., Jr., Ed., CRC Press, Cleveland, 1973, 91–107. (19 ref.)
27. **Waereing, P. F. and Cooper, J. P., Eds.,** *Potential Crop Production. A Case Study*, Heinemann, London, 1971.
28. **Westlake, D. F.,** Comparison of plant productivity, *Biol. Rev.*, 38, 385–425, 1963.
29. **Willey, R. W. and Heath, S. B.,** The quantitative relationship between plant production and crop yield, *Adv. Agron.*, 21, 281–321, 1969.
30. **Wilson, H. K. and Myers, W. M.,** *Field Crop Production. Agronomic Principles and Practices*, Lippincott, Philadelphia, 1954, 674 pp.
31. **Wilson, H. K. and Richter, A. C.,** *Production of Field Crops*, Interstate, Danville, Ill., 1960, 336 pp.
32. **Wolfe, T. K. and Kipps, M. S.,** *Production of Field Crops. A Textbook of Agronomy*, 5th ed., McGraw-Hill, New York, 1959, 538 pp.

2. Plant Breeding

1. **Allard, R. W.,** *Principles of Plant Breeding*, John Wiley & Sons, New York, 1960, 485 pp.
2. **Borlaug, N.,** Genetic improvement of crop foods, *Nutr. Today*, January–February 1972, 20–21, 24–25.
3. **Briggs, F. N. and Knowles, P. F.,** *Introduction to Plant Breeding*, Reinhold, New York, 1967.
4. **Briggs, D. and Walters, S. M.,** *Plant Variation and Evolution*, McGraw-Hill, New York, 1969, 256 pp.
5. **Creech, J. L. and Reitz, L. P.,** Plant germ plasm now and for tomorrow, *Adv. Agron.*, 23, 1–49, 1971.
6. **Day, P. R.,** Plant disease resistance, *Sci. Progr.*, 56, 357–370, 1968.
7. **Day, P. R.,** Crop resistance to pets and pathogens, in *Pest Control Strategies for the Future*, National Academy of Sciences, Washington, D.C., 1972, 257–271.
8. **Elliot, F. C.,** *Plant Breeding and Cytogenetics*, McGraw-Hill, New York, 1958, 395 pp.
9. **Frankel, O. H. and Bennett, E., Eds.,** *Genetic Resources in Plants*, Blackwell, Oxford, 1970.
10. **Frey, K. J., Ed.,** *Plant Breeding. A Symposium*, Iowa State University Press, Ames, 1966, 430 pp.
11. **Frey, K. J.,** Improving crop yields through plant breeding, in *Moving Off the Yield Plateau*, American Society of Agronomy, Madison, Wis., 1971, 15–58.
12. *Genetic Vulnerability of Maior Crops*, National Academy of Sciences, Washington, D.C., 1972, 307 pp.
13. Germ Plasm Resources, AAAS Publ. 66, Washington, D.C., 1961.
14. **Gustafsson, A.,** Mutations in agricultural plants, *Hereditas*, 33, 1–100, 1947.
15. **Hayes, H. K., Immer, F. R., and Smith, D. C.,** *Methods of Plant Breeding*, 2nd ed., McGraw-Hill, New York, 1955, 551 pp.
16. **Hooker, A. L. and Saxena, K. M. S.,** Genetics of diseases resistance in plants, *Annu. Rev. Genet.*, 5, 407–424, 1971.
17. **Jenkins, M. T.,** Genetic improvement of food plants for increased yield, *Am. Phil. Soc. Proc.*, 95, 84–91, 1951.

18. **Kuc, J.,** Resistance of plants to infectious agents, *Annu. Rev. Microbiol.,* 20, 337–370, 1966.
19. **Mangelsdorf, P. C.,** Genetic potentials for increasing yields of food crops and animals, *Proc. Natl. Acad. Sci. U.S.A.,* 56, 370–375, 1966.
20. Manual on Mutation Breeding, Tech. Rep. Ser. No. 119, IAEA, Vienna, 1970, 240 pp.
21. Mutations in Plant Breeding, Proc. Panel on Co-ordination of Research on the Production and Use of Induced Mutations in Plant Breeding, Vienna, January 17–21, 1966, IAEA, Vienna, 1966, 271 pp.; Vol. 2, 1968, 315 pp.
22. **Myers, W. M.,** *Recent Plant Breeding Research Svalof, 1946–1961,* John Wiley & Sons, New York, 1963.
23. New Approaches to Breeding for Improved Plant Protein, Proc. Panel Organized by the Joint FAO/IAEA Division of Atomic Energy in Food and Agriculture, Röstånga, June 1968, IAEA, Vienna, 1969, 195 pp.
24. **Nelson, R. R., Ed.,** *Breeding Plants for Disease Resistance, Concepts and Applications,* Pennsylvania State University Press, University Park, 1972.
25. Nutritional Society Symposium, Improving the nutritional value of cereal grains by the use of gene mutations and selective breeding, *Fed. Proc.,* 25, 1662–1678, 1966.
26. **Painter, A. H.,** *Insect Resistance in Crop Plants,* Macmillan, New York, 1951.
27. **Poehlman, J. M.,** *Breeding Field Crops,* Holt, Rinehart & Winston, New York, 1959, 427 pp.
28. Proc. Plant Breeding Symposium, Iowa State University Press, Ames, 1966.
29. **Simmonds, N. W.,** Variability in crop plants, its use and conservation, *Biol. Rev.,* 37, 422–465, 1962.
30. **Srb, A. M., Own, R. C., and Edgar, R. S.,** *General Genetics,* 2nd ed., W. H. Freeman Publishing, San Francisco, 1965.
31. Statistical Genetics and Plant Breeding, Publ. 982, National Academy of Sciences, Washington, D.C., 1963.
32. U.S. Department of Agriculture, Better Plants and Animals, I, II, The Yearbook of Agriculture 1936, 1937, U.S. Gov. Print. Off., Washington, D.C., 1936, 1937.
33. **Vavilov, N. I.,** The origin, variation, immunity, and breeding of cultivated plants, *Chron. Bot.,* Vol. 13, Waltham, Mass., 1949–1950, 364 pp.
34. **Wright, J. W.,** Genetics of Forest Tree Improvement, FAO, Rome, 1962.

3. Plant Nutrition

1. **Aldrich, D. G.** et al., *The Care and Feeding of Garden Plants,* American Society of Horticultural Science, Washington, D.C., 1954.
2. **Allaway, W. H.,** Agronomic controls over the environmental cycling of trace elements, *Adv. Agron.,* 20, 235–274, 1968.
3. **Anderson, A. J.,** Molybdenum as a fertilizer, *Adv. Agron.,* 8, 163–202, 1956.
4. **Anderson, A. J. and Underwood, E. J.,** Trace-element deserts, *Sci. Am.,* 200, 29–106, 1959.
5. **Berger, K. C.,** Boron in soils and crops, *Adv. Agron.,* 1, 321–351, 1949.
6. **Berger, K. C.,** Micronutrient deficiencies in the United States, *J. Agric. Food Chem.,* 10, 178–181, 1962.
7. **Bollard, E. G.,** Transport in the xylem, *Annu. Rev. Plant Physiol.,* 11, 141–166, 1960.
8. **Beeson, K. C.,** The mineral composition of crops with particular reference to the soils in which they were grown. A review and compilation, U.S. Dep. Agric., Misc. Publ. 369, 1941.
9. **Borys, M. W. and Childers, N. F.,** *The Role of Molybdenum in Plants and Soils,* Rutgers University Horticulture Dept., New Brunswick, N.J., 1961, 344 pp.
10. **Brown, J. C.,** Iron chlorosis in plants, *Adv. Agron.,* 13, 329–369, 1961.
11. **Brown, J. C.,** Interactions involving nutrient elements, *Annu. Rev. Plant Physiol.,* 14, 93–106, 1963.
12. **Camp, A. F. and Fudge, B. R.,** Zinc as a nutrient in plant growth, *Soil Sci.,* 60, 157–164, 1945.
13. **Chapman, H. D., Ed.,** *Diagnostic Criteria for Plants and Soils,* University of California, Riverside, 1966, 793 pp.
14. **Childers, N. F., Ed.,** *Fruit Nutrition,* Somerset Press, Somerville, N.J., 1954.
15. **Cooke, G. W.,** *Control of Soil Fertility,* Hafner Publishing, New York, 1967.
16. **Cooper, H. P.,** Minor plant nutrients, *Commercial Fertilizer,* 74, 14–44, 1947.
17. **De Turk, E. E.,** The physiological basis of plant nutrient deficiency symptoms, *J. Ind. Eng. Chem.,* 33, 648–653, 1941.
18. **Donald, C. M.,** Competition among crop and pasture plants, *Adv. Agron.,* 15, 1–118, 1963.
19. **Eastin, J. D., Ed.,** *Physiological Aspects of Crop Yield,* American Society of Agronomy, Madison, Wis., 1970.
20. **Epstein, E.,** Mineral nutrition of plants: Mechanism of uptake and transport, *Annu. Rev. Plant Physiol.,* 7, 1–24, 1956.
21. **Epstein, E.,** *Mineral Nutrition of Plants. Principles and Perspectives,* John Wiley & Sons, New York, 1972, 412 pp.
22. **Essau, K., Currier, H. B., and Cheadle, V. I.,** Physiology of phloem, *Annu. Rev. Plant Physiol.,* 8, 349–374, 1957.
23. **Evans, H. J.,** Role of molybdenum in plant nutrition, *Soil Sci.,* 81, 199–208, 1956.
24. **Evans, H. J. and Sorger, G. J.,** Role of mineral elements with emphasis on the univalent cations, *Annu. Rev. Plant Physiol.,* 17, 47–76, 1966.
25. **Fried, M. and Broeshart, H.,** *Soil-Plant System in Relation to Inorganic Nutrition,* Academic Press, New York, 1967.
26. **Fussell, G. E.,** *Crop Nutrition: Science and Practice Before Liebig,* Coronado Press, Lawrence, Kan., 1971, 232 pp.

27. **Gauch, H. G.,** *Inorganic Plant Nutrition,* Doroden, Hutchinson and Ross, Stroudsburg, Pa., 1972, 528 pp.
28. **Gerloff, G. C.,** Comparative mineral nutrition of plants, *Annu. Rev. Plant Physiol.,* 14, 107–124, 1963.
29. **Gilbert, F. A.,** *Mineral Nutrition and the Balance of Life,* University of Oklahoma Press, Norman, 1957, 350 pp.
30. **Gilbert, F. A.,** *Mineral Nutrition of Plants and Animals,* University of Oklahoma Press, Norman, 1953, 131 pp.
31. **Gilbert, F. A.,** Copper in nutrition, *Adv. Agron.,* 4, 147–173, 1952.
32. **Goodall, D. W. and Gregory, F. G.,** Chemical composition of plants as an index of their nutritional status, Imp. Bur. Hort. and Plant Crops Tech. Comm. No. 17, Kent, East Malling, 1947, 167 pp.
33. **Hewitt, E. J.,** The role of mineral elements in plant nutrition, *Annu. Rev. Plant Physiol.,* 2, 25–52, 1951.
34. **Hewitt, E. J.,** Essential nutrient elements for plants, in *Plant Physiology,* Vol. 3, Steward, F. C., Ed., Academic Press, New York, 1963, 137–360.
35. **Hoagland, D. R.,** *Inorganic Nutrition of Plants,* Chronica Botanica, Waltham, Mass., 1944.
36. **Hodgson, J. F.,** Chemistry of the micronutrient elements in soils, *Adv. Agron.,* 15, 119–159, 1963.
37. **Hoffer, G. N.,** Potassium plant metabolism, *J. Ind. Eng. Chem.,* 30, 885–889, 1938.
38. **Jacob, A. and Vexhall, O., von,** *Nutrition and Manuring of Tropical Crops,* Hannover, Germany, 1959.
39. **Jackson, W. A.,** Physiological effects of soil acidity, in *Soil Acidity and Liming,* Pearson, R. W. and Adams, F., Eds., *Am. Soc. Agron. Monogr.,* 12, 43–124, 1967.
40. **Jordan, H. V. and Ensminger, L. E.,** The role of sulfur in soil fertility, *Adv. Agron.,* 10, 407–434, 1958.
41. **Lee, S. and Aronoff, S.,** Boron in plants: A biochemical role, *Science,* 158, 798–799, 1967.
42. **Lewin, J. and Reimann, B. E. F.,** Silicon and plant growth, *Annu. Rev. Plant Physiol.,* 20, 289–304, 1969.
43. **Loneragan, J. F.,** Nutrient requirements of plants, *Nature,* 220, 1307–1308, 1968.
44. **Mitchell, R. L.,** Trace elements, in *Chemistry of the Soil,* Bear, F. E., Ed., Reinhold, New York, 1955, 253–285.
45. **Mulder, E. G. and Gerretsen, F. C.,** Soil manganese in relation to plant growth, *Adv. Agron.,* 4, 221–277, 1952.
46. **Mortvedt, J. J., Giordano, P. M., and Lindsay, W. L.,** Eds., *Micronutrients in Agriculture,* Soil Science Society of America, Madison, Wis., 1972, 666 pp. (24 papers)
47. **Nason, A. and McElroy, W. D.,** Modes of action of the essential mineral elements, in *Plant Physiology,* Vol. 3, Steward, F. C., Ed., Academic Press, New York, 1963, 465 – 508.
48. **Nelson, L. G. Berger, K. C., and Andries, H. J.,** Copper requirement and deficiency symptoms of a number of field and vegetable crops, *Soil Sci. Soc. Am. Proc.,* 20, 69 – 72, 1956.
49. **Parker, F. W. and Nelson, L. B.,** More fertilizers for more food, *Proc. Natl. Acad. Sci. U.S.A.,* 56, 382 – 388, 1966.
50. **Pierre, W. H.,** *Soil and Fertilizer Phosphorus in Crop Nutrition,* Academic Press, New York, 1953, 492 pp.
51. **Pillai, K. M.,** *Crop Nutrition,* Asia Publishing, New York, 1968.
52. **Price, C. A.,** Iron compounds and plant nutrition, *Annu. Rev. Plant Physiol.,* 18, 239 – 248, 1968.
53. **Rorison, I. H.,** Ed., *Ecological Aspects of the Mineral Nutrition of Plants, Brit. Ecol. Soc. Symp. No. 9,* Blackwell Scientific Publications, Oxford, 1969, 484 pp.
54. **Russell, R. S.,** Root systems and plant nutrition – Some new approaches, *Endeavour* (Engl. Ed.), 29, 60 – 66, 1970.
55. **Samish, R. M.,** Ed., *Recent Advances in Plant Nutrition. A Colloquium, Tel-Aviv, March 1970,* Gordon & Breach, New York, 1972. (2 vols.)
56. **Sauchelli, V.,** *Trace Elements in Agriculture,* Van Nostrand Reinhold, New York, 1969.
57. **Schrift, A.,** Aspects of selenium metabolism in higher plants, *Annu. Rev. Plant Physiol.,* 20, 475 – 494, 1969.
58. **Shaw, B. T.,** *Soil Physical Conditions and Plant Growth,* Academic Press, New York, 1952.
59. **Skok, J.,** The role of boron in the plant cell, in *Trace Elements,* Lamb, C. A., Bentley, O. G., and Beattie, J. M., Eds., Academic Press, New York, 1958, 227 – 243.
60. **Smith, P. F.,** Mineral analysis of plant tissue, *Annu. Rev. Plant Physiol.,* 13, 81 – 108, 1962.
61. **Sprague, H. B.,** Ed., *Hunger Signs in Crops,* 3rd ed., McKay, New York, 1964, 461 pp. (12 papers)
62. **Stiles, W.,** *Trace Elements in Plants,* 3rd ed., Cambridge University Press, New York, 1961.
63. **Thorne, D. W.,** Zinc deficiency and its control, *Adv. Agron.,* 9, 31 – 65, 1957.
64. **Tisdale, S. L. and Nelson, W. L.,** *Soil Fertility and Fertilizers,* 2nd ed., Macmillan, New York, 1966.
65. **Trugg, E., Ed.,** *Mineral Nutrition of Plants,* University of Wisconsin Press, Madison, 1951.
66. **Ulrich, A.,** Physiological bases for assessing the nutritional requirements of plants, *Annu. Rev. Plant Physiol.,* 3, 207–228, 1952.
67. **Viets, F. G., Jr.,** Zinc deficiency in the soil – plant system, in *Zinc Metabolism,* Prassad, A. S., Ed., Charles C Thomas, Springfield, 1966, 90 – 128.
68. **Wallace, T.,** Diagnosis of Mineral Deficiencies by Visual Symptoms, HMSO, London, 1951, 107 pp.
69. **Wallace, T.,** *The Diagnosis of Mineral Deficiencies in Plants,* 2nd ed., Chemical Publishing, New York, 1961, 125 pp.
70. **Wallace, A. and Lunt, O. R.,** Iron chlorosis in horticultural plants: A review, *Proc. Am. Soc. Hort. Sci.,* 75, 819 – 841, 1960.
71. **Wilkins, M. B.,** *Physiology of Plant Growth and Development,* McGraw-Hill, New York, 1969.
72. **Yoshida, S.,** Physiological aspects of grain yield, *Annu. Rev. Plant Physiol.,* 23, 437 – 464, 1972. (296 ref.)

73. **Younts, S. E.**, Trends in soil fertility and plant nutrition, in *Moving off the Yield Plateau*, Eastin, J. D. and Munson, R. D., Eds., American Society of Agronomy, Madison, Wis., 1971, 69 – 82.
74. **Zimmermann, M. H.**, Transport in the phloem, *Annu. Rev. Plant Physiol.*, 11, 167 – 190, 1960.

4. Plant Diseases and Disease Control

1. **Agrios, G. N.**, *Plant Pathology*, Academic Press, New York, 1969, 629 pp.
2. American Potash Institute, *Diagnostic Techniques for Soil and Crops*, American Potash Institute, Washington, D.C., 1948, 308 pp.
3. **Barnes, E. H.**, *Atlas and Manual of Plant Pathology*, Appleton-Century Crofts, New York, 1968, 325 pp.
4. **Boyce, J. S.**, *Forest Pathology*, 3rd ed., McGraw-Hill, New York, 1961, 572 pp.
5. **Brooks, F. T.**, *Plant Diseases*, Clarendon Press, Oxford, 1953.
6. **Butler, E. and Jones, S. A.**, *Plant Pathology*, Macmillan, London, 1955, 979 pp.
7. **Carter, W.**, *Insects in Relation to Plant Diseases*, John Wiley & Sons, New York, 1963.
8. **Chester, K. S.**, *Nature and Prevention of Plant Diseases*, 2nd ed., McGraw-Hill, New York, 1947, 525 pp.
9. Committee on Phytopathological Reviews, Bibliography of reviews, 1949–1959, *Phytopathology*, 50(8), part 2, B1–B116, 1960.
10. **Couch, H. B.**, *Diseases of Turf Grasses*, Reinhold, New York, 1962, 289 pp.
11. **Dickson, J. G.**, *Disease of Field Crops*, 2nd ed., McGraw-Hill, New York, 1956, 517 pp.
12. **Dowson, W.**, *Plant Diseases Due to Bacteria*, 2nd ed., Cambridge University Press, Cambridge, England, 1957, 231 pp.
13. **Eckert, J. W.**, Chemical Treatments for Control of Postharvest Diseases, *World Rev. Pest Control*, 8, 116–137, 1969. 74 ref.
14. **Evans, E.**, *Plant Diseases and their Chemical Control*, Blackwell Scientific, Oxford, 1968, 288 pp.
15. **Goodman, R. N., Király, Z., and Zaitlin, M.**, *The Biochemistry and Physiology of Infectious Plant Disease*, Van Nostrand, New York, 1967, 354 pp.
16. **Gram, E. and Weber, A.**, *Plant Diseases in Orchard, Nursery and Garden Crops*, Dennis, R. W. G., Ed., Philosophical Library, New York, 1953, 618 pp.
17. **Grant, T. J.**, Role of plant pathology in tropical agriculture, *Phytopathology*, 52, 930–936, 1962.
18. **Heald, F. D.**, *Manual of Plant Diseases*, 2nd ed., McGraw-Hill, New York, 1933, 891 pp.
19. **Hepting, G. H.**, Climate and forest diseases, *Annu. Rev. Phytopathol.*, 1, 31–50, 1963.
20. **Holton, C. S., Fischer, G. W., Fulton, R. W., Hart, H., and McCallan, S. E. A., Eds.**, *Plant Pathology. Problems and Progress, 1908–1958*, University of Wisconsin Press, Madison, 1959, 588 pp. (51 papers)
21. **Horsfall, J. G.**, Selective chemicals for plant disease control, in *Pest Control Strategies for the Future*, National Academy of Sciences, Washington, D.C., 1972, 216–225.
22. **Horsfall, J. G. and Dimmond, A. E., Eds.**, *Plant Pathology. An Advanced Treatise*, Academic Press, New York, 1956–1960. (3 vols.)
23. **Klinkowski, M., Mühle, E., and Reinmuth, E.**, *Phytopathologie und Pflanzenschutz*, Akademie-Verlag, Berlin, 1966. (2 vols.)
24. **Leach, J. G.**, *Insect Transmission of Plant Diseases*, McGraw-Hill, New York, 1940, 615 pp.
25. **Levitt, J.**, *The Hardiness of Plants*, Academic Press, New York, 1956, 278 pp.
26. McNew, G. L., Progress in the battle against plant disease, in *Scientific Aspects of Pest Control*, National Academy of Sciences, Publ. 1402, Washington, D. C., 1966, 73–101.
27. **Melhus, I. E. and Kent, G. C.**, *Elements of Plant Pathology*, Macmillan, New York, 1939, 493 pp.
28. **Metlitsky, L. V. and Ozeretskovskaya, O. K.**, *Plant Immunity. Biochemical Aspects of Plant Resistance to Parasitic Fungi*, Plenum Press, New York, 1968.
29. **Miller, P. R. and O'Brien, M. J.**, Prediction of plant disease epidemics, *Annu. Rev. Microbiol.*, 11, 77–110, 1957.
30. **Paddock, W. C.**, Phytopathology in a hungry world, *Annu. Rev. Phytopathol.*, 5, 375–390, 1967. (25 ref.)
31. **Parker, J.**, Physiological diseases of trees and shrubs, *Adv. Frontiers Plant Sci.*, 12, 97–248, 1965.
32. **Peace, T. R.**, *Pathology of Trees and Shrubs, with Special Reference to Britain*, Oxford University Press, London, 1962, 753 pp.
33. **Pirone, P. P., Dodge, B. O., and Rickett, H. W.**, *Diseases and Pests of Ornamental Plants*, 3rd ed., Ronald Press, New York, 1960, 775 pp.
34. Plant Pathologists's Pocketbook, Commonwealth Mycological Institute, 1968, 267 pp.
35. **Rich, S.**, Current developments in control of pathogenesis of key economic crops, *Dev. Ind. Microbiol.*, 10, 81–87, 1969.
36. **Rubin, B. A. and Artsikhovskaya, Y. V.**, *Biochemistry and Physiology of Plant Immunity*, Pergamon Press, Oxford, 1963, 358 pp.
37. **Shurtleff, M. C.**, *Control of Plant Diseases*, Iowa State University Press, Ames, 1962.
38. **Smith, K. M.**, *Textbook of Plant Virus Diseases*, 2nd ed., Little, Brown & Co., Boston, 1957.
39. **Smith, W. H.**, *Tree Pathology: A Short Introduction*, Academic Press, New York, 1970, 309 pp.
40. **Stakman, E. C.**, Opportunity and obligations in plant pathology, *Annu. Rev. Phytopathol.*, 2, 1–12, 1964.
41. **Stakman, E. C. and Harrar, J. G.**, *Principles of Plant Pathology*, Ronald Press, New York, 1957, 582 pp.

42. **Stevens, N. E. and Stevens, R. B.,** *Diseases in Plants,* Chronica Botanica, Waltham, Mass., 1952, 219 pp.
43. **Streets, R. B.,** The Diagnosis of Plant Diseases, Ariz. Coop. Ext. Serv. Agric. Exp. Sta., 1969.
44. **Thomas, M. D.,** Effects of air pollution on plants, in *Air Pollution,* WHO Monogr. Ser. 46, WHO, Geneva, 1961, 233–278.
45. U.S. Department of Agriculture, Plant Diseases, The Yearbook of Agriculture 1953, U.S. Gov. Print. Off., Washington, D.C., 1953, 940 pp.
46. U.S. Department of Agriculture, Crops Research Division, ARS, Index of Plant Diseases in the United States, Agric. Handb. No. 165, U.S. Gov. Print. Off., Washington, D.C., 1960, 531 pp.
47. **van der Plank, J. E.,** *Plant Diseases: Epidemics and Control,* Academic Press, New York, 1963, 350 pp.
48. **van der Plank, J. E.,** *Disease Resistance in Plants,* Academic Press, New York, 1968, 206 pp.
49. **Walker, J. C.,** *Plant Pathology,* 3rd ed., McGraw-Hill, New York, 1969, 819 pp.
50. **Westcott, C.,** *Plant Disease Handbook,* 2nd ed., Van Nostrand, Princeton, N.J., 1960, 825 pp.
51. **Wood, R. K. S.,** *Physiological Plant Pathology,* Blackwell Scientific, Oxford, 1967, 570 pp.
52. National Research Council, Committee on Plant and Animal Pests, *Plant-Disease Development and Control,* in *Principles of Plant and Animal Pest Control,* Vol. 1, National Academy of Sciences, Publ. 1596, Washington, D.C., 1968, 205 pp.

5. Reproduction and Plant Propagation

1. **Crocker, W.,** Life-span of seeds, *Bot. Rev.,* 4, 235–274, 1938.
2. **Crocker, W. and Barton, L. V.,** *Physiology of Seeds,* Chronica Botanica, Waltham, Mass., 1953, 267 pp.
3. **Hartmann, H. T. and Kester, D. E.,** *Plant Propagation,* 2nd ed., Prentice-Hall, Englewood Cliffs, N.J., 1968.
4. **Salisbury, E.,** *The Reproductive Capacity of Plants,* Bell, London, 1942.
5. U.S. Department of Agriculture, Woody Plant Seed Manual, U.S. Dep. Agric. Misc. Publ. 654, 1948.
6. U.S. Department of Agriculture, Manual for Testing Agricultural and Vegetable Seeds, Agric. Handb. No. 30, Washington, D.C., 1952.
7. U.S. Department of Agriculture, Seeds, The Yearbook of Agriculture 1961, U.S. Gov. Print. Off., Washington, D.C., 1961, 591 pp.

C. Production of Specific Crops

1. Crops In General

1. **Acland, J. D.,** *East African Crops,* FAO, Longmans, London, 1971, 252 pp.
2. **Aiyer, A. K., Yegna Narayan,** *Field Crops of India,* 4th ed., Bangalore Printing and Publishing, Bangalore, 1950, 618 pp.
3. **Bland, B. F.,** *Crop Production: Cereals and Legumes,* Academic Press, New York, 1971, 466 pp.
4. **Janick, J., Schery, R. W., Woods, F. W., and Ruttan, V. W.,** *Plant Science. An Introduction to World Crops,* W. H. Freeman Publishing, San Francisco, 1969, 629 pp.
5. **Northern, H. T.,** *Introductory Plant Science,* 2nd ed., Ronald Press, New York, 1958, 718 pp.
6. **Ochse, J. J., Soule, M. J., Jr., Dijkman, M. J., and Wehlburg, C.,** *Tropical and Subtropical Agriculture,* Macmillan, New York, 1961. (2 vols.)
7. **Percival, J.,** *Agricultural Botany,* 7th ed., Duckworth, London, 1926, 839 pp.
8. **Purseglove, J. W.,** *Tropical Crops: Dicotyledons,* John Wiley & Sons, New York, 1968. (2 vols.)
9. **Rather, H. C. and Harrison, C. M.,** *Field Crops,* 2nd ed., McGraw-Hill, New York, 1951, 446 pp.
10. **Robbins, W. W.,** *Botany of Crop Plants,* 3rd ed., Blakiston's Sons, Philadelphia, 1931, 639 pp.
11. **Wilson, C. L. and Loomis, W. E.,** *Botany,* 3rd ed., Holt, Rinehart & Winston, New York, 1962, 573 pp.

2. Cereals

a. General

1. **Hector, J. M.,** *Introduction to the Botany of Field Crops,* Vol. 1, *Cereals,* Central News Agency, Johannesburg, S. Africa, 1936.
2. **Johnson, D. G. and Gustafson, R. L.,** *Grain Yields and the American Food Supply. An Analysis of Yield Changes and Possibilities,* University of Chicago Press, Chicago, Ill., 1962, 146 pp.
3. **Krantz, B. A. and Melsted, S. W.,** Nutrient deficiences in corn, sorghums, and small grains, in *Hunger Signs in Crops,* Sprague, H. B., Ed., McKay, New York, 1964, 25–46.
4. **Leonard, W. H. and Martin, J. H.,** *Cereal Crops,* Macmillan, New York, 1963, 824 pp.
5. **Milthorpe, F. L. and Ivins, J. D., Eds.,** *The Growth of Cereals and Grasses,* Butterworth, London, 1966.
6. **Sprague, R.,** *Diseases of Cereals and Grasses in North America,* Ronald Press, New York, 1950, 538 pp.
7. **Wilson, H. K.,** *Grain Crops,* 2nd ed., McGraw-Hill, New York, 1955, 396 pp.

b. Barley

1. **Åberg, E. and Wiebe, G. A.,** Taxonomic Value of Characters in Cultivated Barley, U.S. Dep. Agric. Tech. Bull., 942, 1948.

2. **Aufhammer, E. G., Bergal, P., Hagberg, A., Horne, F. R., and van Veldhuizen, H.,** *Barley Varieties,* 2nd ed., American Elsevier, New York, 1958.
3. **Bland, B. F.,** Barley, in *Crop Production: Cereals and Legumes,* Academic Press, New York, 1971, 63–120.
4. **Clark, H. H.,** The origin and early history of cultivated barleys. A botanical and archeological synthesis, *Agric. Hist. Rev.,* 15, 1–18, 1967.
5. **Cook, A. H., Ed.,** *Barley and Malt: Biology, Biochemistry, Technology,* Academic Press, New York, 1962, 740 pp.
6. **Harlan, H. V.,** *One Man's Life with Barley,* Exposition Press, New York, 1957, 223 pp.
7. **Hunter, H.,** *The Barley Crop,* Crosby Lockwood, London, 1952.
8. **Leonard, W. H. and Martin, J. H.,** Barley, in *Cereal Crops,* Macmillan, New York, 1963, 478–543. (127 ref.)
9. **Luekel, R. W. and Tapke, V. F.,** Barley Diseases and their Control, U.S. Dep. Agric. Bull. 2089 Washington, D.C., 1965.
10. **Shands, H. L. and Dickinson, A. D.,** Barley – botany, production, harvesting, processing, utilization and economics, *Econ. Bot.,* 7, 3–26, 1953.
11. **Smith, L.,** Cytology and genetics of barley, *Bot. Rev.,* 17, 1–51, 133–202, 285–355, 1951.
12. **Standt, G.,** The origin of cultivated barleys: A discussion, *Econ. Bot.,* 15, 205–212, 1961.
13. **Takahashi, R.,** The origin and evolution of cultivated barley, *Adv. Genet.,* 7, 227–266, 1955.
14. **Weaver, J. C.,** *American Barley Production,* Burgess Publishing, Minneapolis, 1950, 115 pp.
15. **Wiebe, G. A.,** Barley: Origin, Botany, Culture, Winter Hardiness, Genetics, Utilization, Pests, U.S. Dep. Agric., Agric. Handb. No. 338, Washington, D.C., 1968.

c. Corn or Maize

1. **Aldrich, S. R. and Leng, E. R.,** *Modern Corn Production,* The Farm Quarterly, Cincinnati, 1965, 308 pp.
2. **Blair, R. A.,** Growth rate of maize under field conditions, *Plant Physiol.,* 17, 619–631, 1942.
3. **Bland, B. F.,** Maize, in *Crop Production: Cereals and Legumes,* Academic Press, New York, 1971, 216–263.
4. **Crabb, R. A.,** *Hybrid Corn Makers,* Rutgers University Press, New Brunswick, N.J., 1947.
5. **Cunard, A. C.,** Maize agronomy. I. Nutrition and silage, *World Crops,* 19(1), 20–27, 1967.
6. **Cunard, A. C.,** Maize agronomy. II. Nutrients and nutrient uptake, *World Crops,* 19(3), 24–28, 1967.
7. **Cunard, A. C.,** Maize agronomy. III. Growth and fertilizer usage, *World Crops,* 19(4), 50–60, 1967.
8. **Dungan, G. H., Lang, A. L., and Pendleton, J. W.,** Corn plant population in relation to soil productivity, *Adv. Agron.,* 10, 435–473, 1958.
9. **Erwin, A. T.,** Sweet corn–mutant or historic species, *Econ. Bot.,* 5, 302–306, 1951.
10. **Hageman, R. H., Leng, E. R., and Dudley, J. W.,** A biochemical approach to corn breeding, *Adv. Agron.,* 19, 45–86, 1967.
11. **Huelsen, W. A.,** *Sweet Corn,* World Crop Books, Interscience, New York, 1954.
12. **Inglett, G. E., Ed.,** *Corn: Culture, Processing, Products,* Avi Publishing, Westport, Conn., 1970, 369 pp. (18 papers)
13. **Leonard, W. H. and Martin, J. H.,** Indian corn or maize, in *Cereal Crops,* Macmillan, New York, 1963, 129–274. (85 ref.)
14. **Mangelsdorf, P. C.,** Hybrid corn, *Sci. Am.,* 185(2), 39–47, 1951.
15. **Mangelsdorf, P. C.,** The origin and evolution of maize, *Adv. Genet.,* 1, 161–207, 1947.
16. **Mangelsdorf, P. C., MacNeish, R. S., and Galinat, W. C.,** Domestication of corn, *Science,* 143, 538–545, 1964.
17. **Martin, J. H. and Leonard, W. H.,** Maize (Corn), in *Field Crop Production,* Macmillan, New York, 1950, 331–396.
18. **Miracle, M. P.,** *Maize in Tropical Africa,* University of Wisconsin Press, Madison, 1966.
19. **Montgomery, E. G.,** *The Corn Crops,* Macmillan, New York, 1913, 347 pp.
20. National Fertilizer Development Center, Nutrition of Maize, Bull. Y-26, Tennessee Valley Authority, Muscle Shoals, Alabama, 1971, 107 pp. (653 ref.)
21. **Nelson, L. B.,** The mineral nutrition of corn as related to its growth and culture, *Adv. Agron.,* 8, 321–375, 1956.
22. **Norden, A. J., Rossman, E. C., and Benne, E. J.,** Some factors that affect protein content of corn, *Mich. Agric. Exp. Sta. Q. Bull.,* 35(3), 210–225, 1952.
23. **Pesek, J.,** Potassium nutrition of soybeans and corn, in *Role of Potassium in Agriculture,* American Society of Agronomy, Iowa State University, Ames, 1968, 147–168.
24. **Pierre, W. H., Aldrich, S. A., and Martin, W. P., Eds.,** *Advances in Corn Production,* Iowa State University Press, Ames, 1966, 476 pp.
25. **Rhoades, M. M. and McClintock, B.,** The cytogenetics of maize, *Bot. Rev.,* 1, 292–325, 1935.
26. **Richey, F. D. and Dawson, R. F.,** A survey of the possibilities and methods of breeding high-niacin corn (maize), *Plant Physiol.,* 23, 238–254, 1948.
27. **Richey, F. D.,** Corn breeding, *Adv. Genet.,* 3, 159–192, 1950.
28. **Richey, F. D. and Dawson, R. F.,** Experiments on the inheritance of niacin in corn (maize), *Plant Physiol.,* 26, 474–493, 1951.
29. **Sprague, G. F.,** The experimental basis for hybrid maize, *Biol. Rev.,* 21, 101–120, 1946.
30. **Sprague, G. F., Ed.,** *Corn and Corn Improvement,* Academic Press, New York, 1955, 699 pp.
31. **Sprague, G. F.,** *Corn: Breeding, Production, and Utilization,* Academic Press, New York, 1955.

32. **Stringfield, G. H.**, Objectives in corn improvement, *Adv. Agron.*, 16, 102–137, 1964.
33. **Stringfield, G. H. and Anderson, M. S.**, Corn production, U.S. Dep. Agric. Farmer's Bull., 2073, 1954.
34. **Wallace, H. A. and Bressman, E. N.**, *Corn and Corn Growing*, 5th ed., John Wiley & Sons, New York, 1949, 424 pp.
35. **Wallace, H. A. and Brown, W. L.**, *Corn and Its Early Fathers*, Michigan State University Press, Lansing, 1956, 134 pp.
36. **Weatherwax, P.**, *The Story of the Maize Plant*, University of Chicago Press, Chicago, 1923, 247 pp.
37. **Weijir, J.**, A catalogue of genetic maize types together with a maize bibliography, *Bibl. Genet.*, 14, 189–425, 1952.

d. Oats

1. **Bland, B. F.**, Oats, in *Crop Production: Cereals and Legumes*, Academic Press, New York, 1971, 121–176.
2. **Brownlee, H. J. and Gunderson, F. L.**, Oats and oat production: Culture, botany, seed structure, milling, composition, and uses, *Cereal Chem.*, 15, 257–272, 1938.
3. **Coffman, F. A., Ed.**, Oats and Oat Improvement, ASA Monogr., Vol. 8, American Society of Agronomy, Madison, Wis., 1961, 650 pp.
4. **Findlay, W. M.**, *Oats, Their Cultivation and Use from Ancient Times to the Present Day*, Aberdeen University Studies 137, Oliver & Boyd, Edinburgh, 1956.
5. **Hunter, H.**, *Oats: Their Varieties and Characteristics*, Benn, London, 1924, 131 pp.
6. **Leonard, W. H. and Martin, J. H.**, Oats, in *Cereal Crops*, Macmillan, New York, 1963, 544–603. (152 ref.)
7. **Stanton, T. R.**, Production, harvesting, processing, utilization, and economic importance of oats, *Econ. Bot.*, 7, 43–64, 1953.

e. Rice

1. **Adair, C. R.**, Rice improvement and culture in the United States, *Adv. Agron.*, 14, 61–148, 1962.
2. **Adair, C. R.**, Rice in the United States. Varieties and Production, U.S.Dep.Agric., Agric. Handb. No. 289, U.S. Gov. Print. Off., Washington, D.C., 1966, 124 pp.
3. **Angladette, A. A.**, *Le Riz*, G. P. Maisonneuve & LaRosa, Paris, 1966.
4. **Chandler, R. F., Ed.**, *Rice Genetics and Cytogenetics, Proc. Symposium*, International Rice Research Institute, Los Banos, American Elsevier, New York, 1964.
5. **Chandraratna, M. F.**, *Genetics and Breeding of Rice*, Longmans, London, 1964.
6. **Comhaire, M.**, *Rice Manuring*, Centre International d'Information et de Documentation des Producteurs de Phosphate Thomas, Brussels, 1965, 197 pp.
7. **Copeland, E. B.**, *Rice*, Macmillan, London, 1924, 352 pp.
8. **de Geus, J. G.**, *Means of Increasing Rice Production*, Centre d'Etude de l'Azote, Geneva, 1954.
9. **de Geus, J. G.**, Rice, in *Fertilizer Guide for Tropical and Subtropical Farming*, Centre d'Etude de l'Azote, Zurich, 1967, 31–69.
10. **Dore, W. G.**, Wild Rice, Publ. 1393, Can. Dep. Agric. Res. Branch, 1969, 84 pp.
11. **Doyle, J. J.**, The Response of Rice to Fertilizer, FAO Agric. Stud. No. 70, FAO, Rome, 1966, 69 pp.
12. **Efferson, J. N.**, *The Production and Marketing of Rice*, The Rice Journal, New Orleans, La., 1952, 534 pp.
13. Food and Agriculture Organization, Rice Trade Glossary, FAO, Rome, 1957.
14. Food and Agriculture Organization, Rice. Grain of Life, World Food Problems No. 6, FAO, Rome, 1966, 93 pp.
15. **Grist, D. H.**, *Rice*, 4th ed., Longmans, London, 1965, 548 pp.
16. **Grist, D. H.**, Cultivation of rice, *World Crops*, 21(1), 17–21, 1969.
17. International Rice Research Institute, *International Bibliography of Rice Research, 1951–1961*, Scarecrow Press, Metuchen, N.J., 1964, 881 pp; Annu. Suppl.
18. International Rice Research Institute, *The Mineral Nutrition of the Rice Plant, Proc. Symposium, Los Banos, Philippines, February 1964*, Johns Hopkins Press, Baltimore, 1965, 494 pp.
19. International Rice Research Institute, *Symposium on the Major Insect Pests of the Rice Plants, 1964*, Johns Hopkins Press, Baltimore, 1967.
20. **Ishizuka, Y.**, Physiology of the rice plant, *Adv. Agron.*, 23, 241–315, 1971.
21. **Iso, E.**, Rice and Crops in its Relations in Subtropical Zones, Jap. FAO Assoc., 1954, 611 pp.
22. **Janick, J., Schery, R. W., Woods, F. W., and Ruttan, V. W.**, Rice, in *Plant Science*, W. H. Freeman Publishing, San Francisco, 1969, 352–355.
23. **Juliano, B. O.**, Physicochemical Data on the Rice Grain, International Rice Research Institute, Tech. Bull. No. 6, Manila, 1966.
24. **Kemmler, G.**, The manuring of paddy rice – experiences from South and East Asia, *Tropenlandwirt*, 69, 8–26, 1968 (in German).
25. **Kemmler, G.**, Manuring of rice in Japan, *Fert. News*, 14(4), 11–17, 1969.
26. **Leonard, W. H. and Martin, J. H.**, Rice, in *Cereal Crops*, Macmillan, New York, 1963, 603–678 (137 ref.).
27. **McCall, E. R., Hoffpauir, C. L., and Skau, D. B.**, The chemical composition of rice. A literature review, U.S. Dep. Agric. Publ. AIC-312, U.S. Dep. Agric., Washington, D.C., 1951.
28. **Nagai, I.**, *Japonica Rice, Its Breeding and Culture*, Yokondo, Tokyo, 1959, 843 pp.

29. **Nagao, S.**, Genic analysis and linkage relationship of characters in rice, *Adv. Genet.*, 4, 181–212, 1951.
30. National Fertilizer Development Center, Nutrition of Rice, Bulletin Y-36, Tennessee Valley Authority, Muscle Shoals, Alabama, 1971, 94 pp.
31. **Noguchi, Y. and Sugawara, T.**, Potassium and Japonica Rice: Effects of Potassium on the Growth, Yield, Histological Structure, and Metabolism of Rice Plants, International Potash Institute, Berne, 1966, 102 pp.
32. **Nuttonson, M. Y.**, Global Agroclimatic Analogues for the Rice Regions of the Continental United States, American Institute of Crop Ecology, Washington, D.C., 1965, 9 pp.
33. **Nuttonson, M. Y.**, Rice Culture and Rice-Climate Relationships with Special Reference to the United States Rice Areas and their Latitudinal and Thermal Analogues in Other Countries, American Institute of Crop Ecology, Washington, D.C., 1965, 555 pp.
34. **Padwick, G. W.**, Manual of Rice Diseases, Commonwealth Mycological Institute, Kew, Surrey, 1950, 198 pp.
35. **Shiroshita, T.**, *Theory and Practice of Growing Rice*, Fuji Publishing, Tokyo, 1965.
36. **Stout, B. A.**, Equipment for Rice Production, FAO Agric. Dev. Paper No. 84, FAO, Rome, 1966, 169 pp.
37. **Tanaka, A. and Yoshida, S.**, Nutritional Disorders of the Rice Plant in Asia, IRRI Tech. Bull. 10, International Rice Research Institute, Los Banos, Philippines, 1970, 51 pp.
38. U.S. Department of Agriculture, Rice in the United States: Varieties and Production, U.S. Dep. Agric., Agric. Handb. No. 289, 1966, 124 pp.
39. **Watabe, T. and Kawaguchi, K.**, Increasing the rice yield in South and Southeast Asia, *Asian Survey*, 8(10), 820–828, 1968.
40. **Wickizer, V. D. and Bennett, M. K.**, *The Rice Economy of Monsoon Asia*, Stanford University Food Research Institute, Stanford, Calif., 1941, 358 pp.
41. **Yamada, N. and Lusanandana, B.**, Rice production in the ADB region, in *Asian Agricultural Survey*, University of Tokyo Press, Tokyo, 1969, 141–196.

f. Rye

1. **Bland, B. F.**, Rye, in *Crop Production: Cereals and Legumes*, Academic Press, New York, 1971, 177–215.
2. **Deodikar, G. B.**, Rye. *Secale cereale Linn.*, Indian Council of Agricultural Research, Cereal Crop Ser. No. III, New Delhi, 1963, 152 pp.
3. **Herriot, R. I.**, Rye, *J. Dep. Agric. So. Aust.*, 51(6), 319–326, 1948.
4. **Leonard, W. H. and Martin, J. H.**, Rye, in *Cereal Crops*, Macmillan, New York, 1963, 449–477. (83 ref.)
5. **Nuttonson, M. Y.**, Rye-climate Relationships and the Use of Phenology in Ascertaining the Thermal and Photothermal Requirements of Rye, American Institute of Crop Ecology, Washington, D.C., 1958, 219 pp.
6. **Peterson, R. F.**, Improvement of rye through inbreeding, *Sci. Agric.*, 14, 651–668, 1934.
7. **Taylor, A. E.**, Rye in its relation to wheat, Stanford University Food Research Institute, *Wheat Studies*, 4(5), 181–234, 1928.

g. Sorghum

1. **Anderson, E. and Martin, J. H.**, World production and consumption of millet and sorghum, *Econ. Bot.*, 3, 265–288, 1949.
2. **Doggett, H.**, *Sorghum*, Longmans, Green & Co., London, 1970, 403 pp.
3. **Ganga Prasada Rao, N. and House, L. R., Eds.**, *Sorghum in Seventies*, Oxford & IBH Publishing, New Delhi, 1972, 638 pp.
4. **Jotwani, M. G. and Young, W. R., Eds.**, *Control of Sorghum Shoot Fly*, Oxford & IBH Publishing, New Delhi, 1972, 324 pp.
5. **Karper, R. E. and Quinby, J. R.**, Sorghum – its production, utilization and breeding, *Econ. Bot.*, 1, 355–371, 1947.
6. **Leonard, W. H. and Martin, J. H.**, Grain sorghum, in *Cereal Crops*, Macmillan, New York, 1963, 679–739. (142 ref.)
7. **Martin, J. H.**, Sorghum and pearl millet, in *Handbuch der Pflanzenzuchtung*, II Bd., 2 Aufl., 1959, 565–589.
8. **Moats, R. W.**, *Sorghum: Bibliography of World Literature 1930–1963*, Scarecrow Press, Metuchen, N.J., 1967, 301 pp.
9. **Quinby, J. R. and Martin, J. H.**, Sorghum improvement, *Adv. Agron.*, 6, 305–359, 1954.
10. **Quinby, J. R.**, The natural genes of sorghum, *Adv. Agron.*, 19, 267–305, 1967.
11. **Snowden, J. D.**, *The Cultivated Races of Sorghum*, Adlard & Son, London, 1936, 274 pp.
12. **Vinall, H. N., Stephens, J. C., and Martin, J. H.**, Identification, history, and distribution of common sorghums varieties, U.S. Dep. Agric. Tech. Bull. 506, 1936.
13. **Wall, J. S. and Ross, W. M., Eds.**, *Sorghum Production and Utilization*, Avi Publishing, Westport, Conn., 1970, 702 pp. (19 papers)
14. **Woodle, H. A.**, Grain sorghums, Clemson Agric. Coll., U.S. Dep. Agric. Circ. 285, 1960.

h. Wheat

1. **Alsberg, C. D. and Griffing, E. P.**, Environment, heredity, and wheat quality, Stanford University Food Research Institute, *Wheat Studies*, 10, 229–249, 1934.

2. **Bland, B. F.**, Wheat, in *Crop Production: Cereals and Legumes,* Academic Press, New York, 1971, 2–62.
3. **Callaghan, A. R. and Millington, A. J.**, *The Wheat Industry in Australia,* Angus and Robertson, Sydney, 1956, 486 pp.
4. **Hehn, E. R. and Barmore, M. A.**, Breeding wheat for quality, *Adv. Agron.*, 17, 85–114, 1965.
5. **Johnson, V. A. and Schmidt, W. A.**, Hybrid wheat, *Adv. Agron.*, 20, 199–233, 1968.
6. **Kopetz, L. M.**, *Progressive Wheat Production,* Centre d'etude de l'azote, Geneva, 1960, 338 pp.
7. **Leonard, W. H. and Martin, J. H.**, Wheat, in *Cereal Crops,* Macmillan, New York, 1963, 275–446. (85 ref.)
8. **Mangelsdorf, P. C.**, Wheat, *Sci. Am.*, 189(1), 50–59, 1953.
9. **Pal, B. P.**, Wheat, Indian Council of Agricultural Research, New Delhi, 1966, 370 pp.
10. **Percival, J.**, *The Wheat Plant,* Duckworth, London, 1921, 463 pp.
11. **Percival, J.**, *Wheat in Great Britain,* Duckworth, London, 1934, 126 pp.
12. **Peterson, R. F.**, *Wheat: Botany, Cultivation and Utilization,* Interscience, New York, 1965, 422 pp.
13. **Quisenberry, K. S. and Reitz, L. P., Eds.**, *Wheat and Wheat Improvement,* American Society of Agronomy, Madison, Wis., 1967, 560 pp. (15 papers)
14. **Reitz, L. P.**, Wheat breeding and our food supply, *Econ. Bot.*, 8, 251–268, 1954.
15. **Reitz, L. P.**, Wheat quality components responsive to genic control, *Qual. Plant. Mater. Veg.*, 11(1), 1–16, 1964.
16. **Salmon, S. C., Mathews, O. R., and Leukel, R. W.**, A half century of wheat improvement in the United States, *Adv. Agron.*, 5, 1–151, 1953.
17. **Schlehuber, A. M. and Tucker, B. B.**, Factors affecting the protein content of wheat, *Cereal Sci. Today,* 4, 240–242, 1959.
18. **Shellenberger, J. A.**, Production and utilization of wheat, in *Wheat Chemistry and Technology,* Pomeranz, Y., Ed., American Association of Cereal Chemists, St. Paul, Minn., 1971, 1–18.
19. **Swanson, C. O.**, *Wheat and Flour Quality,* Burgess Publishing, Minneapolis, 1938, 227 pp.
20. **Timoshenko, V. P.**, *Agricultural Russia and the Wheat Problem,* Stanford University Food Research Institute, Stanford, Calif., 1932, 571 pp.
21. Wheat quality improvement. Document FAO Conference on Cereal and Bread Technology for the Near East Region, Cairo, Feb. 1966, FAO, Rome, 1966, NERO CBT/66/7.

i. Other Cereals

1. **Leonard, W. H. and Martin, J. H.**, Millets, in *Cereal Crops,* Macmillian, New York, 1963, 740–769. (59 ref.)
2. **Zilinsky, F. J. and Borlaug, N. E.**, Progress in Developing Triticale as an Economic Crop, CIMMYT Res. Bull. No. 17, February 1971, 27 pp.

3. Pulses and Oilseeds

a. Legumes In General

1. **Bland, B. F.**, Forage legumes, in *Crop Production: Cereals and Legumes,* Academic Press, New York, 1971, 383–449.
2. **Burkart, A.**, *Las leguminosas Argentinas silvestres y cultivadas* (Wild and Cultivated Legumes of Argentina), Istituto de Botanica Darwinión, San Isidoro, 1943, 620 pp.
3. Food and Agriculture Organization, Tabulated Information on Tropical and Subtropical Grain Legumes, FAO, Rome, 1959.
4. **Hallsworth, E. G., Ed.**, *Nutrition of the Legumes,* Butterworth, London, 1958.
5. **Jenkins, B. C.**, Oilseed crops. New developments and possibilities, *Agric. Inst. Rev.*, 8(5), 31–32, 67, 1953.
6. **Nelson, W. L. and Barber, S. A.**, Nutrient deficiencies in legumes for grain and forage, in *Hunger Signs in Crops,* Sprague, H. B., Ed., McKay, New York, 1964, 143–170.
7. **Roberts, L. M.**, The Food Legumes. Recommendations for Expansion and Acceleration of Research, Rockefeller Foundation, New York, November 1970.
8. **Robinson, D. H.**, *Leguminous Forage Plants,* 2nd ed., Edward Arnold, London, 1947.
9. **Stanton, W. R., Ed.**, Grain Legumes in Africa, FAO, Rome, 1966, 183 pp.
10. **Thimann, K. V.**, On the physiology of the formation of nodules on legume roots, *Proc. Natl. Acad. Sci. U.S.A.*, 22, 511–514, 1936.
11. **Whyte, R. O., Nilsson-Leissner, G., and Trumble, H. C.**, Legumes in Agriculture, FAO Agric. Stud. No. 21, FAO, Rome, 1953, 367 pp.

b. Pulses

(1) General

1. Indian Council of Agricultural Research, Ministry of Food and Agriculture, Bibliography of Pulse Crops, New Delhi, 1960.
2. **Stanton, W. R., et al.**, Grain Legumes in Africa, FAO, Rome, 1966.

(2) Beans

1. **Bland, B. F.,** Beans, in *Crop Production: Cereals and Legumes*, Academic Press, New York, 1971, 264–325.

(3) Cowpeas

1. **Morse, W. J.,** *Cowpeas: Culture and Varieties*, U.S.Dep. Agric. Farmer's Bull. 1148, 1947.
2. **Sellschop, J. P. F.,** *Cowpea, Vigna u. (L.) Walp., Field Crop Abstr.*, November 1969.

(4) Lentils

1. **Youngman, V. E.,** Lentils–a pulse of the palouse, *Econ. Bot.*, 22, 135–139, 1968.

(5) Mung-Beans

1. **Ligon, L. L.,** Mung-bean – a Legume for Seed and Forage Production, Okla. Agric. Exp. Sta. Bull. 284, 1945.

(6) Peas

1. **Bland, B. F.,** Peas, in *Crop Production: Cereals and Legumes*, Academic Press, New York, 1971, 326–382.

c. Castor

1. **Weiss, E. A.,** *Castor, Sesame and Safflower*, Leonard Hill, London, 1971, 901 pp.
2. **Zimmerman, L. H.,** Castorbeans: A new oil crop for mechanized production, *Adv. Agron.*, 10, 257–288, 1958.

d. Coconut

1. **Child, R.,** *Coconuts*, Longmans, London, 1964, 216 pp.
2. **Copeland, E. B.,** *The Coconut*, Macmillan, London, 1931.
3. **Menon, K. P. V. and Pandalai, K. M.,** *The Coconut Palm*, Indian Central Coconut Common, Ernakulum, S. India, 1957, 384 pp.
4. **Woodroof, J. G.,** *Coconuts: Production, Processing, Products*, Avi Publishing, Westport, Conn., 1970, 241 pp.

e. Cottonseed

1. **Adams, J. E., Ed.,** Cotton, *Adv. Agron.*, 2, 2–80, 1950.
2. **Bailey, A. E., Ed.,** *Cottonseed and Cottonseed Products. Their Chemistry and Chemical Technology*, Interscience, New York, 1948, 936 pp.
3. **Brown, H. B. and Ware, J. O.,** *Cotton*, 3rd ed., McGraw-Hill, New York, 1958.
4. **Cardozier, V. R.,** *Growing Cotton*, McGraw-Hill, New York, 1957.
5. Cottonseed and its By-products, Proc. Symposium, Hyderabad, Dec. 5–7, 1958, Indian Central Oilseeds Committee, Hyderabad, 1959, 360 pp.
6. **Donald, L.,** Nutrient deficiencies in cotton, in *Hunger Signs in Crops*, Sprague, H. B., Ed., McKay, New York, 1964, 59–90.
7. **Eaton, F. M.,** Physiology of the cotton plant, *Annu. Rev. Plant Physiol.*, 6, 299–328, 1955.
8. **Elliot, F. C., Ed.,** *Advances in Production and Utilization of Quality Cotton*, Iowa State University Press, Ames, 1968, 532 pp.
9. National Cottonseed Product Association, Cottonseed and its Products, Dallas, 1948.

f. Oil Palm

1. **Hartley, C. W. S.,** *The Oil Palm*, Longmans, London, 1970, 706 pp.

g. Peanut

1. **Beattie, J. H.,** *Growing Peanuts*, U.S. Dep. Agric. Farmer's Bull. 2063, 1954.
2. **Cheek, E.,** List of Theses and Dissertations on Peanuts and Peanut Related Research, Ga. Agric. Exp. Sta. Res. Rep. 54, 1969.
3. **Martin, J. H. and Leonard, W. H.,** Peanuts, in *Principles of Field Crop Production*, Macmillan, New York, 1949, 784–800.
4. National Fertilizer Association, The Peanut. The Unpredictable Legume. A Symposium, The National Plant Food Institute, Washington, D.C., 1951.
5. North Carolina Agricultural Extension Service, Peanut Production Guide, N.C. Ext. Circ., 1963.
6. **Oram, P. A.,** Recent developments in groundnut production, with special regard to Africa, *Field Crop Abstr.*, Feb. 1958; Part II, May 1958.
7. *Peanuts – Culture and Uses. A Symposium*, American Peanut Research and Education Association, Stone Printing, Roanoke, Va., 1972, 684 pp.
8. **Seshadri, C. R.,** The Groundnut, Oil Seeds Committee, Indian Council of Agricultural Research, Ministry of Food and Agriculture, New Delhi, 1962.
9. **Woodroof, J. G.,** *Peanuts – Production, Processing, Products*, Avi Publishing, Westport, Conn., 1966, 291 pp.

h. Rapeseed

1. **Appelqvist, L. A. and Ohlson, R., Eds.**, *Rapeseed: Cultivation, Composition, Processing and Utilization*, American Elsevier, New York, 1972, 391 pp.
2. **Youngs, C. G.**, Canadian developments in breeding and utilization of rapeseed, *Oleagineux*, 20, 679–681, 1965.

i. Safflower

1. **Knowles, P. F.**, Safflower – production, processing and utilization, *Econ. Bot.*, 9, 273–299, 1955. (129 ref.)
2. **Knowles, P. F.**, Safflower, *Adv. Agron.*, 10, 289–323, 1958.
3. **Larson, N. G.**, Safflower 1900–1960. A List of Selected References, U.S. Library of Congress, Washington, D.C., 1962.
4. Proc. Safflower Conference, Tucson, Ariz., June 1963.
5. U.S. Department of Agriculture, Safflower – An Oilseed Crop, U.S. Dep. Agric. Farmer's Bull. 2133, 1966.

j. Sesame

1. **Barrett, O. W.**, Sesame, in *The Tropical Crops*, Macmillan, New York, 1928, 360–361.
2. **Langham, D. G.**, Genetics of sesame, *J. Hered.*, 36, 135–142, 245–253, 1945.

k. Soybean

1. **Berg, G. L., Ed.**, New Comprehensive Manual – Modern Soybean Production, Auchem Products, Ambler, PA, 1967.
2. **Cartter, J. L. and Hartwig, E. E.**, The management of soybeans, *Adv. Agron.*, 14, 360–412, 1962.
3. **Cartter, J. L. and Hopper, T. H.**, Factors affecting the chemical composition of soybean seed, U.S. Dep. Agric. Tech. Bull. 787, 1942.
4. **Dies, E. J.**, *Soybeans. Gold from the Soil*, Macmillan, New York, 1942.
5. **Dunleavy, J. M.**, Soybean Diseases, Agric. Handb. No. 302, U.S. Dep. Agric., U.S. Gov. Print. Off., Washington, D.C., 1966, 38 pp.
6. **Holman, L. E. and Carter, D. G.**, Soybean storage, Ill. Agric. Exp. Sta. Bull. 553, 1952.
7. **Howell, R. R.**, Physiology of the soybean, *Adv. Agron.*, 12, 265–310, 1960.
8. **Larson, N. G. and Cowan, J. C.**, Soybeans: A Selected List of References, U.S. Bureau of Agricultural and Industrial Chemistry, Washington, D.C., 1953.
9. **Ling, L.**, Bibliography of soybean diseases, Plant Diseases Report Suppl. 204, U.S. Dep. Agric., U.S. Bureau of Plant Industry, Soils and Agricultural Engineering, 1951.
10. **Morse, W. J. and Cartter, J. L.**, Soybeans: Culture and varieties, U.S. Dep. Agric. Farmer's Bull. 1520, 1949.
11. **Norman, A. G., Ed.**, *The Soybeans. Genetics, Breeding, Physiology, Nutrition, Management.* Academic Press, New York, 1963, 239 pp. (4 papers)
12. **Ohlrogge, A. J.**, Mineral nutrition of soybean, *Adv. Agron.*, 12, 230–263, 1960.
13. **Piper, C. V. and Morse, W. J.**, *The Soybean*, McGraw-Hill, New York, 1923.
14. Soybean Farming, National Soybean Processors Association, Chicago, Ill., 1961.
15. **Viljoen, N. J.**, The Composition of Soybeans in South Africa, Union So. Africa Dep. Agric. For. Sci. Bull. 169, 1935.
16. **Weber, C. R.**, *Vegetable Soybeans*, Iowa Agric. Exp. Sta. F.S. 651, 1956.
17. **Weiss, M. G.**, Soybeans, *Adv. Agron.*, 1, 78–157, 1949.

l. Sunflower

1. National Agricultural Library, Sunflower: A Literature Survey. January 1960 – June 1967, Library List No. 95, U.S. Dep. Agric., National Agricultural Library, Beltsville, MD, 1969, 133 pp.

4. Roots and Tubers

a. General

1. **Leslie, K. A.**, The significance of root crops in the tropics, *Proc. Int. Symposium Trop. Root Crops*, University of West Indies, St. Augustine, Trinidad, 2(5), 1–14, 1967.
2. **Plucknett, D. L., Ed.**, Tropical Root and Tuber Crops Tomorrow. Proc. 2nd International Symposium on Tropical Root and Tuber Crops, University of Hawaii College of Tropical Agriculture, August 1970. (2 vols.)
3. **Rees, A. M. M.**, Some economic aspects of root crop production, Proc. International Symposium Trop. Root Crops, University of West Indies, St. Augustine, Trinidad, 2(5) 18–33, 1967.
4. **Tai, E. A., Charles, W. B., Iton, E. F., Haynes, P. H., and Leslie, K. A., Eds.**, Proc. International Symposium on Tropical Root Crops, University of West Indies, St. Augustine, Trinidad, April 2–8, 1967. (2 vols.)
5. **Whittington, W. J., Ed.**, *Root Growth*, Plenum Publishing, New York, 1969, 450 pp.

b. Cassava

1. **Araque, R.,** Cassava and its cultivation and uses, *Ser. cultivos* (Venezuela), 1961, 1–20.
2. **Doku, E. V.,** *Cassava in Ghana,* Ghana University Press, Ghana, 1969, 57 pp.
3. **Jennings, D. L.,** Cassava in Africa, *Field Crop Abstr.,* 23, 271–278, 1970.
4. **Jones. W. O.,** *Manioc in Africa,* Stanford University Press, Stanford, Calif., 1959, 315 pp.
5. **Krochmal, A.,** Cassava in the new world, *World Crops,* 19(1), 74–75, 1967.
6. **Lamire, M. E.,** Manioc, *J. Geogr.,* 49(9), 376–385, 1950.
7. **Martin, F. W.,** Cassava in the world of tomorrow, *Proc. 2nd Int. Symp. on Trop. Root and Tuber Crops,* 1, 51–58, 1970.
8. University of Georgia Team, A Literature Review and Research Recommendations on Cassava (Manihot esculenta Crantz), AID Contract No. csd/2497, March 1972, 326 pp. (847 ref.)

c. Potato

1. **Burton, W. G.,** *The Potato,* Chapman & Hall, London, 1966.
2. **Chucka, J. A. and Steinmetz, F. A.,** Potatoes, Me. Agric. Exp. Sta. Bull. 438, 1945.
3. **Correll, D. S.,** *The Potato and its Wild Relatives,* Texas Research Foundation, Renner, 1962; Stechert-Hafner, New York, 1966, 606 pp.
4. **Dudley, J. E.,** Control of Potato Insects, U.S. Dep. Agric. Farmer's Bull. 2040, 1952.
5. **Edgar, A. D.,** Potato Storage, U.S. Dep. Agric. Farmer's Bull. 1986, 1947.
6. **Gardner, J. S.,** *Potato Growing,* Ky. Agric. Ext. Serv. Circ. 307, 1951.
7. **Grubb, E. H. and Guilford, W. S.,** *The Potato,* Doubleday Page, New York, 1912, 525 pp.
8. **Hardenburg, E. V.,** *Potato Production,* Comstock, Ithaca, N.Y., 1949.
9. **Hawkes, J. G. and Hjerting, J. P.,** *The Potatoes of Argentina, Brazil, Paraguay, and Uruguay, Ann. Botany Memoirs No. 3,* Oxford University Press, New York, 1969, 525 pp.
10. **Houghland, G. V. C.,** Nutrient deficiencies in the potato, in *Hunger Signs in Crops,* Sprague, H. B., Ed., McKay, New York, 1964, 219–238.
11. **Kykstra, T. P.,** Production of Disease-Free Potatoes, U.S. Dep. Agric. Circ. 674, 1948.
12. Potato Association of America, *Potato Handbook,* Rutgers State University, New Brunswick, N.J., 1967.
13. **Smith, O.,** *Potatoes; Production, Storing, Processing,* Avi Publishing, Westport, Conn., 1968, 642 pp.
14. **Smith, O.,** Potato production, *Adv. Agron.,* 1, 353–390, 1949.
15. **Talburt, W. F. and Smith, O.,** *Potato Processing,* Avi Publishing, Westport, Conn., 1959, 475 pp.

d. Sweet Potato

1. **Boswell, V. R.,** Commercial growing and harvesting of sweet potatoes, U.S. Dep. Agric. Farmer's Bull. 2020, 1950.
2. **Edmond, J. B. and Ammerman, G. R.,** *Sweet Potatoes: Production, Processing, Marketing,* Avi Publishing, Westport, Conn., 1971, 334 pp.
3. **Hildebrand, E. M. and Cook, H. T.,** Sweet potato diseases, U.S. Dep. Agric. Farmer's Bull. 1059, 1959.

e. Yam

1. **Coursey, D. G.,** *Yams,* Longmans, Green & Co., London, 1967, 244 pp.
2. **Harris, R. V. and Oke, O. L.,** *Report on Yam,* Tropical Products Institute, London, 1969.
3. **Martin, F. W.,** *Yam Production Methods,* Production Research Rep. No. 147, USDA, ARS, Washington, D.C., 1972, 17 pp.

5. Sugar Crops

1. **Artschwager, E. and Brandes, E. W.,** *Sugar Cane,* U.S. Dep. Agric. Handb. No. 122, 1958.
2. **Aykroyd, W. R.,** *The Story of Sugar,* Quadrangle, Chicago, 1967.
3. **Barnes, A. C.,** *The Sugar Cane,* Leonard Hill, London, 1964.
4. **Dillewijn, C. van,** *Sugar Cane Breeding,* Int. Inst. of Agric., Rome, 1946.
5. **Dillewijn, C. van** *Handbook of Sugarcane,* Chron. Bot., Waltham, Mass., 1952.
6. **Draycott, A. P.,** *Sugar-Beet Nutrition,* John Wiley & Sons, New York, 1972, 250 pp.
7. **Halliday, D. J.,** *The Manuring of Sugar Cane,* Centre d'Etude de l'Azote, Geneva, 1956.
8. **Humbert, R. P.,** *The Growing of Sugar Cane,* American Elsevier, New York, 1968.
9. International Society of Sugar Cane Technologists, *Sugarcane Diseases of the World,* American Elsevier, New York, 1961, 1964. (2 vols.)
10. **King, N. J., Mungomery, R. W., and Hughes, C. G.,** *Manual of Cane-Growing,* American Elsevier, New York, 1965, 375 pp.
11. **Peay, W. E.,** Sugar Beet Insects. How to Control Them, U.S. Dep. Agric. Farmer's Bull. 2219, July 1968, 25 pp.
12. Pest and Disease Control Program for Sugar Beets, Calif. Agric. Exp. Sta., Ext. Ser. Bull., 1969.
13. **Schmehl, W. R. and Humbert, R. P.,** Nutrient deficiencies in sugar crops, in *Hunger Signs in Crops,* Sprague, H. B., Ed., McKay, New York, 1964, 415–441.

14. **Sherrod, L. B., Campbell, C. M., and Tshizaki, S. M.,** Nutritive value of seed (sugar) cane toppings and mill cane strippings with and without supplemental protein, Hawaii Agric. Exp. Sta. Tech. Progr. Rep. 168, 1968, 9 pp.
15. **Stevenson, G. C.,** *Genetics and Breeding of Sugarcane,* Longmans, Green & Co., London, 1965, 284 pp.
16. **Taggart, W. G. and Simon, E. C.,** *The History of Sugarcane,* La. Sta. Dept. Agric., Baton Rouge, 1939.
17. U.S. Beet Sugar Association, The Beet Sugar Story, Washington, D.C., 1959.

6. Horticultural Crops In General

1. American Society for Horticultural Science, Care and Feeding of Garden Plants, National Plant Food Institute, Washington, D.C., 1954, 184 pp.
2. **Bailey, H.,** *Cyclopedia of Horticulture,* Macmillan, New York, 1947. (3 vols.)
3. **Camp, W. H., Boswell, W. R., and Magness, J. R.,** *The World in Your Garden,* National Geographic Society, Washington, D.C., 1957, 231 pp.
4. **Chandler, W. H.,** *Deciduous Orchards,* Lea & Febiger, Philadelphia, 1947.
5. **Chandler, W. H.,** *Evergreen Orchards,* Lea & Febiger, Philadelphia, 1950.
6. **Cruess, W. V.,** *Commercial Fruit and Vegetable Products,* McGraw-Hill, New York, 1958.
7. **Denison, E. L.,** *Principles of Horticulture,* Macmillan, New York, 1958.
8. **Edmund, J. B., Senn, T. L., and Andrews, F. S.,** *Fundamentals of Horticulture,* 3rd ed., McGraw-Hill, New York, 1964.
9. Food and Agriculture Organization, Fruits and Vegetables in West Africa, FAO, Rome, 1965, 259 pp.
10. **Gardner, V. R.,** *Basic Horticulture,* Macmillan, New York, 1951.
11. **Janick, J.,** *Horticultural Science,* W. H. Freeman Publishing, San Francisco, 1963.
12. **Mack, W. B., Larson, R. E., White, D. G., and Sampson, H. O.,** *Vegetable and Fruit Growing,* Lippincott, Chicago, 1956, 560 pp.
13. **MacMillan, H. F.,** *Tropical Planting and Gardening,* Macmillan, New York, 1954.
14. **Mortensen, E. and Bullard, E. T.,** *Handbook of Tropical and Sub-Tropical Horticulture,* AID, Washington, D.C., 1968. 186 pp.
15. **Nijdam, J. and de Jong, A., Eds.,** *Elsevier's Dictionary of Horticulture,* American Elsevier, New York, 1970, 561 pp.
16. **Ochse, J. J., Soule, M., Jr., Dijkman, M. J., and Wehlbrurg, C.,** *Tropical and Subtropical Horticulture,* Macmillan, New York, 1961. (2 vols.)
17. **Shoemaker, J. S.,** *General Horticulture,* 2nd Ed., Lippincott, 1956.
18. **Shoemaker, J. S. and Teskey, B. J. E.,** *Practical Horticulture,* John Wiley & Sons, New York, 1955.
19. **Talbert, T. J.,** *Growing Fruit and Vegetable Crops,* Lea & Febiger, Philadelphia, 1953.
20. **Thomas, A.,** *Gardening in Hot Countries,* Faber & Faber, London, 1965, 207 pp.
21. **Tindall, H. D.,** Fruits and Vegetables in West Africa, FAO, Rome, 1965, 259 pp.
22. **Wood, R.,** A Notebook of Tropical Horticulture, Imp. Col. Agric. Univ. W. Indies, Trinidad, 1957, 256 pp.

7. Fruit Crops

a. General

1. **Anderson, H. W.,** *Diseases of Fruits Crops,* McGraw-Hill, New York, 1956, 501 pp.
2. **Cain, J. C. and Shear, C. B.,** Nutrient deficiencies in deciduous tree fruits and nuts, in *Hunger Signs in Crops,* Sprague, H. B., Ed., McKay, New York, 1964, 287–313.
3. **Cheema, G.,** *Commercial Fruits of India,* MacMillan, 1954, 422 pp.
4. **Childers, N. F., Ed.,** Nutrition of Fruit Crops, Horticultural Publications, Rutgers University, New Brunswick, N. J., 1966, 888 pp.
5. **Childers, N. F.,** *Modern Fruit Science. Orchard and Small Fruit Culture,* 2nd ed., Horticultural Publications, Rutgers University, New Brunswick, N.J., 1961, 893 pp.
6. **Coursey, D. G., Ed.,** Proceedings of the Conference on Tropical and Subtropical Fruits, Tropical Products Institute, London, 1970.
7. **Cruess, W. V.,** *Commercial Fruit and Vegetable Products,* McGraw-Hill, New York, 1958.
8. **Gardner, V. R.,** *Fundamentals of Fruit Production,* 3rd ed., McGraw-Hill, New York, 1952.
9. **Gourley, J. H. and Howlett, F. S.,** *Modern Fruit Production,* Macmillan, New York, 1941.
10. **Hayes, W. B.,** *Fruit Growing in India,* Alahabad, India, 1954, 283 pp.
11. **Kennard, W. C. and Winters, H. F.,** Some Fruits and Nuts for the Tropics, U.S. Dep. Agric. Misc. Publ. 801, 1960, 135 pp.
12. **Knight, R. L.,** *Abstract Bibliography of Fruit Breeding and Genetics to 1965. Prunus,* Commonwealth Agricultural Bureaux, Farnham Royal, Bucks., England, 1969, 649 pp.
13. **Miller, C. D., Bazore, K., and Bartow, M.,** *Fruits of Hawaii,* 4th ed., University of Hawaii Press, Honolulu, 1965, 229 pp.
14. **Mowry, H., Toy, L. R., and Wolfe, H. S.,** Miscellaneous Tropical and Subtropical Florida Fruits, Fla., Exp. Ser. Bull. 156, 1953.
15. **Naik, K. C.,** *South India Fruits and Their Culture,* Varadachy, Madras, 1949.
16. **Popenoe, W.,** *Manual of Tropical and Subtropical Fruits,* Macmillan, New York, 1920.

17. **Schneider, G. W. and Scarborough, C. C.,** *Fruit Growing,* Prentice-Hall, Englewood Cliffs, N.J., 1960, 307 pp.
18. **Sheer, A. H. and Juergenson, E. M.,** *Approved Practices in Fruit Production,* Interstate, Danville, Ill., 1964.
19. **Shoemaker, J. S.,** *Small-Fruit Culture,* 2nd ed., Blakiston, Philadelphia, 1948.
20. **Shoemaker, J. S., and Teskey, B. J. E.,** *Tree Fruit Production,* John Wiley & Sons, New York, 1959.
21. **Smith, J. R.,** *Tree Crops,* Devin-Adair, New York, 1950.
22. **Smith, C. R., Shaulis, N., and Cook, J. A.,** Nutrient deficiencies in small fruits and grapes, in *Hunger Signs in Crops,* Sprague, H. B., Ed., McKay, New York, 1964, 327–345.
23. **Tukey, H. B.,** *Dwarfed Fruit Trees,* Macmillan, New York, 1964.

b. Berries

1. **Anderson, W.,** *The Strawberry: A World Bibliography 1920–1966,* Scarecrow Press, Metuchen, N.J., 1969.
2. **Darrow, G. M.,** *The Strawberry: History, Breeding and Physiology,* Holt, Rinehart & Winston, New York, 1966, 447 pp.
3. **Eck, P. and Childers, N. F., Eds.,** *Blueberry Culture,* Rutgers University Press, New Brunswick, N.J., 1967.

c. Pomes

1. **Batjer, L. P., Schomer, H. A., Newcomer, E. J., and Coyier, D. L.,** Commercial Pear Growing, Agric. Handb. No. 330, USDA, ARS, Washington, D.C., 1967, 47 pp.
2. **Smock, R. M. and Neubert, A. M.,** *Apples and Apple Products,* Interscience, New York, 1950.
3. **Upshall, W. H., Ed.,** *North American Apples. Varieties, Rootstocks, Outlook,* Michigan State University Press, East Lansing, 1970.
4. **Zielinski, Q. B.,** *Modern Systematic Pomology,* William C. Brown & Co., Dubuque, IA, 1955, 296 pp.

d. Citrus Fruit

(1) General

1. **Fawcett, H. S.,** *Citrus Diseases and Their Control,* 2nd ed., McGraw-Hill, New York, 1936, 656 pp.
2. **Hume, H. H.,** *Citrus Fruits,* Macmillan, New York, 1957, 444 pp.
3. **Jones, W. W. and Smith, P. F.,** Nutrient deficiencies in citrus, in *Hunger Signs in Crops,* Sprague, H. B., Ed., McKay, New York, 1964, 359–401.
4. **Knorr, L. C., Suit, R. F., and Ducharine, E. P.,** Handbook of Citrus Diseases of Florida, Fla. Agric. Exp. Sta. Bull. 587, 1957, 157 pp.
5. **Knorr, L. C.,** Serious Diseases of Citrus Foreign to Florida, Fla. Dep. Agric., Gainesville, 1965, 59 pp.
6. **Reuther, W., Webber, H. J., and Batchelor, L. D., Eds.,** *The Citrus Industry,* 2nd ed., Univ. Calif. Div. Agric. Sci., Berkeley, 1967–1968. (2 vols.)
7. **Ziegler, L. W. and Wolfe, H. S.,** *Citrus Growing in Florida,* University of Florida Press, Gainesville, 1961.

(2) Grapefruit

1. **Robinson, T. R.,** Grapefruit and pummelos, *Econ. Bot.,* 6, 228–245, 1962.
2. **Sinclair, W. B., Ed.,** The Grapefruit – Its Composition, Physiology, and Products, Univ. Calif. Div. Agric. Sci., Berkeley, 1972, 645 pp.

(3) Lemon

1. **Bartholomew, E. T. and Sinclair, W. B.,** *The Lemon Fruit: Its Composition, Physiology, and Products,* University of California Press, Berkeley, 1951, 163 pp.

(4) Orange

1. McPhee, J., *Oranges,* Farrar, Straus & Giroux, New York, 1967.
2. **Sinclair, W. B., Ed.,** *The Orange: Its Biochemistry and Physiology,* University of California Press, Berkeley, 1961, 475 pp.

e. Tropical Fruits (Excluding Citrus Fruit)

(1) Banana

1. **Simmonds, N. W.,** *Bananas,* 2nd ed., Longmans, London, 1966, 512 pp.
2. **Wardlaw, C. W.,** *Banana Diseases,* Longmans, London, 1961.

(2) Date

1. **Dawson, V. H. W. and Aten, A.,** Dates – Handling, Processing and Packing, FAO, Rome, 1962.
2. **Nixon, R. W.,** The date palm – "tree of life" in the subtropical deserts, *Econ. Bot.,* 5, 274–301, 1951.

(3) Fig

1. **Condit, I. J.,** *Ficus: The Exotic Species,* Univ. Calif. Div. Agric. Sci., Berkeley, 1969.
2. **Condit, I. J.,** *The Fig,* Ronald Press, New York, 1947, 222 pp.
3. **Condit, I. J.,** Figs – facts and figures of 1948, *Econ. Bot.,* 2, 403–418, 1948.
4. **Condit, I. J.,** Fig varieties - a monograph, *Hilgardia,* 23, 323–528, 1955.
5. **Condit, I. J. and Enderud, J.,** A bibliography of the fig, *Hilgardia,* 25, 1–663, 1956.

(4) Lychee

1. **Groff, G. W.,** *The Lychee and Longan,* Orange Judd Publishing, New York, 1921, 188 pp.

(5) Mango

1. **Gangolly, S. R., Singh, R., Katyl, S. L., and Singh, D.,** The Mango, Indian Council of Agric. Res., New Delhi, 1957, 530 pp.
2. **Singh, L. B.,** *The Mango: Botany, Cultivation, and Utilization,* Interscience, New York, 1960, 435 pp.

(6) Pineapple

1. **Collins, J. L.,** *The Pineapple: Botany, Cultivation, and Utilization,* 2nd ed., Interscience, New York, 1967.

8. Vegetable Crops

a. General

1. **Brown, H. D. and Hutchison, C. S.,** *Vegetable Science,* Lippincott, Philadelphia, 1949.
2. **Chauhan, D. V. S.,** *Vegetable Production in India,* Agra, Ram Prahad, 1965, 339 pp.
3. **Chupp, C. and Sherf, A. F.,** *Vegetable Diseases and their Control,* Ronald Press, New York, 1960, 693 pp.
4. **Hoare, A. H.,** *Vegetable Crops for Market and Processing,* Crosby Lockwood & Son, London, 1949.
5. **Knott, J. E.,** *Vegetable Growing,* 5th ed., Lea & Febiger, Philadelphia, 1955.
6. **Knott, J. E.,** *Handbook for Vegetable Growers,* John Wiley & Sons, New York, 1960.
7. **Knott, J. E. and Deanon, J. R., Jr.,** *Vegetable Production in Southeast Asia,* University of the Philippines Press, Laguna, 1967, 366 pp.
8. **Knott, J. E. and Lorenz, O. A.,** Vegetable production, *Adv. Agron.,* 2, 114–155, 1950.
9. **MacGillivray, J. H.,** *Vegetable Production,* McGraw-Hill, New York, 1953.
10. **Ochse, J. J.,** *Vegetables of the Dutch East Indies,* Van Riemsdyck, The Hague, 1931, 1005 pp.
11. **Ogilvie, L.,** *Diseases of Vegetables,* Min. of Agric. Fish. Food Bull. 123, HMSO, London, 1961, 101 pp.
12. **Purvis, E. R. and Carolus, R. L.,** Nutrient deficiencies in vegetable crops, in *Hunger Signs in Crops,* Sprague, H. B., Ed., McKay, New York, 1964, 245–275.
13. **Thompson, H. C. and Kelly, W. C.,** *Vegetable Crops,* 5th ed., McGraw-Hill, New York, 1957, 611 pp.
14. **Tindall, H. D.,** *Commercial Vegetable Growing,* Oxford University Press, London, 1968, 300 pp.
15. **Walker, J. C.,** *Diseases of Vegetable Crops,* McGraw-Hill, New York, 1952, 529 pp.
16. **Ware, G. W. and McCollum, J. P.,** *Producing Vegetable Crops,* Interstate, Danville, Ill., 1968, 558 pp.
17. **Winters, H. F. and Miskimen, G. W.,** *Vegetable Gardening in the Caribbean Area,* U.S. Dep. Agric. Handb. No. 323, 1967, 114 pp.
18. **Work, P. and Carew, J.,** *Vegetable Production and Marketing,* 2nd ed., John Wiley & Sons, New York, 1955.

b. Specific Crops

1. **Huelsen, W. A.,** *Sweet Corn,* World Crops Books, Interscience, New York, 1954.
2. **Jones, H. A. and Mann, L. K.,** *Onions and Their Allies: Botany, Cultivation, and Utilization,* Interscience, New York, 1963.
3. **Nieuwhof, M.,** *Cole Crops,* Chemical Rubber Co., Cleveland, 1969, 353 pp.
4. **Whitaker, T. W. and Davis, G. N.,** *Cucurbits: Botany, Cultivation, and Utilization,* Interscience, New York, 1962.

9. Nuts

a. General

1. **Howes, F. N.,** *Nuts: Their Production and Everyday Uses,* Faber & Faber, London, 1948, 264 pp.
2. **Jaynes, R. A.,** *Handbook of North American Nut Trees,* Northern Nut Growers Association, Knoxville, Tenn., 1969, 421 pp.
3. **Reed, C. A. and Davidson, J.,** *The Improved Nut Trees of North America and How to Grow Them,* Devin-Adair, New York, 1958.
4. **Woodroof, J. G.,** *Tree Nuts: Production, Processing, Products,* Avi Publishing, Westport, Conn., 1967. (2 vols.)

b. Specific Nuts

1. **Morton, J. F.,** The cashew's bright future, *Econ. Bot.,* 15, 57–58, 1961.

VI. ANIMAL PRODUCTION

A. The Role of Animals in Food Production

1. General

1. **Blaxter, K. L.,** The animal harvest, *Sci. J.*, 4(5), 53–59, 1968.
2. **Byerly, T. C.,** The role of livestock in food production, *J. Anim. Sci.*, 25, 552–566, 1966.
3. **Byerly, T. C.,** The relation of animal agriculture to world food needs, in *The Role of Animal Agriculture in Meeting World Food Needs,* Proc. 15th Annu. Meeting of the Agricultural Research Institute, Oct. 10–11, 1966; National Academy of Sciences, Washington, D.C., 1966, 31–44.
4. **Catron, D. V. and McRoberts, M. R.,** Animal proteins in the diets of the world's peoples, in *The Role of Animal Agriculture in Meeting World Food Needs,* National Academy of Sciences, Washington, D.C., 1966, 45–63.
5. **Cockrill, W. R., Ed.,** More animal protein for tomorrow's world, *World Rev. Anim. Prod.*, 3(13), 1–107, 1967.
6. **Cox, C. B., Binkerd, E. F., and Crigler, T. P.,** Animal protein – keystone of foods, *Food Technol.*, 25, 808–812, 1971.
7. **Cuthbertson, D.,** Animal production and world needs, in *Nutrition of Animals of Agricultural Importance, International Encyclopaedia of Food and Nutrition,* Vol. 17, Part 2, Pergamon Press, Oxford, 1969, 1311–1329.
8. **De Graff, H., Brown, R. E., Garrigus, U. S., Pritchard, W. R., Broadbent, E. E., and Salisbury, G. W.,** Animal Production and World Food Needs, Univ. Ill. Coll. Agric. Spec. Publ. No. 12, 1968.
9. **Hodgson, R. E.,** Place of animals in world agriculture, *J. Dairy Sci.*, 54, 442–447, 1971.
10. **Holmes, W.,** Animals for food, *Proc. Nutr. Soc.*, 29, 237–244, 1970. (33 ref.)
11. **Kolari, O. E.,** The use of animal proteins for food, *J. Anim. Sci.*, 25, 567–574, 1966.
12. **Lawrie, R. A.,** *Meat Science,* 2nd ed., Pergamon Press, Elmsford, N.Y., 1974, 419 pp.
13. **Leitch, I. and Gorden, W.,** The efficiency of farm animals in the conversion of feedingstuffs to food for man, Tech. Commun. No. 14, Commonwealth Bureaux of Animal Nutrition, Farnham Royal, Slough, 1953.
14. **McMeekan, C. P.,** More food from livestock, *Span,* 4(1), 15–17, 1961.
15. **Mahadevan, P.,** Food resources in the animal kingdom, in *The Food Resources of Mankind,* Agri-World Press, Montreal, 1968, 81–97.
16. **Ovington, J. D., Ed.,** *The Better Use of the World's Fauna for Food,* Symposia of the Institute of Biology No. 11, Institute of Biology, London, 1963, 175 pp. (15 papers)
17. **Phillips, R. W.,** Expansion of livestock production in relation to human needs, *Nutr. Abstr. Rev.*, 21, 241–256, 1951.
18. **Phillips, R. W.,** Animal products in the diets of present and future world populations, *J. Anim. Sci.*, 22, 251–262, 1963.
19. **Phillips, R. W.,** Animal agriculture in the emerging nations, in *Agricultural Sciences for the Developing Nations,* Moseman, A. H., Ed., American Association for the Advancement of Science, Washington, D.C., 1964, 15–32.
20. **Phillips, R. W.,** The necessity of defining needs and establishing priorities for solution of animal production problems, taking into consideration the need of human nutrition, *Proc. World Conference on Animal Production,* Vol. 1, Rome, September 2–7, 1963, European Association for Animal Production, Rome, 1963, 7–34.
21. **Phillips, R. W.,** Animal production and the world's food supply, Int. Agric. Ser. 6, University of Minnesota Institute of Agriculture, 1967, 22 pp.
22. **Phillips, R. W.,** The livestock industry. The scope and potential, in *Introduction to Livestock Production–Including Dairy and Poultry,* Cole, H. H., Ed., W. H. Freeman Publishing, San Francisco, 1966, 1–34.
23. **Phillips, R. W.,** Future programs for increasing food production with reference to the pre-school child, *Proc. Int. Conference on Prevention of Malnutrition in the Pre-School Child,* National Academy of Sciences, Washington, D.C., 1966, 305–316.
24. **Phillips, R. W.,** Increasing output of animal production: Relative potential of specific meat-producing animals, in *Man, Food, and Nutrition. Strategies and Technological Measures for Alleviating the World Food Problem,* Rechcigl, M., Jr., Ed., CRC Press, Cleveland, 1973, 147–162. (23 ref.)
25. **Price, J. F. and Schweigert, B. S., Eds.,** *The Science of Meat and Meat Products,* 2nd ed., W. H. Freeman Publishing, San Francisco, 1971, 660 pp.
26. **Schweigert, B. S.,** Meat in Human Nutrition, American Meat Institute Foundation, Circ. 8, 1953.
27. **Stillings, B. R.,** World supplies of animal protein, in *Proteins in Human Nutrition,* Porter, J. W. G. and Rolls, B. A., Eds., Academic Press, London, 1973, 11–33.
28. *The Role of Animal Agriculture in Meeting World Food Needs. Proc. 15th Annual Meeting of the Agricultural Research Institute,* Oct. 10–11, 1966, National Academy of Sciences, Washington, D.C., 1966, 260 pp. (15 papers)
29. Symposium. The future of animals as human food, *Proc. Nutr. Soc.*, 29, 237–244, 1970. (6 papers)
30. Use of animals in producing foods, *Proc. 2nd World Conference on Animal Production*, American Dairy Science Association, Urbana, Ill., 1969, 15–51. (4 papers with discussion)
31. **Van Horn, H. H., Cunha, T. J., and Harms, R. H.,** The role of livestock in meeting human food needs, *Bioscience,* 22, 710–714, 1972.
32. **Whyte, R. O.,** *Land, Livestock and Human Nutrition in India,* Preager, New York, 1968, 401 pp.

33. **Wilcke, H. L.**, General outlook for animal protein in food supplies of developing countries, in *World Protein Resources, Advances in Chemistry, Ser. No. 57,* American Chemical Society, Washington, D.C., 1966, 16–26. (14 ref.)
34. **Wilcke, H. L.**, Place of food producing animals in providing food for the world population, *Feedstuffs,* 38 (June 11), 6–7, 1966.
35. **Wilson, P. N.**, Biological ceilings and economic efficiencies for the production of animal protein, A.D. 2000, *Chem. Ind.,* July 6, 1968, 899–902.
36. **Zeigler, P. T.**, *The Meat We Eat,* Interstate, Danville, Ill., 1966, 547 pp.

2. Ruminants

a. Meat Production

1. **Blaxter, K. L.**, Domesticated ruminants as sources of human food, *Proc. Nutr. Soc.,* 29, 244–253, 1970.
2. **Cockrill, W. R.**, Meat and milk production from the domestic buffalo, *Bubalus bubalis, World Rev. Anim. Prod.,* 3, 98–107, 1967.
3. **Cuthbertson, D. P.**, Role of the ruminant in world food supply, *World Rev. Nutr. Diet.,* 12, 413–451, 1970. (55 ref.)
4. **Harvey, D.**, Some aspects of the importance of buffaloes as farm stock, *Nutr. Abstr. Rev.,* 33, 931–936, 1963.
5. **Harvey, D. and Rigg, J. C.**, Some aspects of goats as livestock, *Nutr. Abstr. Rev.,* 34, 641–645, 1964.
6. **Moore, L. A., Putnam, P. A., and Bayley, N. D.**, Ruminant livestock. Their role in the world protein deficit, *Agric. Sci. Rev.,* 5(2), 1–7, 1967. (28 ref.)
7. **Reid, J. T.**, The future role of ruminants in animal production, in *Physiology of Digestion and Metabolism in the Ruminant,* Phillipson, A. T., Ed., Oriel Press, Newcastle upon Tyne, 1970, 1–22. (50 ref.)
8. **Warwick, E. J.**, Beef for tomorrow, in *Proc. 18th Annu. Meeting Agric. Res. Inst., Oct. 14–15, 1969,* National Academy of Sciences, Washington, D.C., 1969, 147–162.

b. Milk Production

1. **Eckles, C. H.**, *Milk and Milk Products,* 4th ed., McGraw-Hill, New York, 1951, 454 pp.
2. Increasing the role of milk and milk products in meeting the world food needs, *Proc. 2nd World Conference on Animal Production,* American Dairy Science Association, Urbana, Ill., 1969, 187–219. (5 papers with discussion)
3. International Symposium:Milk and milk products in the world food supply, *J. Dairy Sci.,* 48, 1541–1554, 1965. (4 papers)
4. **Josephson, D. V.**, General outlook for milk protein, in *World Protein Resources, Advances in Chemistry Ser. No. 57,* American Chemical Society, Washington, D.C., 1966, 27–36. (20 ref.)
5. **Judkins, H. F. and Keener, H. A.**, *Milk Production and Processing,* John Wiley & Sons, New York, 1963.
6. **Rusoff, L. L.**, The role of milk in modern nutrition, *Borden's Rev. Nutr. Res.,* 25, 17–49, 1964.
7. **Teply, L. J.**, Role of milk and milk products in nutrition in tropical and sub-tropical countries, *J. Dairy Sci.,* 48, 1547–1550, 1965.
8. **Whyte, R. O.**, *Milk Production in Developing Countries,* Faber & Faber, London, 1967.

3. Swine

1. **Braude, R.**, The future of animals as sources of human food. Pigs–whither? *Proc. Nutr. Soc.,* 29, 262–270, 1970.

4. Poultry

1. **Bolton, W.**, The role of poultry, *Proc. Nutr. Soc.,* 29, 253–262, 1970.
2. **Halnan, E. T.**, Poultry as food converters, *Nature,* 148, 335–336, 1941.

5. Fish

1. **Bogorov, V.**, The ocean as food source, *Curr. Dig. Sov. Press,* 20 (May 22), 13–14, 1968.
2. **Borgstrom, G. A.**, New methods of appraising the role of fisheries in world nutrition, *Fish. News Int.,* 1(1), 12–20, 1961.
3. **Borgstrom, G. A.**, The role of fisheries in the Pacific, *Fish. News Int.,* 1(3), 16–20, 1961.
4. **Borgstrom, G. A., Ed.**, *Fish as Food,* Academic Press, New York, 1961–1965. (4 vols.)
5. **Borgstrom, G. A.**, The sea and our future, *Fish Boat,* Oct. 8, 1963.
6. **Borgstrom, G. A.**, Fish in world trade, *Fish News Int.,* 3(1), 10–14, 1963.
7. **Chapman, W. McL.**, Resources of the ocean and their potentialities for man, *Food Technol.,* 20, 895–901, 1966.
8. **Chapman, W. McL.**, Food from the ocean, in *Proc. 14th Annu. Meeting of the Agricultural Research Institute,* National Academy of Sciences, Washington, D.C., 1965, 65–94. (17 ref.)
9. **Chapman, W. McL.**, Seafood supply and world famine – positive approach, in *Food From the Sea, A Symposium,* American Association for the Advancement of Science. Washington, D.C., 1969.
10. **Christy, F. T. and Scott, A.**, *The Common Wealth in Ocean Fisheries,* Johns Hopkins Press, Baltimore, 1966.
11. **Finn, D. B.**, The sea and world food supplies, *Nutr. Abstr. Rev.,* 24, 487–496, 1954.

12. **Galtsoff, P. S.**, Food resources of the ocean, in *World Population and Future Resources*, Hatt, P. K., Ed., American Book, New York, 1952, 108–118.
13. **Gulland, J. A.**, The ocean reservoir, *Sci. J.*, 4(5), 81–86, 1968.
14. **Hamlisch, R. and Taylor, R. A.**, The demand for fish as human food, in *Fish in Nutrition*, Heen, E. and Kreuzer, R., Ed., Fishing News Books, London, 1962, 385–407.
15. **Heen, E. and Kreuzer, R., Eds.**, *Fish in Nutrition, FAO International Conference on Fish Nutrition*, Fishing News Books, London, 1962, 447 pp.
16. **Holt, S. J.**, The food resources of the ocean, *Sci. Am.*, Sept. 1969, 178–194.
17. **Holt, S. J.**, The contribution of freshwater fish production to human nutrition and well-being, in *The Biological Basis of Freshwater Fish Production*, Gerking, S. D., Ed., Blackwell, Oxford, 1967, 455–467.
18. **Idyll, C. P.**, *The Sea Against Hunger*, Crowell, New York, 1970, 221 pp.
19. **Kashara, H.**, Food production in the ocean, Proc. 7th International Congress of Nutrition, Hamburg, August 1966.
20. **Kreuzer, R., Ed.**, *The Technology of Fish Utilization*, Fishing News Books, London, 1965, 280 pp.
21. **Loftas, T.**, *The Last Resource*, Hamilton, London, 1969, 256 pp.
22. **Meseck, G.**, Importance of fisheries production and utilization in the fish economy, in *Fish in Nutrition*, Heen, E. and Kreuzer, R., Ed., Fishing News Books, London, 1962, 23–38.
23. **Moore, R. W., Ed.**, *Progress in Fishery and Food Science*, University of Washington Press, Seattle, 1972.
24. **Needler, A. H. W.**, Food resources in the sea, in *The Food Resources of Mankind*, Agri-World Press, Montreal, 1968, 105–108.
25. **Olcott, H. S. and Schaefer, M. B.**, Food from the sea, *Food Technol.*, 19, 774–777, 1965.
26. **Pariser, E. R. and Hammerle, O. A.**, Some cultural and economic limitations on the use of fish as food, *Food Technol.*, 20, 629–632, 1966.
27. **Parman, G.**, Food from the sea, *Feedstuffs*, 39(5), 66, 68–69, 1967.
28. **Poppensiek, G. C.**, Aquaculture and marine culture: Protein potentials, *J. Am. Vet. Med. Assoc.*, 161, 1467–1475, 1972.
29. **Ricker, W. E.**, Food from the sea, in *Resources and Man*, W. H. Freeman Publishing, San Francisco, 1969, 87–108.
30. **Sprague, L. M. and Arnold, J. H.**, Trends and prospects for the future harvest of the world fisheries resources, *J. Am. Oil Chem. Soc.*, 49, 345A–350A, 1972. (15 ref.)
31. **Ricker, W. E.**, Food from the sea, in *Resources and Man*, W. H. Freeman Publishing, San Francisco, 1969, 87–108.
32. **Sprague, L. M. and Arnold, J. H.**, Trends and prospects for the future harvest of the world fisheries resources, *J. Am. Oil Chem. Soc.*, 49, 345A–350A, 1972. (15 ref.)
33. Symposium. Food from the sea, in *Proc. 7th International Congress of Nutrition*, Vol. 4, Pergamon Press, Oxford, 1967, 936–1048. (16 papers)
34. **Tressler, D. K. and Lemon, J. McW.**, *Marine Products of Commerce. Their Acquisition, Handling, Biological Aspects, and the Science and Technology of their Properties and Preservation*, 2nd ed., Reinhold, New York, 1951, 782 pp.

6. Wildlife

1. **Ajaji, S. S.**, Wildlife as a source of protein in Nigeria: Some priorities for developement, *Nigerian Field*, 36(3), 115–127, 1971.
2. **Asibey, E. O. A.**, Wildlife as a source of protein in Africa South of the Sahara: African Forestry Commission, Ad hoc Working Party on Wildlife Management, FAO, Rome, 1972.
3. **Beresford-Peirse, H.**, Forest, Food and People, FFHC Basic Study No. 20, FOA, Rome, 1968.
4. **Charter, J. R.**, *Nigeria's Wildlife: A Forgotten National Asset*, International Union for Conservation of Nature and Natural Resources Publ., New Ser., 22 Morges, Switzerland, 1971.
5. **Clottey, St., J. A.**, Wildlife as a source of protein in Ghana, *Food and Nutr. Afr.*, No. 9, 7–11, 1971.
6. **de Vos, A. and Kaittany, K.**, Economic value of wildlife in Africa, Proc. 7th World Forestry Congress, 1972.
7. **French, M. H.**, The scale of contribution of wild herbivores to human welfare in desert lands, in Symp. Large Wild and Farm Herbivores, Aberdeen, Cambridge, 1965; *World Rev. Anim. Prod.*, 3, 89, 1966.
8. **Guggisberg, C. A. W.**, *Man and Wildlife*, ARCO, New York, 1970, 224 pp.
9. **Kay, R. N.**, Meat production from wild herbivores, *Proc. Nutr. Soc.*, 29, 271–278, 1970. (52 ref.)
10. **Ledger, H. P., Sachs, R. and Smith, N. S.**, Wildlife and food production, *World Rev. Anim. Prod.*, 3(11), 13–37, 1967.
11. **Maloiy, G. M. O.**, African game animals as a source of protein, *Nutr. Abstr. Rev.*, 35, 903–908, 1965. (34 ref.)
12. **Riney, T.**, The importance of wild life as a marginal form of land use in developing countries, in Symp. Large Wild and Tame Herbivores, Aberdeen, Cambridge, 1965.
13. **Sachs, R. and Glees, A.**, Report on the Preservation of Wildlife, Utilization of Mammals and Processing of Game Meat, Tanzania, 1967.
14. **Stewart, D. R. M.**, Development of wildlife as an economic asset, *Bull. Epizoot. Dis. Afr.*, 11, 167–171, 1963.
15. **Talbot, L. M.**, Wild animals as sources of food, in *Proc. 6th International Congress Nutrition, Edinburgh, Aug. 9–15, 1963*, Livingstone, Edinburgh, 1964, 243–251.

16. **Talbot, L. M., Payne, W. J. A., Ledger, H. P., Verdcourt, L. D., and Talbot, M. H.,** *The Meat Production Potential of Wild Animals in Africa,* Animal Breed. Tech. Commun. No. 16, Edinburgh, Oliver & Boyd, 1965, Comm. Agric. Bur., Franham Royal, England, 1965, 42 pp.
17. **Talbot, L. M.,** *Wild Animals as a Source of Food,* Bur. Sports Fisheries and Wildlife, Washington, D.C., 1966.

B. Technical Measures for Increasing Livestock Productivity

1. General

1. **Acker, D.,** *Animal Science and Industry,* Prentice-Hall, Englewood Cliffs, N.J., 1971, 604 pp.
2. **Allen, W.,** *The African Husbandman,* Oliver & Boyd, London, 1965.
3. **Anderson, A. L.,** *Introductory Animal Husbandry,* Macmillan, New York, 1958.
4. **Anderson, A. L. and Kiser, J. J.,** *Introductory Animal Science,* 4th ed., Macmillan, New York, 1963, 800 pp.
5. **Blaxter, K. L.,** Increasing output of animal production: Technical measures for increasing productivity, in *Man, Food, and Nutrition. Strategies and Technological Measures for Alleviating the World Food Problem,* Rechcigl, M., Jr., Ed., CRC Press, Cleveland, 1973, 127–146.
6. **Bundy, C. E. and Diggins, R. V.,** *Livestock and Poultry Production. Principles and Practices,* Prentice-Hall, New York, 1954.
7. **Byerly, T. C.,** *Livestock and Livestock Products,* Prentice-Hall, Englewood Cliffs, N.J., 1964, 422 pp.
8. **Byerly, T. C.,** Efficiency of feed conversion, *Science,* 157, 890–895, 1967.
9. **Campbell, J. R. and Lasley, J. F.,** *The Science of Animals that Serve Mankind,* McGraw-Hill, New York, 1969, 771 pp.
10. **Cole, H. H., Ed.,** *Introduction to Livestock Production. Including Dairy and Poultry,* 2nd ed., W.H. Freeman Publishing, San Francisco, 1966, 827 pp.
11. **Cole, H. H. and Ronning, M., Eds.,** *Animal Agriculture. The Biology of Domestic Animals and Their Use by Man,* W. H. Freeman Publishing, San Francisco, 1974, 788 pp.
12. **El-Shazly, K.,** The potential for increasing efficiency of production of farm animals, *World Rev. Anim. Prod.,* 3(11), 45–56, 1967.
13. **Ensminger, M. E.,** *Animal Science,* 6th ed., Interstate, Danville, Ill., 1969, 1253 pp.
14. **Ensminger, M. E.,** *The Stockman's Handbook,* 4th ed., Interstate, Danville, Ill., 1970, 957 pp.
15. **Garrigues, W. P.,** *Introductory Animal Science,* 3rd ed., Lippincott, Chicago, 1960.
16. **Kays, J. M.,** *Basic Animal Husbandry,* Prentice-Hall, Englewood Cliffs, N.J., 1958.
17. Livestock, *Dev. Dig.,* 10(2), 45–80, 1972. (6 papers)
18. **Lodge, G. A. and Lamming, G. E., Eds.,** *Growth and Development of Mammals,* Butterworths, London, 1968.
19. **McDowell, R. E.,** Problems of Cattle Production in Tropical Countries, Cornell Int. Agric. Dev. Monogr. 17, N.Y. State College of Agriculture, Ithaca, N.Y., 1969.
20. **McDowell, R. E.,** Effective Planning for Expanding Livestock Production in Developing Countries, Cornell Int. Dev. Monogr. 32, N.Y. State College of Agriculture, Ithaca, N.Y., 1969.
21. **Morrison, F. B.,** Animal production in an efficient economy, *Chron. Bot.,* 11(4), 247–256, 1948.
22. National Research Council, *Basic Problems and Techniques in Range Research,* National Academy of Sciences, Washington, D.C., 1962, 341 pp.
23. **Peters, W. H. and Deyoe, G. P.,** *Raising Livestock,* McGraw-Hill, New York, 1946.
24. **Peters, W. H. and Grummer, R. H.,** *Livestock Production,* McGraw-Hill, New York, 1954.
25. **Rhodes, D. N., Ed.,** *Meat Production from Entire Male Animals,* Churchill, London, 1969.
26. **Shaw, J. C.,** Problems of increasing animal production, *Proc. Nutr. Soc.,* 21, 99–106, 1962.
27. Techniques and Procedures in Animal Production Research, American Society Animal Production, Beltsville, MD, 1959, 228 pp.
28. *The UFAW Handbook on the Care and Management of Farm Animals,* Churchill Livingstone, Edinburgh, 1971, 283 pp.
29. **Valentine, J. F.,** *Range Development Improvements,* Brigham Young University Press, Provo, UT, 1972, 516 pp.
30. **Widmer, J.,** *Practical Animal Husbandry,* Charles Scribner's Sons, New York, 1949.
31. **Williamson, G. and Payne, W. J. A.,** *An Introduction to Animal Husbandry in the Tropics,* 2nd ed., Longmans, Green & Co., London, 1965, 447 pp.

2. Animal Nutrition

a. General

1. **Abrams, J. T.,** *Animal Nutrition and Veterinary Dietetics,* 4th ed., Williams & Wilkins, Baltimore, 1962, 826 pp.
2. **Abrams, J. T., Ed.,** *Recent Advances in Animal Nutrition,* Churchill, London, 1966, 261 pp. (7 papers)
3. **Allman, R. T. and Hamilton, T. S.,** Nutritional Deficiencies in Livestock, FAO Stud. No. 5, FAO, Washington, D.C., 1949, 102 pp.
4. Animal nutrition and food production, (7 papers) Proc. Fifth International Congress on Nutrition, Washington, D.C., Sept. 1–7, 1960, *Fed. Proc.,* 20(1), III, Suppl. 7, 261–313, 1961.

5. **Ashton, W. M.,** *Elements of Animal Nutrition,* Griffin, London, 1950, 208 pp.
6. **Bigwood, E. J.,** Problems of animal nutrition in underdeveloped areas, *Fed. Proc.,* 20, 261–267, 1961.
7. **Bisschop, J. H. R. and Groenwald, J. W.,** Limitations to efficiency imposed by inadequate nutrition with particular reference to feed resources, *Proc. World Conference on Animal Production,* Vol. 1, Rome, Sept. 2–7, 1963, European Association for Animal Production, Rome 1963, 47–72.
8. **Breireim, K.,** Increasing efficiency of animal production by nutrition and feeding research, *Proc. 2nd World Conference on Animal Production,* American Dairy Science Association, Urbana, Ill., 1969, 109–124. (124 ref.)
9. **Charton, A. and Lesbouyries, G.,** *Nutrition des mammiferes domestiques,* Vigot Freres, Paris, 1957, 479 pp.
10. **Cuthbertson, D. P., Ed.,** *Progress in Nutrition and Allied Sciences,* Oliver & Boyd, Edinburgh, 1963, 452 pp. (39 papers)
11. **Cuthbertson, D. P., Ed.,** *Nutrition of Animals of Agricultural Importance, International Encyclopaedia of Food and Nutrition,* Vol. 17, Pergamon Press, Oxford, 1969, 1443 pp. (35 papers)
12. **Fědorović-Tome, M., Obradović, M., and Stošić, D.,** *Norms and Tables for Animal Nutrition,* Zadružna knjiga, Beograd, 1967, 319 pp.
13. **Guilbert, H. R. and Loosli, J. K.,** Comparative nutrition of farm animals, *J. Anim. Sci.,* 10, 22–41, 1951.
14. **Hafez, E. S. E. and Dyer, I. A., Eds.,** *Animal Growth and Nutrition,* Lea & Febiger, Philadelphia, 1969, 402 pp. (20 papers)
15. **Hock, A.,** *Vergleichende Ernährungslehre des Menschen und seiner Haustiere,* G. Fischer, Jena, 1966, 968 pp.
16. **King, J. O. L.,** *Veterinary Dietetics,* Williams & Wilkins, Baltimore, 1961, 230 pp.
17. **Kirchgessner, M.,** *Tierernährung,* DLG-Verlag, Frankfurt/Main, 1970, 406 pp.
18. **Kolb, E. and Gürtler, H.,** *Ernährungsphysiologie der landwirthschaftlichen Nutztiere,* VEB Gustav Fischer Verlag, Jena, 1971, 957 pp.
19. **Lenkeit, W., Breirem, K., and Crasemann, E.,** *Handbuch der Tierernährung,* Verlag Paul Parey, Hamburg, 1969. (2 vols.)
20. **Lewis, D.,** The potential for increasing efficiency of feed utilization through newer knowledge of animal nutrition. F. Basic factors influencing the efficiency of utilization of energy and protein with special reference to their interrelationship, *Proc. World Conference on Animal Production,* Vol. 1, Rome, Sept. 2–7, 1963, European Association for Animal Production, Rome, 1963, 305–329.
21. **McDonald, P., Edwards, R. A., and Greenhalgh, J. F. D.,** *Animal Nutrition,* 2nd ed., Hafner Press, New York, 1973, 487 pp.
22. **Maynard, L. A. and Loosli, J. K.,** *Animal Nutrition,* 6th ed., McGraw-Hill, New York, 1969, 613 pp.
23. **Maynard, L. A.,** The role and efficiency of animals in utilising feed to produce human food, *J. Nutr.,* 32, 345–360, 1946.
24. **Mitchell, H. H.,** *Comparative Nutrition of Man and Domestic Animals,* Academic Press, New York, 1963, 1964. (2 vols).
25. **Nehring, K.,** *Lehrbuch der Tierernährung und Futtermittelkunde,* Neumann Verlag, Radebeul, 1964.
26. Nutrition Symposium. Animal nutrition research, a key to some world food and nutrition problems, *Fed. Proc.,* 23, 851–867, 1964. (4 papers)
27. **Oyenuga, V. A.,** Problems of livestock nutrition in Nigeria, *Nutr. Abstr. Rev.,* 28, 985–1000, 1958.
28. **Tyler, C. T.,** *Animal Nutrition,* Chapman & Hall, London, 1964, 253 pp.

b. Feeds

1. **Bixler, W. B.,** *Feed Additive Compendium,* Miller Publishing, Minneapolis, 1971.
2. **Butterworth, M. H.,** The digestibility of tropical grasses, *Nutr. Abstr. Rev.,* 37, 349–368, 1967.
3. **Cassard, D. W.,** *Approved Practices in Feeds and Feeding,* Interstate, Danville, Ill., 1963.
4. **Crampton, E. W. and Harris, L. E.,** *Applied Animal Nutrition. The Use of Feedstuffs in the Formulation of Livestock Rations,* W. H. Freeman, San Francisco, Calif., 1969, 753 pp.
5. **Evans, R. E.,** *Rations for Livestock,* Min. Agric. Fish. Foods, Bull. No. 148, HMSO, 1960.
6. **Gutcho, M.,** *Animal Feeds,* 1970, Noyes Data Corp., Park Ridge, N.J., 1970, 353 pp.
7. **Gutcho, M.,** *Feeds for Livestock, Poultry and Pets,* Noyes Data Corp., Park Ridge, N.J., 1973.
8. **Kolb, E.,** *Wirkstoff-Vademekum. Chemische and biologishche Präparata in der Tierernährung,* Fisher, Jena, 1970, 319 pp.
9. **Morrison, F. B.,** *Feeds and Feeding,* 23rd ed., Morrison Publishing, Clinton, IA, 1967.
10. National Research Council, Composition of Concentrated By-Product Feeding Stuffs, Publ. No. 449, National Academy of Sciences, Washington, D.C., 1956.
11. National Research Council, Composition of Cereal Grains and Forages, Publ. No. 585, National Academy of Sciences, Washington, D.C., 1958.
12. National Research Council, *U.S.-Canadian Tables of Feed Composition,* National Academy of Sciences, Washington, D.C., 1969.

13. National Research Council, *Atlas of Nutritional Data on United States and Canadian Feeds,* National Academy of Sciences, Washington, D.C., 1971, 772 pp.
14: **Perry, T. W.,** *Feed Formulation Handbook,* Interstate, Danville, Ill., 1966.
15. **Raymond, W. F.,** The nutritive value of forage crops, *Adv. Agron.,* 21, 1–108, 1969.
16. **Schneider, B. H.,** Feeds of the World. Their Digestibility and Composition, Agric. Exp. Sta., West Virginia University, Morgantown, 1947.
17. **Seiden, R. and Pfander, W. H.,** *The Handbook of Feedstuffs. Production, Formulation, Medication,* Springer Publishing, New York, 1957, 591 pp.
18. **Sherwood, R. M.,** *Feed Mixers' Handbook,* Interstate, Danville, Ill., 1951.
19. Symposium. Problems in providing rations for Britain's livestock, *Proc. Nutr. Soc.,* 13, 1–40, 1954. (11 papers)
20. Use of land and crop resources in animal production, *Proc. 2nd World Conference on Animal Production,* American Dairy Science Association, Urbana, Ill., 1969, 52–95. (4 papers with discussion)
21. **Woodman, H. E.,** The Composition and Nutritive Value of Feeding Stuffs, Min. Agric. Fish. Bull. 8, HMSO, London, 1950.

3. Genetics and Animal Breeding

1. **Bogart, R.,** *Improvement of Livestock,* Macmillan, New York, 1959.
2. **Briggs, H. M. and Briggs, D. M.,** *Modern Breeds of Livestock,* 3rd ed., Macmillan, London, 1969, 714 pp.
3. **Donald, H. P.,** Opportunities and directions for research in genetics and breeding, *Proc. 2nd World Conference on Animal Production,* American Dairy Science Association, Urbana, Ill., 1969, 96–102. (8 ref.)
4. **Hagedoorn, A. L.,** *Animal Breeding,* 5th ed., Crosby Lockwood, London, 1954, 364 pp.
5. **Hammond, J.,** *Farm Animals. Their Growth, Breeding and Inheritance,* 3rd ed., Edward Arnold & Co., London, 1960, 322 pp.
6. **Hutt, F. B.,** *Genetic Resistance of Disease in Domestic Animals,* Comstock Publishing, Ithaca, N.Y., 1958.
7. **Hutt, F. B.,** *Animal Genetics,* Ronald Press, New York, 1964.
8. **Johansson, I.,** Hereditary defects in farm animals, *World Rev. Anim. Prod.,* 3, 19–30, 1965.
9. **Johansson, I. and Rendel, J.,** *Genetics and Animal Breeding,* Oliver & Boyd, Edinburgh, 1968, 489 pp.
10. **Kelley, R. B.,** *Native and Adapted Cattle,* Angus & Robertson, Sydney, 1959.
11. **Kelley, R. B.,** *Principles and Methods of Animal Breeding,* Angus & Robertson, Sydney, 1960, 350 pp.
12. **Lasley, J. F.,** *Genetics of Livestock Improvement,* 2nd ed., Prentice-Hall, Englewood Cliffs, N.J., 1963, 429 pp.
13. **Lerner, I. M.,** *Population Genetics and Animal Improvement,* Cambridge University Press, London, 1951.
14. **Lerner, I. M. and Donald, H. P.,** *Modern Developments in Animal Breeding,* Academic Press, New York, 1966, 294 pp.
15. **Lush, J. L.,** *Animal Breeding Plans,* Iowa State University Press, Ames, 1970, 443 pp.
16. **McDowell, R. E.,** *Improvement of Livestock Production in Warm Climates,* W. H. Freeman Publishing, San Francisco, 1972, 712 pp.
17. **Nichols, J. E.,** *Livestock Improvement in Relation to Heredity and Environment,* Oliver & Boyd, Edinburgh, 1957, 240 pp.
18. **Phillips, R. W.,** Breeding Livestock Adapted to Unfavorable Environments, FAO, Washington, D.C., 1949.
19. **Phillips, R. W., Ed.,** Improving Livestock under Tropical and Sub-Tropical Conditions, FAO Dev. Paper No. 6, FAO, Washington, D.C., 1950, 55 pp.
20. **Phillips, R. W.,** Untapped sources of animal germ plasm, in Germ Plasm Resources, AAAS Publ. No. 66, American Association for the Advancement of Science, Washington, D.C., 1961, 43–75.
21. **Phillips, R. W.,** Animal genetic resources, *World Rev. Anim. Prod.,* 3, 28–31, 1967.
22. **Phillips, R. W., Johnson, R. G., and Moyer, R. T.,** The Livestock of China, U.S. Dep. of State Publ. 2249 Far Eastern Ser. 9, 174 pp.
23. **Rice, V. A., Andrews, F. N., and Warwick, E. J.,** *Breeding Better Livestock,* McGraw-Hill, New York, 1953, 465 pp.
24. **Rice, V. A., Andrews, F. N., Warwick, E. J., and Legates, J. E.,** *Breeding and Improvement of Farm Animals,* McGraw-Hill, New York, 1967, 477 pp.
25. **Robertson, A.,** Influence of breeding on production efficiency, *Proc. World Conference on Animal Production,* Vol. 1, Rome, Sept. 2–7, 1963, European Association for Animal Production, 1963, 99–110.
26. A symposium on the effects of climate on animals, *J. Hered.,* 49(2), 46–81, 1958.
27. **Winters, L. M.,** *Animal Breeding,* 5th ed., John Wiley & Sons, New York, 1954.

4. Reproduction

1. **Arthur, G. H.,** *Wright's Veterinary Obstetrics Including Diseases of Reproduction,* 3rd ed., Williams & Wilkins, Baltimore, 1964, 549 pp.
2. **Asdell, S. A.,** *Patterns of Mammalian Reproduction,* 2nd ed., Cornell University Press, Ithaca, N.Y., 1964.
3. **Asdell, S. A.,** *Cattle Fertility and Sterility,* 2nd ed., Little, Brown & Co., Boston, 1968, 276 pp.
4. **Cole, H. H. and Cupps, P. T., Eds.,** *Reproduction in Domestic Animals,* 2nd ed., Academic Press, New York, 1969, 657 pp.

5. **Faulkner, L. C., Ed.**, *Abortion Diseases of Livestock*, Charles C Thomas, Springfield, 1968, 207 pp. (15 papers)
6. **Hafez, E. S. E., Eds.**, *Reproduction and Breeding Techniques for Laboratory Animals*, Lea & Febiger, Philadelphia, 1970, 375 pp. (18 papers)
7. **Hafez, E. S. E., Ed.**, *Reproduction in Farm Animals*, 2nd ed., Lea & Febiger, Philadelphia, 1968, 440 pp. (23 papers)
8. **Holmes, R. L.**, *Reproduction and Environment*, Oliver & Boyd, Edinburgh, 1968.
9. **Karg, H.**, Opportunities for improving reproduction as means to increasing efficiency of animal production, *Proc. 2nd World Conference on Animal Production*, American Dairy Science Association, Urbana, Ill., 1969, 102–109. (44 ref.)
10. **Laing, J. A.**, *Fertility and Infertility in Domestic Animals*, 2nd ed., Williams & Wilkins, Baltimore, 1971.
11. **Laing, J. A., Ed.**, *Fertility and Infertility in the Domestic Animals. Aetiology, Diagnosis and Treatment*, 2nd ed., Baillière, Tindall & Cassell, London, 1970, 480 pp. (12 papers)
12. **Nalbandov, A. V.**, *Reproductive Physiology*, 2nd ed., W. H. Freeman Publishing, San Francisco, 1964.
13. **Perry, E. J., Ed.**, *The Artificial Insemination of Farm Animals*, 4th ed., Rutgers University Press, New Brunswick, N.J., 1968, 473 pp.
14. **Rheinbold, H. L., Ed.**, *Maternal Behavior in Mammals*, John Wiley & Sons, New York, 1963.
15. **Rowlands, I. W., Ed.**, *Comparative Biology of Reproduction in Mammals*, Academic Press, New York, 1966.
16. **Salisbury, G. W. and Van Denmark, N. L.**, *Physiology of Reproduction and Artificial Insemination of Cattle*, W. H. Freeman Publishing, San Francisco, 1961.
17. **Young, W. C., Ed.**, *Sex and Internal Secretions*, Williams & Wilkins, Baltimore, 1961.
18. **Zemjanis, R.**, *Diagnostic and Therapeutic Techniques in Animal Reproduction*, 2nd ed., Williams & Wilkins, Baltimore, 1970, 242 pp.

5. Animal Diseases and Disease Control

1. *Advances of Veterinary Science*, Vol. 1–17–, Academic Press, New York, 1953–1973–.
2. **Andrews, P. N.**, Animal and veterinary science programs in developing countries, in *Agricultural Sciences for the Developing Nations*, American Association for the Advancement of Science, Washington, D.C., 1964, 69–79.
3. **Benbrook, E. A. and Sloss, M. W.**, *Veterinary Clinical Parasitology*, 2nd ed., Iowa State College Press, Ames, 1961.
4. **Benjamin, M. M.**, *Outline of Veterinary Clinical Pathology*, 2nd ed., Iowa State University Press, Ames, 1965, 186 pp.
5. **Blood, D. C. and Henderson, J. A.**, *Veterinary Medicine*, 3rd ed., Williams & Wilkins, Baltimore, 1968, 927 pp.
6. **Boddie, G. F.**, *Introduction to Veterinary Therapeutics*, Oliver & Boyd, London, 1952, 196 pp.
7. **Cheng, T. C., Ed.**, *Current Topics in Comparative Pathology*, Vol. 1, Academic Press, New York, 1971.
8. **Cockrill, W. R.**, The veterinarian in the developing countries, *Vet. Rec.*, 75, 722–729, 1963.
9. **Cockrill, W. R.**, The profession and the science. International trends in veterinary medicine, *Adv. Vet. Sci.*, 9, 251–325, 1964.
10. **Cottereau, P.**, *La pathologie animale*, Presses universitaires de France, Paris, 1970, 119 pp.
11. **Doxey, D. L.**, *Veterinary Clinical Pathology*, Williams & Wilkins, Baltimore, 1971, 356 pp.
12. **Dykstra, R. R.**, *Animal Sanitation and Disease Control*, 6th ed., Interstate, Danville, Ill., 1961, 858 pp.
13. **Galloway, J. H.**, *Farm Animal Health and Disease Control*, Lea & Febiger, Philadelphia, 1972, 373 pp.
14. **Gibbons, W. J.**, *Clinical Diagnosis of Diseases of Large Animals*, Lea & Febiger, Philadelphia, 1966, 259 pp.
15. **Gresham, G. A. and Jennings, A. R.**, *An Introduction to Comparative Pathology*, Academic Press, New York, 1962, 412 pp.
16. **Hagan, W. A. and Brunner, D. W.**, *The Infectious Diseases of Domestic Animals*, 4th ed., Comstock Publishing, Ithaca, N.Y., 1961, 1033 pp.
17. *Handbook of Animal Diseases in the Tropics*, British Veterinary Association, London, 1967, 236 pp.
18. **Henning, M. H.**, *Animal Diseases in South Africa*, 2nd ed., Central News Agency, Johannesburg, 1949.
19. **Hungerford, T. G.**, *Diseases of Livestock*, 7th ed., Angus & Robertson, 1970, 1035 pp.
20. **Hussel, L.**, *Lehrbuch der Veterinärhygiene*, Hirzel, Leipzig, 1963, 820 pp.
21. **Hutyra, F., Marek, J., and Manninger, R.**, *Special Pathology and Therapeutics of the Domestic Animals*, Alexander Eger, Chicago, 1949.
22. **Jennings, A. R.**, *Animal Pathology*, Baillière, Tindall & Cassell, London, 1970, 262 pp.
23. **Jones, L. M.**, *Veterinary Pharmacology and Therapeutics*, 3rd ed., Iowa State University Press, Ames, 1965, 1037 pp.
24. **Judd, K. V. F. and Kennedy, P. C.**, *Pathology of Domestic Animals*, Academic Press, New York, 1970. (2 vols.)
25. **Konnerup, N. M.**, Diseases of livestock and game in tropical Africa, in *Symposium on Protein and Population in Africa*, Advisory Committee on Africa, National Academy of Sciences, Washington, D.C., 1964.
26. **Lieberman, J., Ed.**, Animal disease and human health, *Ann. N.Y. Acad. Sci.*, 70, 277–762, 1958.
27. **Medway, W., Prier, J. E., and Wilkinson, J. S., Eds.**, *A Textbook of Veterinary Clinical Pathology*, Williams & Wilkins, Baltimore, 1969, 522 pp.
28. **Merchant, I. A. and Packer, R. A.**, *Veterinary Bacteriology and Virology*, 6th ed., Iowa State College Press, Ames, 1961.

29. *The Merck Veterinary Manual. A Reference Handbook of Diagnosis and Therapy for the Veterinarian*, 3rd ed., Merck, Rahway, N.J., 1961, 168 pp.
30. National Research Council, Report of Study Group on Animal Diseases in Africa, Dec. 8, 1965.
31. Pan American Health Organization, Economic aspects of certain animal diseases transmissable to man, Meeting of the Inter-American Economic and Social Council, Costa Rica, July 1963.
32. **Parker, W. H.**, *Health and Disease in Farm Animals for Those Concerned with Animal Husbandry*, Oxford, Pergamon Press, 1970, 301 pp.
33. **Pritchard, W. R.**, Increasing protein foods through improving animal health, *Proc. Natl. Acad. Sci. U.S.A.*, 56, 360–369, 1966.
34. **Radeleff, R. D.**, *Veterinary Toxicology*, Lea & Febiger, Philadelphia, 1970.
35. **Runnels, R. A.**, *Animal Pathology*, 5th ed., Iowa State College Press, Ames, 1954, 718 pp.
36. **Schwabe, C. W.**, *Veterinary Medicine and Human Health*, Williams & Wilkins, Baltimore, 1965.
37. **Seiden, R.**, *Livestock Health Encyclopedia*, 2nd ed., Springer Publishing, New York, 1961.
38. **Smith, H. A. and Jones, T. C.**, *Veterinary Pathology*, Lea & Febiger, Philadelphia, 1961.
39. **Smithcors, J. F.**, *Evolution of Veterinary Art*, Veterinary Medical Publishing, Kansas City, Mo., 1957.
40. **Soulsby, E. J. L.**, *Textbook of Veterinary Clinical Parasitology*, Davis, Philadelphia, 1965.
41. **Stableforth, A. W. and Galloway, I. A., Eds.**, *Infectious Diseases of Animals: Diseases Due to Bacteria*, Academic Press, New York, 1959. (2 vols.)
42. **Steele, J. H.**, Veterinary medicine and the public health, *Proc. 17th World Vet. Congr.*, 2, 1551–1555, 1963.
43. **Steele, J. H.**, Animal Diseases and Human Health, FFHC Basic Stud. No. 3, FAO, Rome, 1962.
44. **Stevenson, D. E. and Wilson, A. A.**, *Metabolic Disorders of Domestic Animals*, Davis, Philadelphia, 1963, 198 pp.
45. **Udall, D. H.**, *The Practice of Veterinary Medicine*, 6th ed., Udall, Ithaca, N.Y., 1954, 811 pp.
46. U.S. Department of Agriculture, Keeping Livestock Healthy. The Yearbook of Agriculture 1942, U.S. Gov. Print. Off., Washington, D.C., 1942, 1271 pp.
47. U.S. Department of Agriculture, Animal Diseases. The Yearbook of Agriculture 1956, U.S. Gov. Print. Off., Washington, D.C., 1956, 591 pp.
48. U.S. Livestock Sanitary Association, Foreign Animal Diseases – Their Prevention, Diagnosis and Control, Washington, D.C., 1964, 307 pp.
49. Veterinary Research Laboratory, Emerging Diseases of Animals, FAO, Rome, 1963, 241 pp.
50. **Weinman, D. and Ristic, M., Ed.**, *Infectious Blood Animals of Man and Animals*, Academic Press, New York, 1968. (2 vols.)
51. **White, E. G. and Jordan, F. T. W.**, *Veterinary Preventive Medicine*, Williams & Wilkins, Baltimore, 1963, 334 pp.
52. **Wooldridge, W. R.**, *Farm Animals in Health and Diseases*, Crosby Lockwood, London, 1960, 533 pp.

C. Production of Specific Animals

1. Ruminants

a. General

1. Agricultural Research Council, The Nutrient Requirements of Livestock No. 2. Ruminants, ARC, London, 1965.
2. **Balch, C. C. and Campling, R. C.**, Regulation of voluntary food intake in ruminants, *Nutr. Abstr. Rev.*, 32, 669–686, 1962.
3. **Beeson, W. M.**, Recent developments in ruminant nutrition, Pfizer, Chas. Co., Inc., Agric. Res. Dev. Dep., Conf. Ser. 1962, 31–41. (41 ref.)
4. **Blaxter, K. L.**, *The Energy Metabolism of Ruminants*, 2nd ed., Charles C Thomas, Springfield, Ill., 1967, 332 pp.
5. **Church, D. C.**, *Digestive Physiology and Nutrition of Ruminants*, Oregon State University, Dep. of Animal Science, Corvallis, 1969–1972. (3 vols.)
6. **Dougherty, R. W.**, *Physiology of Digestion in the Ruminant*, Butterworth, London, 1965.
7. **Hale, W. H.**, Recent findings in ruminant nutrition, Pfizer, Chas. Co., Inc., Agric. Res. Dev. Dep., Conf. Ser. 1961, 33–38. (55 ref.)
8. **Larson, B. L. and Smith, V. R., Eds.**, *Lactation: A Comprehensive Treatise*, Academic Press, New York, 1974. (3 vols.)
9. **Lewis, D., Ed.**, *Digestive Physiology and Nutrition of the Ruminant*, Butterworths, London, 1961, 297 pp. (23 papers)
10. **McDonald, I. W.**, The nutrition of grazing ruminants, *Nutr. Abstr. Rev.*, 38, 381–400, 1968.
11. Nutritional disorders of ruminants, *Proc. Nutr. Soc.*, 28, 171–212, 1969. (6 papers)
12. **Payne, W. J. A.**, Nutrition of ruminants in the tropics, *Nutr. Abstr. Rev.*, 36, 23–67, 1966.
13. Symposium, The utilization of grass by ruminants, *Proc. Nutr. Soc.*, 29, 54–107, 1964.

b. Beef Cattle

(1) General

1. **Cockrill, W. R.**, The water buffalo, *Sci. Am.*, 217(6), 118–125, 1967.
2. **De Graff, H.**, *Beef Production and Distribution*, University of Oklahoma Press, Norman, 1960.

3. **Diggins, R. V. and Bundy, C. E.,** *Beef Production,* Prentice-Hall, Englewood Cliffs, N.J., 1962.
4. **Ensminger, M. E.,** *Beef Cattle Science,* Interstate, Danville, Ill., 1969.
5. **Neumann, A. L. and Snapp, R. R.,** *Beef Cattle,* 6th ed., John Wiley & Sons, New York, 1969, 767 pp.
6. **Payne, W. J. A.,** *Cattle Production in the Tropics,* Longmans, London, 1970. (2 vols.)
7. **Roy, J. H. B.,** *The Calf. Nutrition and Health,* Pennsylvania State University Press, University Park, 1970, 164 pp.
8. **Snapp, R. R. and Neumann, A. L.,** *Beef Cattle,* John Wiley & Sons, New York, 1960, 684 pp.
9. **Wagnon, K. A., Albaugh, R., and Hart, G. H.,** *Beef Cattle Production,* Macmillan, New York, 1960.
10. **Williams, W.,** *Beef Cattle Production in the South,* 2nd ed., Interstate, Danville, Ill., 1955.

(2) Breeding

1. **Beattie, W. A.,** *Beef Cattle Breeding and Management,* Pastoral Review, Melbourne, 1954.
2. *Breeding Beef Cattle for Unfavorable Environments, A Symposium,. . . .,* University of Texas Press, Austin, 1955, 248 pp.
3. **Cunha, T. J., Koger, M., and Warnick, A. C., Eds.,** *Cross Breeding Beef Cattle,* University of Florida Press, Gainesville, 1963.
4. **Gregory, K. E.,** *Beef Cattle Breeding,* Agric. Inf. Bull. 286, U.S. Gov. Print. Off., Washington, D.C., 1969, 55 pp.
5. **Rhoad, A. O.,** *Breeding of Beef Cattle for Unfavorable Environments. A Symposium,* King Ranch Centennial Conference, University of Texas Press, Austin, 1955.
6. **Warwick, W.,** Fifty years of progress in breeding beef cattle, *J. Anim. Sci.,* 17, 922–943, 1958.

(3)Nutrition

1. **Beeson, W. M.,** 50 years of scientific development in cattle feeding, *Proc. Annu. Indiana Cattle Feeders Day,* April 24, 1959.
2. **Dyer, I. A. and O'Mary, C. C., Eds.,** *Feedlot,* Lea & Febiger, Philadelphia, 1972, 224 pp. (15 papers)
3. **McClure, T. J.,** A review of developments in nutrition as it is related to fertility in cattle: 1964–9, *N. Z. Vet. J.,* 18, 61–68, 1970. (81 ref.)
4. National Research Council, *Nutrient Requirements of Domestic Animals. No. 4. Nutrient Requirements of Beef Cattle,* 4th ed., National Academy of Sciences, Washington, D.C., 1971.
5. **Richter, K.,** Möglichkeiten zur Steigerung der Futterverwertung und Grund neuer Erkenntnisse auf dem Gebiet der Tirernährung. A. Rinder (Fleischerzeugung) (Potential for increasing effciency of feed utilization through newer knowledge of animal nutrition. A. Cattle [Meat Production]), *Proc. World Conference on Animal Production, Rome Sept. 1–2, 1963,* European Association for Animal Production, Rome, 1963, 147–168. (134 ref.)

(4) Diseases

1. **Gibbons, W. J., Ed.,** *Diseases of Cattle,* American Veterinary Publications, Santa Barbara, 1963.

c. Dairy Cattle

(1) General

1. **Craplet, C.,** *The Dairy Cow,* Arnold, London, 1963.
2. **Davis, J. G.,** *A Dictionary of Dairying,* Leonard Hill, London, 1955; Suppl., 1965.
3. **Davis, R. F.,** *Modern Cattle Management,* Prentice-Hall, Englewood Cliffs, N.J., 1962, 264 pp.
4. **Eckles, C. H. and Anthony, E. L.,** *Dairy Cattle and Production,* Macmillan, New York, 1956.
5. **Foley, R. C., Bath, D. L., Dickinson, F. N., and Tucker, H. A.,** *Dairy Cattle: Principles, Practices, Problems, Profits,* Lea & Febiger, Philadelphia, 1972, 693 pp.
6. **Henderson, H. O. and Reaves, P. M.,** *Dairy Cattle. Feeding and Management,* 4th ed., John Wiley & Sons, New York, 1963.
7. **Knodt, C. B.,** *Successful Dairying,* McGraw-Hill, New York, 1954, 381 pp.
8. **Lamond, D. R.,** *Dairy Cattle Husbandry,* Angus & Robertson, Sydney, 1961.
9. **Peterson, W. E.,** *Dairy Science,* Lippincott, Philadelphia, 1950.
10. **Porter, A. R., Sims, J. A., and Freeman, C. F.,** *Dairy Cattle in American Agriculture,* Iowa State University Press, Ames, 1965.
11. **Reaves, P. M. and Henderson, H. O.,** *Dairy Cattle Feeding and Management,* John Wiley & Sons, New York, 1960.
12. **Roy, J. H. B.,** *The Calf, Its Management, Feeding and Health,* Farmer & Stock-Breeder Publications, London, 1959.
13. **Tucker, H. A.,** *Dairy Cattle: Principles, Practices, Problems, Profits,* Lea & Febiger, Philadelphia, 1972, 693 pp.
14. **Wing, J. M.,** *Dairy Cattle Management. Principles and Applications,* Reinhold, New York, 1963.
15. **Yapp, W. W. and Nevens, W. B.,** *Dairy Cattle Selection, Feeding and Management,* 4th ed., John Wiley & Sons, New York, 1955.

(2) Breeding

1. **Gilmore, L. O.,** *Dairy Cattle Breeding,* Lippincott, Philadelphia, 1952.
2. **Johansson, I.,** *Genetic Aspects of Dairy Cattle Breeding,* University of Illinois Press, Urbana, 1961.

(3) Nutrition
1. **Bartley, E. E.**, Recent developments in dairy cattle nutrition, Pfizer, Chas. Co., Inc., Agric. Res. Dev. Dep., Conf. Ser. 1961, 41–51. (65 ref.)
2. **Breireim, K.**, Nutrition and lactation in domestic animals and particularly in the cow, *Proc 4th International Congress of Nutrition,* Paris, 1957, A3–A32.
3. **Hoglund, C. R., Johnson, G. L., Lassister, C. A., and McGilliard, L. G., Eds.,** *Nutritional and Economic Aspects of Feed Utilization by Dairy Cows,* Iowa State College Press, Ames, 1949, 287 pp.
4. **Iwema, S.**, The potential for increasing efficiency of feed utilization through newer knowledge of animal nutrition. B. Cattle (Milk production), *Proc. World Conference on Animal Production,* Rome, Sept. 1–2, 1963, European Association for Animal Production, Rome, 1963, 147–168. (134 ref.)
5. **Loosli, J. K.**, Recent advances in dairy cattle nutrition, Pfizer, Chas. Co., Inc., Agric. Res. Dev. Dep., Conf. Ser. 1962, 21–29. (96 ref.)
6. **McCullough, M. E.**, *Optimum Feeding of Dairy Animals,* University of Georgia Press, Athens, 1969, 194 pp.
7. National Research Council, Committee on Animal Nutrition, Nutrient Requirements of Domestic Animals, No. 3. *Nutrient Requirements of Dairy Cattle,* 4th ed., National Academy of Sciences, Washington, D.C., 1971, 54 pp.
8. **Reeves, P. M. and Henderson, H. O., Eds.,** *Dairy Cattle, Feeding and Management,* 5th ed., John Wiley & Sons, New York, 1963.

d. Sheep

(1) General
1. **Coffey, W. C.**, *Productive Sheep Husbandry,* Lippincott, New York, 1937.
2. **Collins, S. B.**, *Profitable Sheep,* Macmillan, New York, 1956.
3. **Diggins, R. V. and Bundy, C. E.**, *Sheep Production,* Prentice-Hall, Englewood Cliffs, N.J., 1958.
4. **Ensminger, M. E.**, *Sheep and Wool Science,* Interstate, Danville, Ill., 1964.
5. **Horlacher, L. J.**, *Sheep Production,* McGraw-Hill, New York, 1927.
6. **Juergenson, E. M.**, *Approved Practices in Sheep Production,* Interstate, Danville, Ill., 1963.
7. **Kammlade, W. G. and Kammlade, W. G., Jr.**, *Sheep Science,* Lippincott, Philadelphia, 1955, 536 pp.
8. **McKinney, J.**, *The Sheep Book,* John Wiley & Sons, New York, 1959.
9. **Marsh, H.**, *Newson's Sheep Diseases,* 3rd ed., Williams & Wilkins, Baltimore, 1965.
10. **Mason, I. L.**, *Sheep Breeds of Mediterranean,* Constable, Edinburgh, 1967.
11. **Moule, G. R.**, Sheep (Pastotal Regions), in *The UFAW Handbook on the Care and Management of Farm Animals,* Churchill Livingstone, Edinburgh, 1971, 77–89.
12. National Sheep Breeders' Association, *British Sheep,* 3rd ed., Horseheath, Cambridge, NSBA, 1968.
13. **Spedding, C. R. W.**, *Sheep Production and Grazing Management,* Ballière, Tindall & Cox, London, 1965.

(2) Breeding
1. **Mason, I. L.**, *Sheep Breeds of Mediterranean,* Constable, Edinburgh, 1967.
2. **Rae, A. L.**, The genetics of the sheep, *Adv. Genet.,* 8, 190–266, 1956.
3. **Terrill, C. E.**, Fifty years progress in sheep breeding, *J. Anim. Sci.,* 17, 944–959, 1958.
4. **Turner, H. N. and Young, S. S. Y.**, *Quantitative Genetics in Sheep Breeding,* Cornell University Press, Ithaca, N.Y., 1969, 332 pp.

(3) Nutrition
1. **Allden, W. G.**, The effects of nutritional deprivation on the subsequent productivity of sheep and cattle, *Nutr. Abstr. Rev.,* 40, 1167–1184, 1970.
2. **Allen, D. M. and Lamming, G. E.**, Nutrition and reproduction in the ewe, *J. Agric. Sci.,* 56, 69–79, 1961.
3. National Research Council, Committee on Animal Nutrition, Nutrient Requirements of Domestic Animals No. 5, *Nutrient Requirements of Sheep,* 4th ed., National Academy of Sciences, Washington, D.C., 1968, 64 pp.
4. Nutrition of hill cattle and sheep, *Proc. Nutr. Soc.,* 30, 195–229, 1971. (6 papers)
5. **Schinckel, P. G.**, The potential for increasing efficiency of feed utilization through newer knowledge of animal nutrition. C. Sheep and goat, *Proc. World Conference on Animal Production, Rome, Sept. 1–2, 1963,* European Association for Animal Production, Rome, 1963, 199–239.

(4) Diseases
1. **Newsom, I. E.**, *Sheep Diseases,* 3rd ed., Williams & Wilkins, Baltimore, 1964.

e. Goats

1. **Abbey, A.**, *Practical Goat Keeping,* Cassell, London, 1953, 124 pp.
2. **Davies, C. J.**, *Goat-keeping for Milk Production,* Country Life, London, 1920, 219 pp.
3. **French, M. H.**, Observations on the Goat, FAO, Rome, 1970, 204 pp.
4. *The Husbandry and Diseases of Goats,* British Veterinary Association, London, 1950.
5. **Jeffery, H. E.**, *Goats,* Cassell, London, 1970, 112 pp.

6. **Linzell, J. L.,** Goats, in *The UFAW Handbook on the Care and Management of Farm Animals,* Churchill Livingstone, Edinburgh, 1971, 90–98.
7. **Mackenzie, D.,** *Goat Husbandry,* Faber & Faber, London, 1967, 349 pp.
8. **Potts, C. G., and Simons, V. L.,** *Milk Goats,* U.S. Dep. Agric. Farmer's Bull. No. 920, Washington, D.C., 1946, 36 pp.
9. **Rogers, A. L.,** *Goat-keeping in the United States,* Admire, Kansas, Dairy Goat Association, 1965.

2. Swine

a. General

1. **Anderson, A. L.,** *Swine Management, Including Feeding and Breeding,* Lippincott, Chicago, 1957, 531 pp.
2. **Baker, J. K. and Juergenson, E. M.,** *Approved Practices in Swine Production,* Interstate, Danville, Ill., 1971.
3. **Bundy, C. E. and Diggins, R. V.,** *Swine Production,* Prentice-Hall, Englewood Cliffs, N.J., 1956.
4. **Carroll, W. E., Krider, J. L., and Andrews, F. N.,** *Swine Production,* McGraw-Hill, New York, 1962, 433 pp.
5. **Cole, D. J. A., Ed.,** *Pig Production,* Butterworths, London, 1972.
6. **Cook, G. C. and Juergenson, E. M.,** *Approved Practices in Swine Production,* Interstate, Danville, Ill., 1962.
7. **Davidson, H. R.,** *The Production and Marketing of Pigs,* 3rd ed., Longmans, London, 1966.
8. **Deyoe, G. P. and Krider, J. L.,** *Raising Swine,* Interstate, Danville, Ill., 1961.
9. **Ensminger, M. E.,** *Swine Production,* 4th ed., Interstate, Danville, Ill., 1970, 881 pp.
10. **Johnson, G.,** *Profitable Pig Farming,* 3rd ed., Farming Press, Ipswich, Suffolk, England, 1968.
11. **Krider, J. L. and Carroll, W. E.,** *Swine Production,* 4th ed., McGraw-Hill, New York, 1971, 528 pp.
12. **Mount, L. E.,** *The Climatic Physiology of Pig,* Arnold, London, 1968.
13. **Pond, W. G. and Maner, J. H.,** *Swine Production in Temperate and Tropical Environments,* W. H. Freeman Publishing, San Francisco, 1974, 646 pp.
14. **Smith, W. W. and Hutchings, L. M.,** *Pork Production,* Macmillan, New York, 1952.
15. **Towne, C. W. and Wentworth, E. N.,** *Pigs from Cave to Cornbelt,* University of Oklahoma Press, Norman, 1952.

b. Nutrition

1. Agricultural Research Council, The Nutrient Requirements of Farm Livestock. No. 3 Pigs, ARC, London, 1966.
2. **Braude, R.,** Problems in pig feeding, *Agriculture* (London), 73, 377–381, 1966.
3. **Cunha, T. J.,** *Swine Feeding and Nutrition,* Interscience, New York, 1957, 296 pp.
4. **Lucas, I. A. M.,** *Feeding Pigs,* Pig Industrial Development Authority, London, 1961.
5. **Lucas, I. A. M. and Lodge, G. A.,** *The Nutrition of the Young Pig, A Review,* Technical Communication No. 22, Commonwealth Agric. Bureaux, Farnham Royal, Slough, Bucks, 1961, 119 pp.
6. **Morgan, J. T. and Lewis, D. Eds.,** *Nutrition of Pigs and Poultry,* Butterworths, London, 1962.
7. National Research Council, Nutrient Requirements of Domestic Animals, No. 2. Nutrient Requirements of Swine, 6th ed., Publ. No. 1599, National Academy of Sciences, Washington, D.C., 1968, 69 pp.

c. Diseases

1. **Anthony, D. J. and Lewis, E. F.,** *A Handbook of the Diseases of the Pig with Introduction to its Husbandry,* 5th ed., Williams & Wilkins, Baltimore, 1961.
2. British Veterinary Association, *The Husbandry and Diseases of Pigs,* 4th ed., British Veterinary Association, London, 1967.
3. **Dunne, H. W., Ed.,** *Diseases of Swine,* 2nd ed., Iowa State University Press, Ames, 1964, 897 pp.
4. **Heard, T. W.,** Preventive medicine in pig practice, in *Veterinary Annual,* 10th ed., Wright, Bristol, 1969, 72–82.

3. Poultry

a. General

1. **Benjamin, C. W., Gwin, J. M., Faber, F. L., and Termohlen, W. D.,** *Marketing Poultry Products,* John Wiley & Sons, New York, 1960.
2. **Card, L. E. and Nesheim, M. C.,** *Poultry Production,* 10th ed., Lea & Febiger, Philadelphia, 1966, 400 pp.
3. **Ensminger, M. E.,** *Poultry Science,* Interstate, Danville, Ill., 1971, 276 pp.
4. Food and Agriculture Organization, Report of the Technical Meeting on Poultry Production in Asia and the Far East, FAO, Rome, 1956.
5. **Freeman, B. M. and Lake, P. E., Eds.,** *Egg Formation and Production,* Constable, Edinburgh, 1971.
6. **Jull, M. A.,** *Poultry Husbandry,* 3rd ed., McGraw-Hill, New York, 1951, 526 pp.
7. **Marsden, S. J. and Martin, J. H.,** *Turkey Management,* Interstate, Danville, Ill., 1955.
8. **North, M. O.,** *Commercial Chicken Production Manual,* Avi Publishing, Westport, Conn., 1972, 645 pp.
9. Report of Technical Meeting on Poultry Production in Asia and the Far East, Poona, India, Oct. 17–27, 1955, FAO, Rome, 1955, 123 pp.
10. Report of the Technical Meeting on Poultry Production in the Near East and North Africa, Cairo, Dec. 8–12, 1957, FAO, Rome, 1957, 92 pp.
11. **Rice, J. E. and Botsford, H. E.,** *Practical Poultry Management,* John Wiley & Sons, New York, 1956.

12. **Romanoff, A. L. and Romanoff, A.,** *The Avian Egg,* John Wiley & Sons, New York, 1963.
13. **Winter, A. R. and Funk, E. M.,** *Poultry Science and Practice,* Lippincott, Chicago, 1960.

b. Nutrition

1. **Almquist, H. H.,** Amino acid requirements of chickens and turkeys – a review, *Poult. Sci.,* 31, 966–981, 1952.
2. **Bell, D. J. and Freeman, B. M.,** *Physiology and Biochemistry of the Domestic Fowl,* Academic Press, New York 1971. (3 vols.)
3. **Bolton, W.,** Digestion, absorption and metabolism in poultry, in The Science of Nutrition of Farm Livestock, Cuthbertson, D., Ed., in *International Encyclopaedia of Food and Nutrition,* Vol. 17, part 1, Pergamon Press, Oxford, 1969, 183–240.
4. **Bolton, W.,** *Poultry Nutrition,* 3rd ed., Gr. Brit. Min. Agric., Fish. Food, Bull. No. 174, HMSO, London, 1967, 102 pp.
5. **Engler, H.,** Chicken Feeding in Europe, FAO, Rome, 1955, 56 pp.
6. **Ewing, W. R.,** *Poultry Nutrition,* 5th ed., W. R. Ewing, Pasadena, Calif., 1963.
7. **Farner, D. S.,** Digestion and the digestive system, in *Biology and Comparative Physiology of Birds,* Vol. 1, Marshall, A. J., Ed., Academic Press, New York, 1960, 411–467.
8. **Ferrando, R.,** *Alimentation du poulet et de la poule pondeuse. Bases et Applications,* Vigot, Paris, 1969, 197 pp.
9. **Heuser, G. F.,** *Feeding Poultry,* 2nd ed., John Wiley & Sons, New York, 1955.
10. **Hill, F. W.,** Poultry nutrition and nutrient requirements, in *Nutrition of Animals of Agricultural Importance,* Cuthbertson, D., Ed., *International Encyclopaedia of Food and Nutrition,* Vol. 17, Pergamon Press, Oxford, 1969, 1137–1179.
11. **Horton-Smith, C. and Amorose, E. C., Eds.,** *Physiology of the Domestic Fowl,* Oliver & Boyd, Edinburgh, 1966, 328 pp.
12. **Morton, R. A. and Amorose, E. C., Eds.,** *Protein Utilization by Poultry,* Oliver & Boyd, Edinburgh, 1967, 215 pp. (12 papers)
13. National Research council, *Nutrient Requirements of Domestic Animals. No. 1. Nutrient Requirements of Poultry,* 6th ed., National Academy of Sciences, Washington, D.C., 1971, 54 pp.
14. *Poultry Feeding in Tropical and Subtropical Countries,* FAO Agric. Deve. Paper 82, FAO, Rome, 1965.
15. **Schaible, P. J.,** *Poultry: Feeds and Nutrition,* Avi Publishing, Westport, Conn., 1970.
16. **Scott, M. L.,** Broiler nutrition, *Feedstuffs,* 35(47), 50–54, 1963.
17. **Scott, M. L., Nesheim, M. C., and Young, R. J.,** *Nutrition of the Chicken,* M. L. Scott, Ithaca, N.Y., 1969. 511 pp.
18. **Sturkie, P. D.,** *Avian Physiology,* 2nd ed., Comstock Publishing, Ithaca, 1965, 766 pp.
19. Symposium on Pacific area. Feedstuffs for poultry nutrition, *World's Poult. Sci. J.,* 18(4), 386–431, 1962.
20. **Titus, H. W.,** *The Scientific Feeding of Chickens,* 5th ed., Interstate, Danville, Ill., 1971.
21. **Vohra, P.,** *Poultry Feed Mixing Manual,* U.S. AID Mission to India, New Delhi, 1962, 90 pp.

c. Diseases

1. **Biester, H. E. and Schwarte, L. H., Eds.,** *Diseases of Poultry,* 5th ed., Iowa State University Press, Ames, 1965, 1382 pp.
2. **Holfstad, M. S., Ed.,** *Diseases of Poultry,* Iowa State University Press, Ames, 1972, 1176 pp.
3. **Pomeroy, B. S.,** Protection of poultry product quality through preventive health measures, in *Food Quality,* Irving, G. W., Jr. and Hoover, S. R., Eds., Publ. No. 77, American Association for the Advancement of Science, Washington, D.C., 1965, 213–221.

d. Reproduction

1. **Taylor, L. W., Ed.,** *Fertility and Hatchability of Chicken and Turkey,* John Wiley & Sons, New York, 1949.

4. Fish

a. General

1. **Bardach, J. E.,** *Harvest of the Sea,* Harper & Row, New York, 1968, 301 pp.
2. **Bardach, J. E. and Ryther, J. H.,** *The Status and Potential of Aquaculture,* American Institute of Biological Sciences, Vol. 1, part 1, Washington, D.C., 1968.
3. **Bardach, J. E., Ryther, J. H., and McLarney, W. O.,** *Aquaculture. The Farming and Husbandry of Freshwater and Marine Organisms,* Interscience, New York, 1972, 861 pp.
4. **Bogorov, V. G.,** Problems concerning the productivity of the ocean (in Russian), *Gidrobiol. Zh.,* 3(5), 12–21, 1967.
5. **Chapman, W. M.,** Potential Resources of the Ocean, Van Camp Sea Food Co., San Diego, 1965.
6. **Christy, F. T. and Scott, A.,** *The Common Wealth in Ocean Fisheries,* Johns Hopkins Press, Baltimore, 1965.
7. **Costlow, J. D., Jr., Ed.,** *Fertility of the Sea,* Gordon & Breach, New York, 1971. (2 vols.)
8. *CRC Handbook of Marine Science,* Vol. 1, Walton Smith, F. G., Ed., CRC Press, Cleveland, 1973, 600 pp.; Vol. 2., Kalber, F. A. and Walton Smith, F. G., Eds. 1973; Vol. 3, Kalber, F. A., Ed., 1974.

9. **Crutchfield, J. A.**, Valuation of fishery resources, *Land Econ.*, 38, 145–154, 1962.
10. **Davis, H. S.**, *Culture and Diseases of Gamefish*, University of California Press, Berkeley, 1956.
11. **Dean, B.**, *A Bibliography of Fishes*, Russell & Russell, New York, 1962. (3 vols.,)
12. **Dietrich, G. and Kalle, K.**, *Allgemeine Meereskunde*, Borntraeger, Berlin, 1957.
13. **Gulland, J. A. and Carroz, J. E.**, The management of fisheries resources, *Adv. Mar. Biol.*, 6, 1–71, 1968.
14. **Hickling, C. F.**, *Tropical Inland Fisheries*, Longmans Green, London, 1960.
15. **Hickling, C. F.**, *The Farming of Fish*, Pergamon Press, Oxford, 1963, 88 pp.
16. **Hickling, C. F.**, *Fish Culture*, Faber & Faber, London, 1971, 317 pp.
17. **Huet, M.**, *Traité de pisciculture*, 3rd ed., de Wyngaert, Bruxelles, 1960, 369 pp.
18. **Idyll, C. P.**, Marine sciences, in Contributions of the biological sciences to human welfare, *Fed. Proc.*, 31, TF121–TF130, 1972.
19. **Iversen, E. S.**, *Farming the Edge of the Sea*, Fishing News Books, London, 1968.
20. **Ivlev, V. S.**, The biological productivity of waters, *J. Fish. Res. Bd. Can.*, 23(11), 1727–1759, 1966.
21. **Keeler, C. R.**, *Books on Aquatic Biology: Freshwater and Marine*, Facile Press, Tallahassee, 1965, 223 pp.
22. **McKee, A.**, *Farming the Sea*, Crowell, New York, 1969, 198 pp.
23. **McNeil, W. J., Ed.**, *Marine Aquaculture*, Oregon State University Press, Corvallis, 1968, 172 pp. (10 papers)
24. **Olson, T. A. and Burgess, F. J., Eds.**, *Pollution and Marine Ecology*, Interscience, New York, 1967, 364 pp.
25. **Ricker, W. E., Ed.**, *Methods for Assessment of Fish Production in Fresh Waters*, Blackwell, Oxford, 1968, 313 pp.
26. **Rounsefell, G. A. and Everhart, W. H.**, *Fishery Science, Its Methods and Applications*, John Wiley & Sons, New York, 1953, 444 pp.
27. **Royce, W. F.**, *Introduction to the Fishery Sciences*, Academic Press, New York, 1972, 351 pp.
28. **Russell-Hunter, W. D.**, *Aquatic Productivity. An Introduction to Some Basic Aspects of Biological Oceanography and Limnology*, Macmillan, New York, 1970, 306 pp.
29. **Ryther, J. H.**, Photosynthesis and fish production in the sea, *Science*, 166, 72–76, 1969.
30. **Ryther, J. H. and Bardach, J. E.**, *The Status and Potential of Aquaculture, American Institute of Biological Sciences*, Washington, D.C., 1968. (2 vols.)
31. **Shelbourne, J. E.**, The artificial propagation of marine fish, *Adv. Mar. Biol.*, 2, 1–84, 1967.
32. **Sterba, G.**, *Freshwater Fishes of the World*, Viking Press, New York, 1963, 878 pp.
33. **Troebst, C. C.**, *Conquest of the Sea*, Harper & Row, New York, 1962, 270 pp.
34. **Waters, J. F.**, *The Sea Farmers*, Hastings House, New York, 1970, 120 pp.

b. Nutrition

1. **Barrington, E. J. W.**, The alimentary canal and digestion, in *The Physiology of Fishes*, Vol. 1, Brown, M. E., Ed., Academic Press, New York, 1957, 109–161.
2. **Cowey, C. B. and Sargent, J. R.**, Fish nutrition, *Adv. Mar. Biol.*, 10, 383–492, 1972.
3. **Halver, J. E., Ed.**, *Fish Nutrition*, Academic Press, New York, 1972.
4. **Jφrgensen, C. B.**, *Biology of Suspension Feeding*, Pergamon Press, Oxford, 1966, 357 pp.
5. **Love, R. M.**, *The Chemical Biology of Fishes. With a Key to the Chemical Literature*, Academic Press, New York, 1970, 547 pp.
6. **Neuhaus, O. W. and Halver, J. E.**, *Fish in Research*, Academic Press, New York, 1969, 314 pp.
7. **Pearson, W. E.**, *La nutrition des poissons*, Hoffmann-La Roche, Paris, 1969, 52 pp.
8. **Phillips, A. M., Jr.**, Nutrition, digestion, and energy metabolism, in *Fish Physiology*, Vol. 1, Hoar, W. S. and Randall, D. J., Eds., Academic Press, New York, 1969, 391–432.
9. **Raymont, J. E. G.**, *Plankton and Productivity in the Oceans*, Pergamon Press, New York, 1963, 660 pp.
10. **Steele, J. H., Ed.**, *Marine Food Chains*, University of California Press, Berkeley, 1970, 552 pp.
11. **Wimpenny, R. S.**, *The Plankton of the Sea*, Faber & Faber, London, 1966, 426 pp.

c. Reproduction

1. **Breder, C. M., Jr. and Rosen, D. E.**, *Modes of Reproduction in Fishes*, Natural History Press, Garden City, N.Y., 1966, 941 pp.

d. Diseases

1. **Amlacher, E.**, *Taschenbuch der Fischkrankheiten*, Fischer, Jena, 1961, 286 pp.
2. **Amlacher, E.**, *Textbook of Fish Diseases*, T.F.H. Publications, Jersey City, N.J., 1970, 302 pp.
3. **Dogiel, V. A., Petrushevski, G. K., and Polyanski, Yu.I., Eds.**, *Parasitology of Fishes*, Oliver & Boyd, Edinburgh, 1961, 384 pp.
4. **Duijn, C. van, Jr.**, *Diseases of Fishes*, 2nd ed., Charles C Thomas, Springfield, Ill., 1967, 309 pp.
5. **Hoffman, G. L.**, *Parasites of North American Freshwater Fishes*, University of California Press, Berkeley, 1967, 486 pp.
6. **Reichenbach-Klinke, H. H. and Elkan, E.**, *The Principal Diseases of Lower Vertebrates*, Academic Press, New York, 1965, 600 pp.
7. **Reichenbach-Klinke, H. H.**, *Krankheiten und Schädigungen*, Fischer, Jena, 1966, 389 pp.

8. **Schäperlaus, W.,** *Fishkrankeiten,* Akademie Verlag, Berlin, 1954, 708 pp.
9. **Schubert, G.,** *Krankheiten der Fische,* 3rd ed., Kosmos, Stuttgart, 1968, 63 pp.
10. **Sindermann, C. J.,** *The Principal Diseases of Marine Fish and Shellfish,* Academic Press, New York, 1970, 369 pp.
11. **Snieszko, S. F.,** *A Symposium on Diseases of Fishes and Shellfishes,* American Fisheries Society, Washington, D.C., 1970, 526 pp.

5. Wildlife

a. General

1. **Allen, D. L.,** The Farmer and Wildlife, Wildl. Manage. Inst. Bull., 1949, 84 pp.
2. **Allen, D. L.,** Wildlife and the business of farming, *J. Soil Water Conserv.,* 7(5), 223–226, 245, 1952.
3. **Allen, D.,** *Our Wildlife Legacy,* Funk & Wagnalls, New York, 1962, 402 pp.
4. **Anderson, W. L.,** Making Land Produce Useful Wildlife, U.S. Dep. Agric. Farmer's Bull. 2035, 30 pp.
5. **Black, J. D.,** *The Management and Conservation of Biological Resources,* Davis, Philadelphia, 1968, 339 pp.
6. **Darling, F. F.,** Wildlife husbandry in Africa, *Sci. Am.,* 203(5), 123–134, 1960.
7. **Darling, F. F.,** *Wildlife in African Territory,* Oxford University Press, London, 1960, 160 pp.
8. **Dasmann, R. F.,** *Wildlife Biology,* John Wiley & Sons, New York, 1964, 231 pp.
9. **Dasmann, R. F.,** *African Game Ranching,* Pergamon Press, Oxford, 1964.
10. **Graham, E. H.,** *The Land and Wildlife,* Oxford University Press, New York, 1947, 232 pp.
11. **Harnaday, W. T.,** *Our Vanishing Wildlife,* Scribners, New York, 1913.
12. **Hey, D.,** *Wildlife Heritage of South Africa,* Oxford University Press, Cape Town, 1966, 246 pp.
13. **Ledger, H. P.,** Animal husbandry research and wildlife in East Africa, *E. Afr. Wildl. J.,* 1, 18–29, 1963.
14. **Leopold, A. S.,** *Game Management,* Scribners, New York, 1933, 481 pp.
15. **Leopold, A. S. and Darling, F. F.,** *Wildlife in Alaska,* Ronald Press, New York, 1953.
16. National Research Council, *Land Use and Wildlife,* National Academy of Sciences, Washington, D.C., 1970, 262 pp.
17. **Riney, T.,** Development of the wildlife resource in Africa, *Unasylva,* 18(4), 30–38, 1964.
18. **Schaller, G. B.,** *The Deer and The Tiger. The Study of Wildlife in India,* University of Chicago Press, Chicago, 1967, 370 pp.
19. **Worthington, E. B.,** The Wild Resorces of East and Central Africa, Colonial No. 352, HMSO, London, 1961, 26 pp.

b. Nutrition

1. **Crawford, M. A., Ed.,** *Comparative Nutrition of Wild Animals, Symposia of the Zoological Society of London No. 21,* Academic Press, New York, 1968.

c. Diseases

1. **Davis, J. W. and Anderson, R. C., Eds.,** *Parasitic Diseases of Wild Animals,* Iowa State University Press, Ames, 1971, 364 pp. (17 papers)
2. **Davis, J. W., Anderson, R. C., Karstad, L., and Trainer, D. O., Eds.,** *Infection and Parasitic Diseases of Wild Birds,* Iowa State University Press, Ames, 171, 344 pp.
3. **Davis, J. W., Karstad, L. H., and Trainer, D. O., Eds.,** *Infectious Diseases of Wild Animals,* Iowa State University Press, Ames, 1970, 421 pp. (43 papers)
4. **Kerschagl, W.,** *Wildkrankheiten* Osterreichischer Jagd- und Fischerei-Verlag des N.-O-Landes jagdverbandes, Wien, 1965, 328 pp.
5. **McDiarmid, A.,** Diseases of Free-Living Wild Animals, FAO Agric. Stud. No. 57, FAO, Rome, 1962, 119 pp.
6. **McDiarmid, A., Ed.,** *Diseases in Free-Living Wild Animals, Symposia of the Zoological Society of London No. 24,* Academic Press, New York, 1969, 332 pp.
7. **Schulze, H.,** *Die Krankheiten des Wildes,* Mayer Verlag, Munchen, 1965, 223 pp.
8. **Wetzel, R. and Rieck, W.,** *Krankheiten des Wildes,* Verlag Paul Parey, Hamburg, 1962, 223 pp.

VII. REDUCING FOOD WASTE

A. Food and Nutrient Losses

1. General

1. **Bunting, H. H.**, Pests, population and poverty, *Trop. Sci.*, 14, 35–46, 1972.
2. Food and Agriculture Organization, War on Waste, FAO, Rome, 1969, 15 pp.
3. Food and Agriculture Organization, Food Losses, The Tragedy and Some Solutions, FAO, Rome, 1970.
4. **Hulse, J. H.**, Increasing food by reducing waste, in *Proc. SOS/70. 3rd International Congress Food Science and Technology*, Institute of Food Technologists, Chicago, 1971, 532–537. (9 ref.)
5. **Majumder, S. K.**, Recent progress in the protection of foods, in *Proc. 1st Asian Congress of Nutrition*, Hyderabad 1971, Tulpule, P. G. and Jaya Rao, K. S., Eds., Nutrition Society of India, Hyderabad, 1972, 725–741. (24 ref.)
6. **Parpia, H. A. B.**, Waste and the protein gap – they can both be reduced, *Ceres*, 2(5), 20–24, 1969.
7. Symposium, Food losses in field and store, *Proc. Nutr. Soc.* 20, 1-24, 1961. (5 papers)
8. U.S. Department of Agriculture, Protecting Our Food. The Yearbook of Agriculture 1966, U.S. Gov. Print. Off., Washington, D.C., 1966, 386 pp.
9. U.S. Department of Agriculture, Losses in Agriculture, Agric. Handb. No. 291, USDA, ARS, Washington, D.C., 1965, 120 pp.

2. Crop Losses

1. **Carlson, G. A.**, The Microeconomics of Crop Losses. Symposium on economic research on pesticides for policy decision making, USDA, ERS, Washington, D.C., 1970.
2. **Chester, K. S.**, Plant disease losses: Their appraisal and interpretation, *Plant Dis. Rep. Suppl.*, 193, 190–362, 1950.
3. **Chiarappa, L.**, International assistance in plant pathology in developing countries with particular reference to FAO programmes, *FAO Plant Prot. Bull.*, 17(1), 1–8, 1969.
4. **Chiarappa, L.**, FAO international collaborative programme for the development of reproducible methods for the assessment of crop losses, *FAO Plant Prot. Bull.*, 18(2), 29–30, 1970.
5. **Chiarappa, L., Ed.**, Crop Loss Assessment Methods, FAO, Rome, 1970, 221 pp.
6. **Chiarappa, L.**, Reducing crop losses, in *Man, Food, and Nutrition. Strategies and Technological Measures for Alleviating the World Food Problem*, Rechcigl, M., Jr., Ed., CRC Press, Cleveland, Ohio 1973, 163–176. (33 ref.)
7. **Chiarappa, L., Chiang, H. C., and Smith, R. F.**, Plant pests and diseases: Assessment of crop losses, *Science*, 176, 769–773, 1972. (14 ref.)
8. **Crowdy, S. H. and Manners, J. G.**, Microbial disease and plant productivity, in *Microbes and Biological Productivity, 21st Symp. Soc. Gen. Microbiol.*, Cambridge University Press, New York, 1971, 103–123.
9. Estimated Crop Losses Due to Plant-Parasitic Nematodes in the United States, Soc. of Nematol., Spec. Publ. 1, 1971.
10. Estimates of Crop Losses and Diseases – Control Costs in California, 1963, Univ. Calif. Agric. Exp. Sta. Agric. Ext. Serv., 1965.
11. Food and Agriculture Organization, Report of the FAO Symposium on Crop Losses, FAO Rep. PL/M/6, Rome, 1967.
12. **Haeussler, G. J.**, Losses caused by insects, in Insects, The Yearbook of Agriculture 1952, U.S. Gov. Print. Off., Washington, D.C., 1952, 141–146.
13. **Haskell, P. T.**, The hungry locust, *Sci. J.*, 6(1), 61–67, 1970.
14. **Hawthorne, R.**, Estimated Damage and Crop Losses Caused by Insect/Mite Pests – 1969, Calif. Dep. Agric., Mimeo. E-82-12.
15. **Holmes, E.**, The contribution of crop protection to increased world food supplies, *Outlook Agric.*, 3, 23–38, 1960. (39 ref.)
16. **LeClerg, E. L.**, Impact of pre-harvest crop losses due to diseases and insects on production in the Near East, in Indicative World Plan, FAO, Plant Production and Protection Division, Rome, 1966.
17. **Padwick, G. W.**, Losses Caused by Plant Diseases in the Colonies, Phytopathol. Paper No. 1, Commonwealth Mycological Institute, Kew, England, 1956, 60 pp.
18. **Tainsh, A. R.**, The gross national waste of food grain, Proc. Symp. Agricultural College of Norway, 1967.
19. U.S. Department of Agriculture, A Survey of Extent and Cost of Weed Control and Specific Weed Problems, USDA, ARS 34-23-1, 1965.
20. **Vallega, J. and Chiarappa, L.**, Plant disease losses as they occur world-wide, *Phytopathology*, 54(11), 1305–1313, 1964.
21. **Way, M. J.**, Crop losses by insects and the problem of control, *Proc. Nutr. Soc.*, 20, 5–11, 1961.

3. Livestock Losses

1. **Aczel, J. A.**, Effect of animal disease on human food supplies. A review of the world's position, *Vet. Rec.*, 77, 596–598, 1965.
2. **Beveridge, W. I. B.**, Economics of animal health, *Vet. Rec.*, 72, 810–815, 1960.

3. **Cockrill, W. R.**, Dead loss, the impact of disease on world livestock, *World Crops*, Sept. 1968, 31–34.
4. **Cunningham, I. J.**, Veterinary science and man's food, *Aust. J. Sci.*, 28, 194–200, 1965.
5. **Derbyshire, J. B.**, Microbial diseases and animal productivity, *Symp. Soc. Gen. Microbiol.*, 21, 125–147, 1971.
6. **Faust, E. C.**, Diseases of domestic animals and human welfare, *Acta Trop.*, 7, 289–297, 1950.
7. Food and Agriculture Organization, The economic losses caused by animal diseases, in Animal Health Yearbook, FAO, Rome, 1962, 284–313.
8. **Grunsell, C. S., Penny, R. H. C., Wragg, S. R., and Allcock, J.**, The practicability and economics of veterinary preventive medicine, *Vet. Rec.*, 84, 26–41, 1969.
9. **Hardie, J. D.**, The economics of veterinary practice, Misc. Publ., School of Agriculture, Aberdeen, Scotland, 1969.
10. **Harnett, P.**, The significance of veterinary science in the national economy, *Ir. Vet. J.*, 10, 130–142, 164–172, 185–188, 1956.
11. **Kaplan, M. M.**, Economic and social aspects of animal diseases in developing countries, in *Biotechnology and Bioengineering Symposium No. 1*, Gaden, E. L., Jr., Ed., 2nd International Conference on Global Impacts of Applied Microbiology, Addis Ababa, Ethiopia, Nov. 6–11, 1967, Interscience, New York, 1969, 211–234.
12. **Klarman, H. E.**, *The Economics of Health*, Columbia University Press, New York, 1965.
13. **Leech, F. B.**, Food losses through animal disease, *Proc. Nutr. Soc.*, 20, 20–24, 1961.
14. Livestock, and poultry losses, in Losses in Agriculture, Agric. Handb. No. 291, USDA, ARS, Washington, D.C., 1965, 72–84.
15. **Mann, I.**, Freedom from protein hunger. A challenge to the veterinary profession in underdeveloped countries, Proc. 17th World Veterinary Congress, Hannover, Germany, 1963.
16. **Meyer, K. F.**, Animal diseases and human welfare, *Adv. Vet. Sci.*, 1, 1–48, 1953.
17. **Morris, R. S. and Blood, D. C.**, The economic basis of planned veterinary services to individual farms, *Aust. Vet. J.*, 45, 337–341, 1969.
18. **Morris, R. S.**, Assessing the economic value of veterinary services to primary industries, *Aust. Vet. J.*, 45, 295–300, 1969.
19. National Research Council, A Historical Survey of Animal – Disease Morbidity and Mortality Reporting, Publ. No. 1346, National Academy of Sciences, Washington, D.C., 1966.
20. National Research Council, *Prenatal and Postnatal Mortality in Cattle*, Publ. No. 1685, National Academy of Sciences, Washington, D.C., 1968.
21. **Nielsen, S. W.**, Animal disease surveys reveal the effects of environmental pollution, *Milestones. Conn. Agric. Home Econ.*, 13, 14–16, 1970.
22. **Nordquist, A. V.**, Estimating livestock losses, Proc. 50th Ann. Meet. U.S. Livestock Sanit. Assoc., 1946, 1947, 199–208.
23. **Pritchard, W. R.**, Increasing protein foods through improving animal health, *Proc. Natl. Acad. Sci. U.S.A.*, 56, 360–369, 1966.
24. **Pritchard, W. R.**, Importance of animal disease in the economic development of countries of the hemisphere, PAHO Inter-American Meeting on Foot-and-Mouth Disease and Zoonoses Control, 1968.
25. **Schwabe, C. W.**, Animal diseases and world health, *J. Am. Vet. Med. Assoc.*, 153, 1859–1863, 1968.
26. **Schwabe, C. W.**, *Veterinary Medicine and Human Health*, 2nd ed., Williams & Wilkins, Baltimore, 1969.
27. **Schwabe, C. W. and Ruppanner, R.**, Animal diseases as contributors to human hunger problems of control, *World Rev. Nutr. Diet.*, 15, 185–224, 1972.
28. **Short, G. V.**, Veterinary science and the world's food, *Outlook Agric.*, 3, 39–50, 1960.
29. **Weisbrod, B. A.**, *Economics of Public Health. Measuring the Economic Impact of Diseases*, University of Pennsylvania Press, Philadelphia, 1962.

4. Food Losses During Handling and Storage

1. **Christensen, C. M.**, Deterioration of stored grain by fungi, *Bot. Rev.*, 23, 108–134, 1957.
2. **Christensen, C. M.**, Effect of fungi on the quality of cereal grains for food, in *Food Quality*, Irving, G. W., Jr. and Hoover, S. R., Eds., American Association for the Advancement of Science, Washington, D.C., 1965, 115–120.
3. **Christensen, C. M. and Kaufmann, H. H.**, *Grain Storage: The Role of Fungi in the Quality Loss*, University of Minnesota Press, Minneapolis, 1969, 153 pp.
4. Food and Agriculture Organization, Destruction of Food in Storage by Insects, Mites, Rodents and Mold Fungi, NU/WaWPs, FAO, Washington, D.C., 1946.
5. **Geddes, W. F.**, The chemistry, microbiology and physics of grain storage, *Food Technol.*, 12(11), 7–14, 1958.
6. Harvesting and storage losses, in Losses in Agriculture, Agric. Handb. No. 291, USDA, ARS, Washington, D.C., 1965, 67–71.
7. **Herford, G. V. B.**, Food lost in store by insect attack, *Span*, 4, 40–42, 1961.
8. **Hill, D. W.**, Losses of stored food, in Handling and Storage of Food Grains in Tropical and Subtropical Areas, FAO Agric. Dev. Paper No. 90, FAO, Rome, 1970, 13–37.
9. **Howe, R. W.**, Losses caused by insects and mites in stored foods and feedingstuffs, *Nutr. Abstr. Rev.*, 35, 285–303, 1965.

10. **Livingston, G. E., Aug, C. Y. W., and Chang, C. M.,** Effects of food service handling, *Food Technol.*, 27(1), 28–34, 1973.
11. **Lyons, J. M.,** Physical and biological considerations relating to storage and handling of harvested crops, in *Engineering of Unconventional Protein Production* Vol. 65, Bieber, H., Ed., American Institute of Chemical Engineers, New York, Chemical Engineering Progress Symposium Ser. No. 93, 1969, 106–111.
12. Marketing and processing losses, in Losses in Agriculture, Agric. Handb. 291, USDA, ARS, Washington, D.C., 1965, 85–94.
13. **Parkin, E. A.,** Insect and stored food: World losses and control measures surveyed, *Food Manuf.*, 34, 164–168, 1959.
14. **Pomeranz, Y.,** Biochemical and functional changes in stored cereal grains, *CRC Crit. Rev. Food Technol.*, 2, 45–80, 1971. (63 ref.)
15. **Seemink, G., Anderson, J. S., and Alcock, A. W.,** Microflora in storage of cereal grains and their products, Am. Assoc. Cereal Chem. Monogr. Ser. No. 2, 1945, 77–151.
16. **Wolpert, V.,** Rats in store, India, *Far Eastern Econ. Rev.*, 53(10), 439, 1966.
17. **Wolpert, V.,** Needless losses, *Far Eastern Econ. Rev.*, 55(9), 411–412, 1967.

5. Nutrient Losses During Handling and Processing

a. General

1. **Allouf, R. and Thomas, G.,** Nutritive value of preserved food, *Bull. Soc. Sci. Hyg. Aliment. Ration. Homme*, 49, 179–204, 1961. (76 ref.)
2. **Barnell, H. R. and Hollingsworth, D. F.,** Food processing and nutritive value for man, *Proc. Nutr. Soc.*, 15, 142–147, 1956.
3. **Bender, A. E.,** Nutritional effects of food processing, *J. Food Tech.*, 1, 261–289, 1966.
4. **Bender, A. E.,** Nutritional effects of food processing. 1. Vitamin losses. 2. Mineral salts, protein and specific commodities. 3. Equipment and methods, *Rev. Nutr. Food Sci.*, 11, 2–6, April 1968; 12, 10–12, July 1968; 13, 6–8, October 1968.
5. **Berk, Z.,** Processing and storage damage to nutritional value of foods, in *Proc. SOS/70. 3rd International Congress Food Science and Technology*, Institute of Food Technologists, Chicago, 1971, 189–191. (13 ref.)
6. **Bramsnaes, F.,** Influence of refrigeration and canning on the nutritive value of fish, *Proc. Conf. Fish Nutr.*, Washington, 1961, 1:1.13.1–1.13.15.
7. **Campbell, J. A. and Morrison, A. B.,** Nutritional impact of modern food processing, *Fed. Proc.*, 25, 130–136, 1966. (78 ref.)
8. **Clifcorn, L. E. and Peterson, G. T.,** Effect of canning and dehydration on the nutritive values of foods, in *Handbook of Food and Agriculture*, Blanck, F. C., Ed., Reinhold, New York, 1955, 365–388.
9. **Cutting, C. L.,** Influence of drying, salting, and smoking on the nutritive value of fish, *Proc. Conf. Fish Nutr.*, Washington, 1961, 1:1.14.1–1.14.19.
10. **Davidson, J. and Woodham, A. A.,** The effects of processing on the nutritive values of foodstuffs, in *Recent Advances in Animal Nutrition*, Abrams, J. T., Ed., Churchill, London, 1966, 1–50.
11. Effects of processing and additives on foods, *Proc. 5th Int. Congr. Nutr.*, Washington, D.C., 1960, *Fed. Proc.*, 20(1), III, Suppl. 7, 209–257, 1960.
12. **Esselen, W. B.,** Food preservation and its contribution to nutrition, *Borden's Rev. Nutr. Res.*, 18, 29–44, 1957.
13. **Harris, R. S. and Von Loesecke, H.,** *Nutritional Evaluation of Food Processing*, John Wiley & Sons, New York, 1960.
14. **Hein, R. E. and Hutchings, I. J.,** Influence of processing on vitamin-mineral content and biological availability in processed foods, in Symposium on Vitamins and Minerals in Processed Foods, AMA Council on Foods and Nutrition and Food Industry Liaison Committee, New Orleans, La., 1971.
15. **Higashi, H.,** Relation between processing techniques and amount of vitamins and minerals in processed fish, *Proc. Conf. Fish Nutr.*, Washington, 1961, 1:1.11.1–1.11.14. (21 ref.)
16. **Hollingsworth, D. F.,** Effects of some new production and processing methods on nutritive values, *J. Am. Diet. Assoc.*, 57, 246–249, 1970.
17. **Hollingsworth, D. F. and Martin, P. E.,** Some aspects of the effects of different methods of production and processing on the nutritive value of food, *World Rev. Nutr. Diet.*, 15, 1–34, 1972.
18. **Jelliffe, D. B.,** Commerciogenic malnutrition? Time of a dialogue, *Food Technol.*, 25(2), 55–56, 1971.
19. **Lepkovsky, S.,** Nutritional stress factors in food processing, *Adv. Food Res.*, 4, 105–132, 1953.
20. **Moore, J. A.,** Nutrient losses involved in cooking, processing, and storage of food. A review, *Food Nutr. Notes Rev.*, 16, 82–87, 1959. (19 ref.)
21. **Muelenaare, H. J. H.,** Malnutrition as a result of food processing, *Food Ind. S. Afr.*, 41–46, 1969.
22. National Research Council, *Effect of Processing on the Nutritional Value of Feeds*, National Academy of Sciences, Washington, D.C., 1973, 494 pp.
23. Proc. Symposium on Nutritive Aspects of Preserved Food, Swedish Institute for Preservation Research, Göteborg, 1956, 173 pp.

24. **Sabry, Z. I.**, The nutritional consequences of developments in food processing, *Can. J. Public Health*, 59, 471–474, 1969.
25. **Schroeder, H. A.**, Losses of vitamins and trace minerals resulting from processing and preparation of foods, *Am. J. Clin. Nutr.*, 24, 562–573, 1971. (26 ref.)
26. **Shewan, J. M.**, Influences of irradiation preservation on the nutritive value of fish and fishery products, *Proc. Conf. Fish Nutr.*, Washington, 1961,1:1.17.1–1.17.22. (39 ref.)
27. **Somogyi, J. C., Ed.**, *The Influence of Industrial and Household Handling on the Composition of Food, Bibl. Nutr. Dieta*, No. 7, S. Karger, Basel, 1965, 242 pp.
28. Symposium: Effects of processing, storage, and handling on nutrient retention in foods, *Food Technol.*, 27(1), 16–38, 1973. (4 papers)
29. **Tape, N. W.**, Food processing. A nutritional problem? *Can. Food Ind.*, 40, 36–38, 1969.
30. **Vail, G. E.**, The effects of processing upon the nutritive value of food, *J. Am. Diet. Assoc.*, 18, 569–574, 1942. (70 ref.)
31. **White, P. L., Ed.**, *Nutrients in Processed Foods*, Publishing Sciences Group, Inc., Acton, Mass., 1974. (3 vols.)

b. Specific Processes

(1) Blanching

1. **Lee F. A.**, The blanching process, *Adv. Food Res.*, 8, 63–109, 1958.

(2) Canning

1. **Cameron, E. J. and Esty, J. R.**, *Canned Foods in Human Nutrition*, National Canners Association Res. Lab., Washington, D.C., 1950, 264 pp.
2. **Cameron, E. J., Clifcorn, L. E., Esty, J. R., Feaster, J. F., Lamp, F. C., Monroe, K. H., and Royce, R.**, *Retention of Nutrients During Canning*, National Canners Association, Washington, D.C., 1955, 93 pp.
3. **Cameron, E. J., Pilcher, R. W., and Clifcorn, L. E.**, Nutrient retention during canned food production, *Am. J. Public Health*, 39, 756–763, 1949.
4. **Ives, M.**, Nutrient, retention during canning, *Food in Canada*, 10(10), 30–32, 34, 36, 1950; 10(11), 18–21, 22–24, 1950.
5. **Kohman, E. F.**, Vitamins in Canned Food, National Canners Assoc. Bull. 19-L, National Canners Association, Washington, D.C., 1922.

(3) Cooking

1. **Callow, A. B.**, *Cooking and Nutritive Value*, Clarendon Press, Oxford, 1945, 156 pp.
2. **Hewston, L. M., Dawson, E. H., Alexander, L. M., and Orent-Keiles, E.**, Vitamin and Mineral Content of Certain Foods as Affected by Home Preparation, Misc. Publ. 628, U.S. Dep. Agric., Washington, D.C., 1948.
3. **Lang, K.**, Influence of cooking on foodstuffs, *World Rev. Nutr. Diet.*, 12, 267–317, 1970.
4. Losses in nutrients during large-scale preparation for direct feeding, in *Nutritional Evaluation of Food Processing*, Harris, R. S. and von Loesecke, H., Eds., Avi Publishing, Westport, Conn., 1960, 391–441. (2 papers)
5. Losses in nutrients during home preparation of foods, in *Nutritional Evaluation of Food Processing*, Harris, R. S. and von Loesecke, H., Eds., Avi Publishing, Westport, Conn., 1960, 442–461. (2 papers)

(4) Dehydration

1. **Labuza, T. P.**, Nutritional losses during drying and storage of dehydrated foods, *CRC Crit. Rev. Food Technol.*, 3(2), 217–240, 1972.
2. **Labuza**, T. P., Effect of dehydration and storage, *Food Technol.*, 27(1), 20–26, 51, 1973.
3. **Salunkhe, D. K., Do, J. Y., and Bolin, H. R.**, Developments in technology and nutritive value of dehydrated fruits, vegetables, and their products, *CRC Crit. Rev. Food Technol.*, 4(2), 153–192, 1973.
4. **Thomas, M. H. and Calloway, D. H.**, Nutritional value of dehydrated foods, *J. Am. Diet. Assoc.*, 39, 105–116, 1961.

(5) Freezing

1. **Burger, M., Hein, L. W., Teply, L. J., Derse, P. H., and Krieger, C. H.**, Nutritional composition of frozen foods, *Quick Frozen Foods*, 3, 223, 1959.
2. **Watt, B. K.**, The nutritive value of frozen foods, in *The Freezing Preservation of Foods*, Vol. 2, 4th ed., Tressler, D. K., Van Arsdell, W. B., and Copley, M. J., Eds., Avi Publishing, Westport, Conn., 1968, 327–362.

(6) Freeze-drying

1. **Goldblith, S. A. and Tannenbaum, S. R.**, The nutritional aspects of the freeze-drying of foods, in *Proc. 7th International Congress of Nutrition*, Pergamon Press, Oxford, 1966, 432–445.

(7) Harvesting of Unprocessed Food

1. Effect of harvesting and handling practices on composition of unprocessed foods, in *Nutritional Evaluation of Food Processing*, Harris, R. S. and von Loesecke, H., Eds., Avi Publishing, Westport, Conn. 1960, 58–108. (2 papers)
2. **Lee, T. C. and Chichester, C. O.,** The influence of harvest time on nutritional values, *Proc. Symposium on Nutritional Qualities of Fresh Fruits and Vegetables*, American Medical Association, Chicago, 1973.

(8) Heat Processing

1. **Goldblith, S. A.,** Thermal processing of foods. A review, *World Rev. Nutr. Diet.*, 13, 165–193, 1971.
2. **Lachance, P. A., Ranadive, A. S., and Matas, J.,** Effects of reheating convenience foods, *Food Technol.*, 27(1), 36–38, 1973.
3. **Lund, D. B.,** Effects of heat processing, *Food Technol.*, 27(1), 16–18, 1973.

(9) Irradiation

1. **Decareau, R. V.,** The effect of microwave energy on the nutrients in foods, *Microwave Energy Applic. Newsletter*, 2(2), 3, 1969.
2. Food and Agriculture Organization, Report of the Meeting on the Wholesomeness of Irradiated Foods with Exclusive Reference to the Evaluation of the Nutritional Adequacy and Safety for Consumption, Brussels, 1961, FAO, Rome, 1962, 205 pp.
3. **Kraybill, H. F.,** The effect of ionizing radiation on vitamins and other physiologically active compounds, Report of Meeting on Wholesomeness of Irradiated Foods, FAO, Rome, 1962, 44–60.
4. **Read, M. S.,** The effect of ionizing radiations on the nutritive value of foods, *Proc. International Conference on Preservation of Foods by Ionizing Radiations*, M.I.T. Press, Cambridge, Mass., July 27–30, 1959, 138–152.
5. Symposium on nutritional and toxicological studies on irradiated foods, *Fed. Proc.*, 15, 905–937, 1956. (9 papers)
6. Symposium on new aspects of nutrition uncovered with irradiated foods, *Fed. Proc.*, 19, 1023–1059, 1960. (9 papers)

(10) Packaging

1. **Karel, M.,** Effects of packaging on maintenance of nutrients in food products, in *Nutritional Evaluation of Food Processing*, Harris, R. S. and von Loesecke, H., Eds., Avi Publishing, Westport, Conn., 1960, 305–336.

(11) Storage

1. Effect of commercial storage on the nutrient content of processed foods, in *Nutritional Evaluation of Food Processing*, Harris, R. S. and von Loesecke, H., Eds., Avi Publishing, Westport, Conn., 1960, 337–390. (7 papers)

c. Specific Nutrients

(1) Lipids

1. **Dugan, L. R., Jr.,** Processing and other stress effects on the nutritive value of lipids, *World Rev. Nutr. Diet*, 9, 181–205, 1968.

(2) Protein

1. **Carpenter, K. J.,** Damage to lysine in food processing: Its measurement and its significance, *Nutr. Abstr. Rev.*, 43, 423–451, 1973.
2. **Dworschak, E.,** Change in the biological value of proteins due to thermal processing of foods. A review, *Elemez. Ipar.*, 22, 257–263, 1968; *Chem. Abstr.*, 70, 36449, 1969.
3. **Ford, J. E.,** Some effects of processing on nutritive value, in *Proteins in Human Nutrition*, Porter, J. W. G. and Rolls, B. A., Eds., Academic Press, London, 1973, 515–529.
4. **Osner, R. G. and Johnson, R. M.,** Nutritional changes in protein during heat processing, *J. Food Technol.*, 3, 81–86, 1968.
5. **Zimmermann, G., Ben-Gera, I., Weissmann, S., and Yannai, S.,** Storage under controlled conditions of dry staple foods and its influence on their protein nutritive value, *Dechema Monogr.*, 63, 347–380, 1969.

(3) Vitamins

1. **Bender, A. E.,** Loss of vitamins in industrially handled foods and its prevention, *Bibl. Nutr. Dieta*, 18, 92–113, 1973. (73 ref.)
2. **Cain, R. F.,** Water soluble vitamins: Changes during processing and storage of fruits and vegetables, *Food Technol.*, 21, 998–1007, 1967.
3. **Farrar, K. T. H.,** The thermal destruction of vitamin B_1 in foods, *Adv. Food Res.*, 6, 257–311, 1955.
4. **Feaster, J. F., Tompkins, M. D., and Ives, M.,** Retention of vitamins in low acid canned foods, *Food Ind.*, 20, 82–85, 218–222, 1948. (14 ref.)
5. **Fixsen, M. A. B.,** The vitamin content of human foods as affected by processes of cooking and canning, *Nutr. Abstr. Rev.*, 8, 281–307, 1938. (156 ref.)

6. **Greenwood, D. A., Kraybill, H. R., Feaster, J. F., and Jackson, J. M.,** Vitamin retention in processed meat: Effect of thermal processing, *Ind. Eng. Chem.*, 36, 922–927, 1944.
7. **Mapson, L. W.,** Effect of processing on the vitamin content of foods, *Br. Med. Bull.*, 12, 73–77, 1956.

d. Specific Foods

(1) Cereals

1. Effects of commercial processing of cereals on nutrient content, in *Nutritional Evaluation of Food Processing,* Harris, R. S. and von Loesecke, H., Eds., Avi Publishing, Westport, Conn., 1960, 197–237. (8 papers)

(2) Fruits and Vegetables

1. Effect of commercial processing on the composition of fruits and vegetables, in *Nutritional Evaluation of Food Processing,* Harris, R. S. and von Loesecke, H., Eds., Avi Publishing, Westport, Conn., 1960, 109–172. (11 papers)
2. **Salunkhe, D. K., Pao, S. K., and Dull, G. G.,** Assessment of nutritive value, quality, and stability of cruciferous vegetables during storage and subsequent to processing, *CRC Crit. Rev. Food Technol.*, 4(1), 1–38, 1973.

(3) Oilseeds

1. Effect of commercial processing of oilseeds and oils on their composition, *Nutritional Evaluation of Food Processing,* Harris, R. S. and von Loesecke, H., Eds., Avi Publishing, Westport, Conn., 1960, 238–260. (2 papers)

(4) Meat

1. Effects of commercial processing on the nutrient composition of meat, poultry, and fish products, in *Nutritional Evaluation of Food Processing,* Harris, R. S. and von Loesecke, H., Eds., Avi Publishing, Westport, Conn., 1960, 261–304. (3 papers)
2. **Leverton, R. M. and Odell, G. V.,** The Nutritive Value of Cooked Meat, Oklahoma Agric. Exp. Sta., Misc. Publ. MP-49, Oklahoma State University, 1958, 90 pp.

(5) Milk

1. **Wanner, R. L.,** Effects of commercial processing of milk and milk products on their nutrient content, in *Nutritional Evaluation of Food Processing,* Harris, R. S. and von Loesecke, H., Eds., Avi Publishing, Westport, Conn., 1960, 173–196.

B. Pests and Pest Control

1. General

1. *Advances in Pest Control Research,* Vol. 1, Interscience, New York, 1957.
2. **Bailey, S. F. and Smith, L. M.,** *Handbook of Agricultural Pest Control,* Industry Publications, New York, 1951.
3. **Beirne, B. P.,** *Pest Management,* Leonard Hill, London, 1966, 123 pp.
4. **de Bach, P., Ed.,** *Biological Control of Insect Pests and Weeds,* Reinhold, New York, 1964, 844 pp.
5. **de Ong, E. R.,** *Chemical and Natural Control of Pests,* Reinhold, New York, 1960, 244 pp.
6. **Hall, S. A., Ed.,** *New Approaches to Pest Control and Eradication,* Advances in Chemistry, Ser. No. 41, ACS, Washington, D.C., 1963, 74 pp.
7. Improved pest control practices, in President's Science Advisory Committee, Restoring the Quality of Our Environment. Report of the Environmental Pollution Panel, U.S. Gov. Print. Off., Washington, D.C., 1965, 227–291.
8. **Kilgore, W. W. and Doutt, R. L.,** *Pest Control: Biological, Physical and Selected Chemical Methods,* Academic Press, New York, 1967, 477 pp. (11 papers)
9. **Mallis, A.,** *Handbook of Pest Control. The Behavior, Life History, and Control of Household Pests,* 4th ed., MacNair-Dorland Co., New York, 1964, 1148 pp.
10. National Research Council. Committee on Plant and Animal Pests, *Principles of Plant and Animal Pest Control,* National Academy of Sciences, Washington, D.C., 1968–1970. (6 vols.)
11. National Research Council, *Scientific Aspects of Pest Control.* A Symposium, National Academy of Sciences, Publ. 1402, Washington, D.C., 1966, 470 pp. (28 papers)
12. National Research Council, *Pest Control Strategies for the Future,* National Academy of Sciences, Washington, D.C., 1972, 376 pp. (24 papers)
13. **Rietz, L. P., Ed.,** *Biological and Chemical Control of Plant and Animal Pests,* American Association for the Advancement of Science, Washington, D.C., 1960, 273 pp.
14. **Sweetman, H. L.,** *The Principles of Biological Control: Interrelation of Hosts and Pests and Utilization in Regulation of Animal and Plant Populations,* W. C. Brown & Co., Dubuque, IA, 1958, 560 pp.
15. **Truman, L. C. and Butts, W. L.,** *Scientific Guide to Pest Control Technology,* 2nd ed., Pest Control Magazine, Cleveland, Ohio, 1967, 187 pp. *World Review of Pest Control,* Vol. 1–6, Harston, London, 1962–1967.

2. Pests and Crop Protection

a. General

1. **Cramer, H. H.,** *Plant Protection and World Crop Production,* Bayer, Leverkusen, Germany, 1967, 524 pp.
2. **Ennis, W. B., Jr. and Miller, P. R.,** Disease, weed and nematode control. Pesticides – Potential and problems, Int. Agric. Dev., No. 40, February 1968.
3. Field Crop Pests, Plant Protection Field Problem Unit, Agric. Protection Program, University of Tennessee, U.S. AID nesa-318, 105 pp.
4. Food and Agriculture Organization, Manual on the Use of FAO Specifications for Plant Protection Products, FAO, Rome, 1970, 64 pp.
5. **Fröhlich, G. and Rodewald, W.,** *Pests and Diseases of Tropical Crops and Their Control,* Pergamon Press, Toronto, 1970, 371 pp.
6. **Goodman, R. N.,** Trends in plant protection against microorganisms, in *Global Impacts of Applied Microbiology,* Starr, M. P., Ed., Almqvist & Wiksell, Stockholm, 1964, 248–255.
7. **Gregory, J. G. and Rawlings, C. J.,** *Technological Economics of Crop Protection and Pest Control,* Soc. Chem. Ind., London, 1970, 255 pp.
8. **Hassall, K. A.,** *World Crop Protection,* Vol. 2, *Pesticides,* Chemical Rubber Co., Cleveland, 1969, 249 pp.
9. **Hussey, N. W., Read, W. H., and Hesling, J. J.,** *Pests of Protected Cultivation,* American Elsevier, New York, 1970, 404 pp.
10. **Jones, F. G. W. and Jones, M. G.,** *Pests of Field Crops,* St. Martin's Press, New York, 1964, 406 pp.
11. **Martin, H., Ed.,** *Insecticide and Fungicide Handbook for Crop Protection,* British Crop Protection Council, Blackwell, London, 1972, 416 pp.
12. **Martin, H.,** *The Scientific Principles of Crop Protection,* 6th ed., Arnold, London, 1973, 424 pp.
13. **Merino-Rodriguez, M., Ed.,** *Elsevier's Lexicon of Plant Pests and Diseases,* Elsevier, Amsterdam, 1966, 351 pp.
14. **Miller, P. R., Hoffmann, C. H., Ennis, W. B., Jr., and Jansen, L. L.,** Improved food production by use of crop-protecting chemicals, in *Proc. SOS/70. 3rd International Congress Food Science and Technology,* Institute of Food Technologists, Chicago, 1971, 544–550.
15. **Nene, Y. L.,** Plant protection: A key to maintaining present gains in food production, in *Some Issues Emerging from Recent Breakthroughs in Production,* Turk, K. L., Ed., New York State College of Agriculture, Cornell University, 1971, 365–377.
16. **Ordish, G. V. B.,** More food by crop protection, *Span,* 4, 36–39, 1961.
17. **Ordish, G.,** *Man, Crops and Pests in Central America,* Pergamon Press, Oxford, 1964.
18. **Ordish, G.,** 150 years of crop pest control, *World Rev. Pest Contr.,* 7, 204–213, 1968.
19. **Pyenson, L.,** *Elements of Plant Protection,* John Wiley & Sons, New York, 1951, 538 pp.
20. **Rivnay, E.,** Field Crop Pests in the Near East, in *Monographiae Biologie,* Vol. 10, Den Haag, 1962, 450 pp.
21. **Rose, G. J.,** *Crop Protection,* Chemical Publishing, New York, 1963, 490 pp.
22. **Stakman, E. C.,** Pest, pathogen, and weed control for increased food production, *Proc. Natl. Acad. Sci. U.S.A.,* 56, 376–381, 1966.
23. **Stapley, J. H. and Gayner, F. C. H.,** *World Crop Protection,* Vol. 1, *Pests and Diseases,* Chemical Rubber Co., Cleveland, Ohio, 1969, 272 pp.
24. **Trought, T. E.,** *Farm Pests: An Aid to Their Recognition,* Blackwell, Oxford, 1965, 32 pp.
25. **Woodford, E. K., Ed.,** Crop Production in a Weed Free Environment. Symposium of the British Weed Control Council No. 2, 1963.
26. **Wyniger, R.,** *Pests of Crops in Warm Climates and Their Control,* Verlag Recht und Gesellschaft, Basel, 1962.

b. Cereal Crops

1. **Murphy, H. C.,** Protection of oats and other cereal crops during production in *Food Quality,* Irving, G. W., Jr. and Hoover, S. R., Eds., American Association for the Advancement of Science, Washington, D.C., 1965, 99–113.

c. Stored Grain

1. **Cotton, R. T.,** *Pests of Stored Grain and Grain Products,* Burgess Publishing, Minneapolis, Minn., 1963, 318 pp.
2. **Hinton, H. E. and Corbert, A. S.,** *Common Insect Pests of Stored Food Products. A Guide to Their Identification,* British Museum, London, 1963, 61 pp.
3. **Semenink, G., Anderson, J. S., and Alcock, A. W.,** Microflora in storage of cereal grains and their products, American Association of Cereal Chemists, Monogr. Ser. No. 2, 1945, 77–151.
4. Stored–Grain Pests, U.S. Dep. Agric. Farmer's Bull. No. 1260, 1958.

d. Fruits and Vegetables

1. **Ebeling, W.,** *Subtropical Fruit Pests,* University of California Div. Agric. Sci., Berkeley, 1959, 436 pp.
2. **Pepper, B. B.,** Protection of fruits and vegetables during production, in *Food Quality,* Irving, G. W., Jr. and Hoover, S. R., Eds., American Association for the Advancement of Science, Washington, D.C., 1965, 35–45.

3. Livestock Pests and Their Control

1. **Furman, D. P.,** Poultry insects and related pests, in *Fundamentals of Applied Entomology,* 2nd ed., Macmillan, New York, 1971, 589–633.
2. **Pfadt, R. E.,** Livestock insects and related pests, in *Fundamentals of Applied Entomology,* 2nd ed., Macmillan, New York, 1971, 351–587.
3. **Seiden, R.,** *Insect Pests of Livestock, Poultry and Pets and Their Control,* Springer Publishing, New York, 1964, 162 pp.

4. Weeds and Weed Control

a. General

1. **Ahlgren, G. H., Klingman, G. C., and Wolf, D. E.,** *Principles of Weed Control,* John Wiley & Sons, New York, 1951, 386 pp.
2. **Ashby, D. G. and Pfeiffer, R. K.,** Weeds a limiting factor in tropical agriculture, *World Crops,* 8, 227–229, 1956.
3. **Bates, G. H.,** *Weed Control,* Spon, London, 1955, 235 pp.
4. **Buchholtz, K. P.,** Weed control – A record of achievement, *Weeds,* 10, 167–170, 1962.
5. **Crafts, A. S.,** Weed control research – past, present, and future, *Weeds,* 10, 535–540, 1960.
6. **Crafts, A. S. and Harvey, W. A.,** Weed control, *Adv. Agron.,* 1, 289–320, 1949.
7. **Crafts, A. S. and Robbins, W. W.,** *Weed Control,* 3rd ed., McGraw-Hill, New York, 1962, 660 pp.
8. **Day, B. E.,** The scientific basis of weed control, in *Scientific Aspects of Pest Control,* Publ. 1402, National Academy of Sciences, Washington, D.C., 1966, 102–114.
9. **Day, B. E.,** Nonchemical weed control, in *Pest Control Strategies for the Future,* National Academy of Sciences, Washington, D.C., 1972, 330–338.
10. **Ennis, W. B., Jr.,** The challenges of modern weed control, *Weeds,* 6, 363–369, 1958.
11. **Ennis, W. B., Shaw, W. C., Danielson, L. L., Klingman, D. L., and Timmons, F. L.,** Impact of chemical weed control on farm management practices, *Adv. Agron.,* 15, 161–210, 1963.
12. **Fryer, J. and Makepeace, R., Eds.,** *Weed Control Handbook,* 6th ed., F. A. Davis & Co., Philadelphia, 1971, 348 pp.
13. **Georgia, A.,** *A Manual of Weeds,* Macmillan, New York, 1914, 593 pp.
14. **Harper, J. L.,** Ecological aspects of weed control, *Outlook Agric.,* 1, 197–205, 1957.
15. **Harper, J. L., Ed.,** *Biology of Weeds,* Blackwell, Oxford, 1960, 256 pp.
16. **Huffaker, C. B.,** Fundamentals of biological control of weeds, *Hilgardia,* 27, 101–157, 1957.
17. **Huffaker, C. B.,** Biological control of weeds with insects, *Annu. Rev. Entomol.,* 4, 251–276, 1959.
18. **Isely, D.,** *Weed Identification and Control,* Iowa State University Press, Ames, 1960.
19. **Kasasian, L.,** *Weed Control in the Tropics,* Leonard Hill, London, 1971, 307 pp.
20. **King, L. J.,** *Weeds of the World. Biology and Control,* Interscience, New York, 1966, 526 pp.
21. **Klingman, G. C.,** *Weed Control: As a Science,* John Wiley & Sons, New York, 1961, 421 pp.
22. **Korsmo, E.,** *Anatomy of Weeds,* Grondahl, Oslo, 1954, 413 pp.
23. **Muenscher, W. C.,** *Weeds,* 2nd ed., Macmillan, New York, 1955, 560 pp.
24. National Research Council, Committee on Plant and Animal Pests, *Weed Control, Principles of Plant and Animal Pest Control,* Vol. 2, National Academy of Sciences, Washington, D.C., 1968, 471 pp.
25. **Robbins, W. W., Crafts, A. S., and Raynor, R. N.,** *Weed Control,* 2nd ed., McGraw-Hill, New York, 1952, 503 pp.
26. **Shaw, W. C.,** Weed science – revolution in agricultural technology, *Weeds,* 12, 153–162, 1964.
27. **Shaw, W. C. and Jansen, L. L.,** Chemical weed control strategies for the future, in *Pest Control Strategies for the Future,* National Academy of Sciences, Washington, D.C., 1972, 197–215.
28. **Simmonds, F. J., Ed.,** *Proc. 1st International Symposium on Biological Control of Weeds, March 1969,* Agricultural Bureaux, Slough, England, 1970, 110 pp.
29. **Wilson, F.,** The biological control of weeds, *Annu. Rev. Entomol.,* 9, 225–243, 1964.
30. **Woodford, E. K., Ed.,** *Weed Control Handbook,* 2nd ed., Blackwell, Oxford, 1960, 264 pp.
31. **Woodford, E. K. and Evans, S. A., Eds.,** *Weed Control Handbook,* 3rd ed., Blackwell, Oxford, 1963, 356 pp.

b. Aquatic Weeds

1. **Greenwald, M.,** List of references on control of aquatic plants including algae, Bull. Chipman Chem. Co., 1956; Suppl., 1957.
2. **Hall, T. F.,** Principles of aquatic plant control, *Adv. Pest Contr.,* 4, 211–247, 1961.
3. **Holm, L. G., Weldon, L. W., and Blackburn, R. D.,** Aquatic weeds, *Science,* 166, 699–709, 1969.
4. **Little, E. C. S.,** The control of water weeds, *Weed Res.,* 8, 79–105, 1968.

5. Fungi and Their Control

1. **Ainsworth, G. C.,** *Ainsworth and Bisby's Dictionary of the Fungi,* 6th ed., Commonwealth Mycological Institute, Kew, Surrey, 1971, 663 pp.
2. **Ainsworth, G. C. and Sussman, A. E., Eds.,** *The Fungi. An Advanced Treatise,* Academic Press, New York, 1965–1968. (3 vols.)

3. **Alexopoulos, C. D.,** *Introductory Mycology,* 2nd ed., John Wiley & Sons, New York, 1962, 613 pp.
4. **Arthur, J. C.,** *Manual of the Rusts in the United States and Canada,* Hafner Publishing, New York, 1962, 438 pp.
5. **Bessey, E. A.,** *Morphology and Taxonomy of Fungi,* Blakiston, Philadelphia, 1950, 791 pp.
6. **Clements, F. E. and Shear, C. L.,** *The Genera of Fungi,* Wilson, New York, 1931, 496 pp.
7. **Cochrane, V. W.,** *Physiology of Fungi,* John Wiley & Sons, New York, 1931, 524 pp.
8. **Fischer, G. W. and Holton, C. S.,** *Biology and Control of the Smut Fungi,* Ronald Press, New York, 1957, 622 pp.
9. **Garrett, S. D.,** *Biology of Root-Infecting Fungi,* Cambridge University Press, Cambridge, 1956, 292 pp.
10. **Garrett, S. D.,** *Pathogenic Root-Infecting Fungi,* Cambridge University Press, London, 1970.
11. **Gilman, J. C.,** *A Manual of Soil Fungi,* Iowa State College Press, Ames, 1957.
12. **Ingold, C. T.,** *Dispersal in Fungi,* Oxford University Press, London, 1953, 198 pp.
13. **Moore, W. C.,** *British Parasitic Fungi,* Cambridge University Press, Cambridge, 1959, 430 pp.
14. **Wolf, F. A. and Wolf, F. T.,** *The Fungi,* John Wiley & Sons, New York, 1947. (2 vols.)

6. Viruses

a. General

1. **Burnet, F. M. and Stanley, W. M., Eds.,** *Viruses. Biochemical, Biological and Biophysical Properties,* Academic Press, New York, 1959. (3 vols.)
2. **Fraenkel-Conrat, H.,** *The Chemistry and Biology of Viruses,* Academic Press, New York, 1969, 293 pp.
3. **Holmes, F. O.,** The filterable viruses, in *Bergy's Manual of Determinative Bacteriology,* Suppl. 2, 6th ed., Williams & Wilkins, Baltimore, 1948, 1127–1186.
4. **Maramorosch, K. and Kurstak, E., Eds.,** *Comparative Virology,* Academic Press, New York, 1971, 552 pp.

b. Animal Viruses

1. **Betts, A. O. and York, C. J., Eds.,** *Viral and Rickettsial Infections of Animals,* Vol. 1, Academic Press, New York, 1967, 388 pp.
2. **Fenner, F.,** *Biology of Animal Viruses,* Academic Press, New York, 1968. (2 vols.)
3. **Fenner, F. and White, D. O.,** *Medical Virology,* Academic Press, New York, 1970, 390 pp.
4. **Smith, K. M.,** *Insect Virology,* Academic Press, New York, 1967, 256 pp.

c. Plant Viruses

1. **Bawden, F. C.,** *Plant Viruses and Virus Diseases,* 4th ed., Ronald Press, 1964, 361 pp.
2. **Corbett, M. K. and Sisler, H. D.,** *Plant Virology,* University of Florida Press, Gainesville, 1964, 527 pp.
3. **Maramorosch, K.,** *Viruses, Vectors, and Vegetation,* Interscience, New York, 1969, 666 pp.
4. **Matthews, R. E. F.,** *Principles of Plant Virology,* Academic Press, New York, 1970, 778 pp.
5. **Smith, K. M.,** *A Textbook of Plant Virus Diseases,* 2nd ed., Brown & Co., Boston, 1957, 652 pp.
6. **Smith, K. M.,** *Plant Viruses,* 4th ed., Methuen, London, 1968, 166 pp.
7. U.S. Department of Agriculture, Virus Diseases and Other Disorders with Viruslike Symptoms of Stone Fruits in North America, U.S. Dep. Agric. Handb. No. 10, Washington, D.C., 1951, 276 pp.

7. Nematodes and Their Control

1. **Chitwood, B. G. and Chitwood, M. B.,** *An Introduction to Nematology,* Monumental Printing, Baltimore, MD, 1950, 213 pp.
2. **Christie, J. R.,** *Plant Nematodes, Their Bionomics and Control,* Drew Co., Jacksonville, Florida, 1959, 256 pp.
3. **Good, J. M. and Taylor, A. L.,** Chemical Control of Plant-Parasitic Nematodes, U.S. Dep. Agric. Handb. No. 286, Washington, D.C., 1965.
4. **Goodey, T.,** *Soil and Freshwater Nematodes,* Methuen, London, 1963, 544 pp.
5. **Hare, W. W.,** The inheritance of resistance of plants to nematodes, *Phytopathology,* 55, 1162–1167, 1965.
6. **Jenkins, W. R. and Taylor, D. P.,** *Plant Nematology,* Reinhold, New York, 1967, 270 pp.
7. **Jones, F. G. W.,** Ecological relationships of nematodes, in *Plant Pathology, Problems and Progress,* Holton, C. S., Fischer, G. W., Fulton, R. W., Hart, H., and McCallan, S. E. A., Eds., University of Wisconsin Press, Madison, 1959, 395–411.
8. **Kehr, A. E.,** Current status and opportunities for the control of nematodes by plant breeding, in Pest Control by Chemical, Biological, Genetic, and Physical Means. A Symposium, USDA, ARS 33-110, 1966, 126–138.
9. **Kirkpatrick, J. D., Van Gundy, S. D., and Mai, W. F.,** Interrelationships of plant nutrition, growth and parasitic nematodes, *Plant Anal. Fert. Probl.,* 4, 189–225, 1964.
10. **Krushberg, L. R.,** Host response to nematode infection, *Annu. Rev. Phytopathol.,* 1, 219–240, 1963.
11. **Lee, D. L.,** *The Physiology of Nematodes,* Oliver & Boyd, London, 1965, 154 pp.
12. **Mai, W. F., Lyon, H. H., and Kruk, T. H.,** *Pictorial Key to Genera of Plant Parasitic Nematodes,* Art Craft of Ithaca, Ithaca, N.Y., 1964, 55 pp.

13. National Research Council, Committee on Plant and Animal Pests, Control of Plant-Parasitic Nematodes, in *Principles of Plant and Animal Pest Control,* Vol. 4, National Academy of Sciences, Washington, D.C., 1968, 166 pp.
14. **Oostenbrink, M.,** Major characteristics of the relation between nematodes and plants, *Meded. Landb. Woogesch. Wageningen,* 66, 3–46, 1966.
15. **Pitcher, R. S.,** Interrelationships of nematodes and other pathogens of plants, *Helminthol Abstr.,* 34, 1–17, 1965.
16. **Pitcher, R. S.,** Role of plant-parasitic nematodes in bacterial diseases, *Phytopathology,* 53, 35–39, 1963.
17. **Powell, N. T.,** The role of plant-parasitic nematodes in fungus diseases, *Phytopathology,* 53, 28–35, 1963.
18. **Pramer, D.,** Nematode-trapping fungi, *Science,* 144, 382-388, 1964.
19. **Raski, D. J. and Hewitt, W. B.,** Plant-parasitic nematodes as vectors of plant viruses, *Phytopathology,* 53, 39–47, 1963.
20. **Rohde, R. A.,** The nature of resistance in plants to nematodes, *Phytopathology,* 55, 1159–1162, 1965.
21. **Sasser, J. N. and Jenkins, W. R., Eds.,** *Nematology, Fundamentals and Recent Advances,* University of North Carolina Press, Chapel Hill, 1960, 480 pp.
22. **Southey, J. F.,** Plant Nematology, Min. Agric. Fish. Food, Tech. Bull. No. 7, London, 1965, 282 pp.
23. **Thorne, G.,** *Principles of Nematology,* McGraw-Hill, New York, 1961, 553 pp.
24. **Van Gundy, S. D.,** Factors in survival of nematodes, *Annu. Rev. Phytopathol.,* 3, 43–68, 1965.
25. **Wallace, H. R.,** *The Biology of Plant-Parasitic Nematodes,* Arnold, London, 1963, 280 pp.

8. Insects and Their Control

1. **Baron, S.,** *The Desert Locust,* Methuen, London, 1971, 228 pp.
2. **Brown, A. W.,** *Insect Control by Chemicals,* John Wiley & Sons, New York, 1951, 817 pp.
3. **Burges, H. D. and Hussey, N. W., Eds.,** *Microbial Control of Insects,* Academic Press, New York, 1971, 862 pp.
4. **Busvine, J. R.,** *Insects and Hygiene. The Biology and Control of Insect Pests of Medical and Domestic Importance,* 2nd ed., Methuen, London, 1966, 467 pp.
5. **Davidson, R. H. and Peairs, L. M.,** *Insect Pests of Farm, Garden and Orchard,* 6th ed., John Wiley & Sons, New York, 1966, 675 pp.
6. **Ebeling, W.,** *Subtropical Entomology,* Lithotype Process Co., San Francisco, 1950, 747 pp.
7. **Edwards, C. A. and Heath, G. W.,** *The Principles of Agricultural Entomology,* Chapman & Hall, London, 1964.
8. **Fernald, H. T. and Shepard, H. H.,** *Applied Entomology,* 5th ed., McGraw-Hill, New York, 1955, 385 pp.
9. **Geier, P.,** Management of insect pests, *Annu. Rev. Entomol.,* 11, 471-490, 1966.
10. **Graham, K.,** *Concepts of Forest Entomology,* Reinhold, New York, 1963, 388 pp.
11. **Harris, W. V.,** *Termites. Their Regulation and Control,* 2nd ed., Longmans, London, 1971, 186 pp.
12. **Herford, G. V. B.,** Recent developments in the protection of foods from insect pests with special reference to the Commonwealth countries, *J. R. Soc. Arts,* 110, 423–435, 1962. (8 ref.)
13. **Hicklin, N. E.,** *Household Insect Pests,* Hutchinson, London, 1964, 172 pp.
14. **Isely, D.,** *Methods of Insect Control,* Burgess, Minneapolis, 1946–47. (2 vols.)
15. **LaBrecque, G. C. and Smith, C. N., Eds.,** *Principles of Insect Chemosterilization,* Appleton-Century Crofts, New York, 1968, 354 pp.
16. **Little, V. A.,** *General and Applied Entomology,* 2nd ed., Harper & Row, New York, 1963, 543 pp.
17. **Metcalf, C. L.,** *Destructive and Useful Insects: Their Habits and Control,* 4th ed., McGraw-Hill, New York, 1962, 1087 pp.
18. **Monro, H. A. U.,** Manual of Fumigation for Insect Control, FAO, Rome, 1961, 289 pp.
19. National Research Council, Committee on Plant and Animal Pests, *Insect-Pest Management and Control,* in *Principles of Plant and Animal Pest Control,* Vol. 3, National Academy of Sciences, Publ. 1695, Washington, D.C., 1969, 508 pp.
20. **Newcomer, E. J.,** Insect Pests of Deciduous Fruits in the West, U.S. Dep. Agric. Handb. No. 306, 1966, 57 pp.
21. **Newman, L. H.,** *Man and Insects. Insect Allies and Enemies,* Natural History Press, New York, 1967, 252 pp.
22. **Oldroyd, H.,** *Elements of Entomology,* Universe Books, New York, 1970, 312 pp.
23. **Peairs, L. M. and Davidson, R. H.,** *Insect Pests of Farm, Garden, and Orchard,* 5th ed., John Wiley & Sons, New York, 1956.
24. **Pfadt, R. E., Ed.,** *Fundamentals of Applied Entomology,* 2nd ed., Macmillan, New York, 1971, 693 pp.
25. **Pradhan, S.,** *Insect Pests of Crops,* National Book Trust, New Delhi, India, 1969, 208 pp.
26. **Riley, W. A. and Johannsen, O. A.,** *Medical Entomology. A Survey of Insects and Allied Forms which Affect the Health of Man and Animals,* 2nd ed., McGraw-Hill, New York, 1938, 483 pp.
27. **Ross, H. H.,** *A Textbook of Entomology,* 3rd ed., John Wiley & Sons, New York, 1965, 539 pp.
28. **Smart, J.,** *A Handbook for the Identification of Insects of Medical Importance,* 4th ed., British Museum, London 1965, 303 pp.
29. U.S. Department of Agriculture, Insects, The Yearbook of Agriculture 1952, U.S. Gov. Print. Off., Washington D.C., 1952, 780 pp.
30. U.S. Department of Agriculture, Locust and Other Insect Control, U.S. Dep. Agric. Misc. Publ. 770, 147 pp.

31. **Uvarov, B. P.,** Locusts and Grasshoppers, Imperial Bureau of Entomology, London, 1928, 352 pp.
32. **Uvarov, B. P.,** *Insects and Climate, Trans. Entomol. Soc. Lond.,* 79(I), 1–247, 1931.
33. **West, T. F. and Hardy, J. E.,** *Chemical Control of Insects,* Chapman & Hall, London, 1961, 206 pp.

9. Vertebrate Pests and Their Control

a. General

1. **Dykstra, W. W. and Lennon, R. E.,** The role of chemicals for the control of vertebrate pests, in Pest Control by Chemical, Biological, Genetic, and Physical Means, A Symposium, Knipling, E. R., Ed., USDA, ARS 33-110, Washington, D.C., 1966, 29–34.
2. **Howard, W. E.,** Vertebrate pests: Biocontrol and chemosterilants, in *Pest Control: Biological, Physical and Selected Chemical Methods,* Kilgore, W. W. and Doutt, R. L., Eds., Academic Press, New York, 1967, 346–386.
3. **McCabe, R. A.,** Vertebrates as pests: A point of view, in Scientific Aspects of Pest Control, National Academy of Sciences, Publ. 1402, Washington, D.C., 1966, 115–134.
4. National Research Council, Committee on Plant and Animal Pests, *Vertebrate Pests: Problems and Control,* in *Principles of Plant and Animal Pest Control,* Vol. 5, National Academy of Sciences, Washington, D.C., 1970, 153 pp.
5. Proc. Vertebrate Pest Control Conference, Sacramento, Calif., 1962, National Pest Control Association, Elizabeth, N.J., 1962, 391 pp.
6. Proc. 2nd Vertebrate Pest Control Conference, Anaheim, Calif., 1964, Cummings, M., Ed., University of California, Davis, 1964, 160 pp.
7. Proc. 3rd Vertebrate Pest Control Conference, University of California, Davis, 1967, 177 pp.
8. Rodents, Bats and Birds Damage to Food Supplies in the Diet-Deficient Regions of the World, Wildlife Research Laboratory, Denver, 1966, 34 pp.
9. *The Control of Vermin,* 5th ed., Gilbertson and Page, Hertford, Eng., 1968, 42 pp.

b. Rodents

1. **Barnett, S. A.,** Infestation Control: Rats and Mice, Ministry of Food..., London, 1946, 36 pp.
2. **Barnett, S. A.,** *The Rat,* Aldine Publishing, Chicago, 1966, 288 pp.
3. **Barnett, S. A.,** Rats, *Sci. Am.,* 216(1), 78–85, 1967.
4. **Bjornson, B. F., Pratt, H. D., and Litting, K. S.,** Control of Domestic Rats and Mice. Training Guide, Rodent Contr. Ser., U.S. Gov. Print. Off., Washington, D.C., 1970, 41 pp.
5. **Buell, N. E.,** Rodent control, *Pest Cont.,* 26(10), 16–20, 1958.
6. **Calhoun, J. B.,** *The Ecology and Sociology of the Norway Rat,* Public Health Service publ. 1008, U.S. Gov. Print. Off. Washington, D.C., 1963, 288 pp.
7. **Chitty, D. and Southern, H. N.,** *Control of Rats and Mice,* Oxford University Press, London, 1954. (3 vols.)
8. **Christian, J. J, and Davis,** D. E., The biological basis of rodent control, *Proc. 9th Pacific Science Conference 1957,* 19, 25–33, 1966.
9. Communicable Disease Center, Rat-borne Disease Prevention and Control, Fed. Sec. Agency, Public Health Service, Atlanta, Ga., 1949, 293 pp.
10. **Crowcroft, P.,** *Mice All Over,* G. T. Foulis, London, 1966, 121 pp.
11. **Davis, D. E.,** The characteristics of rat populations, *Q. Rev. Biol.,* 28, 373–401, 1953.
12. **Davis, D. E.,** Rodent control strategy, in *Pest Control Strategies for the Future,* National Academy of Sciences, Washington, D.C., 1972, 157–171.
13. **Eadie, W. R.,** *Animal Control in Field, Farm and Forest,* Macmillan, New York, 1954, 257 pp.
14. **Elton, C.,** *Voles, Mice and Lemmings: Problems in Population Dynamics,* Clarendon Press, Oxford, 1942.
15. **Hirst, L. T.,** *The Conquest of Plague,* Clarendon Press, Oxford, 1953, 478 pp.
16. **Johnson, W. H. and Bjornson, B. F.,** Rodent Eradication and Poisoning Programs, U.S. Dep. Health, Educ. & Welfare, Public Health Service, Atlanta, Ga., 1964, 84 pp.
17. **Krebs, C. J., Gaines, M. S., Keller, B. L., Myers, J. H., and Tamarin, R. H.,** Population cycles in small rodents. Demographic and genetic events are closely coupled in fluctuating populations of field mice, *Science,* 179, 35–41, 1972.
18. Proc. Rodents as Factors in Disease and Economic Loss, Asia-Pacific Interchange, June 17–27, 1968, Institute for Technical Interchange, East-West Center, Honolulu, Hawaii, 1968, 285 pp.
19. Seminar on Rodents and Rodents' Parasites, Geneva, October 24–28, 1966.
20. **Weinburgh, H. B.,** Field Rodents, Rabbits, and Hares, U.S. Dep. Health, Educ. Welfare, Public Health Service, Atlanta, Ga., 1966, 87 pp.
21. **Welch, J. F.,** Rodent control. A review of chemical repellents for rodents, *J. Agric. Food Chem.,* 2, 142–149, 1954. (30 ref.)

c. Birds

1. **Beck, J. R. and Jackson, W. B., Eds.,** *Proc. 2nd Bird Control Seminar,* Bowling Green State University, Bowling Green, Ohio, 1964, 140 pp.

2. **Murton, R. K., Ed.,** *The Problems of Birds as Pests,* Academic Press, New York, 1969, 270 pp.
3. Proc. 1st Bird Control Seminar, Bowling Green State University, Bowling Green, 1962.
4. Proc. N. Am. Conf. on Blackbird Depredation in Agriculture, Ohio State University, Columbus, Ohio, 1967, 62 pp.
5. **Schneider, D. E. and Jackson, W. B., Eds.,** Proc. 3rd Bird Control Seminar, Bowling Green State University, Bowling Green, 1966, 172 pp.
6. **Stefferud, A., Ed.,** Birds in Our Lives, U.S. Dep. Interior, Washington, D.C., 1966, 561 pp.

C. Food Protection Technology

1. Food Preservation In General

1. **Ball, C. and Olson, F. C. W.,** *Sterilization in Food Technology. Theory, Practice and Calculation,* McGraw-Hill, New York, 1957, 654 pp.
2. **Cambell, C. H., Isker, R. A., and Maclinn, W. A.,** *Cambell's Book. Canning, Pickling, Preserving,* 3rd ed., Vance Publishing, Chicago, 1950, 222 pp.
3. **Desrosier, N. W.,** *The Technology of Food Preservation,* 3rd ed., Avi Publishing, Westport, Conn., 1970, 488 pp.
4. **Esselen, W. B.,** Food preservation and its contribution to nutrition, *Borden's Rev. Nutr. Res.,* 18, 29–44, 1957.
5. **Fellers, C. R.,** Food preservation, in *Handbook of Food and Agriculture,* Blanck, F. C., Ed., Reinhold, New York, 1955, 331–363.
6. **Lechtman, S. C. and Fanning, O., Eds.,** *The Future of Food Preservation,* Midwest Research Institute, Kansas City, Mo., 1957.
7. **Lynch, L. J., Mitchell, R. S., and Casimir, D. J.,** The chemistry and technology of the preservation of green peas, *Adv. Food Res.,* 9, 61–152, 1959.
8. **Mathur, P. B.,** Storage and preservation of food, *Annu. Rev. Biochem. Res. India,* 27, 110–116, 1956.
9. **Scott, J. K.,** The effect of technical advancement on the preservation of food, in *Global Impact of Applied Microbiology,* Starr, M. P., Ed., Almqvist & Wiksell, Stockholm, 1964, 173–200.

2. Cold Preservation

1. **Anderson, O. E., Jr.,** *Refrigeration in America,* Princeton University Press, Princeton, N.J., 1953.
2. **Brody, A. L. and Bedrosian, K.,** Effect of room temperature vs. refrigerated storage on quality of canned fruit and vegetable products, *Food Technol.,* 15, 367–370, 1961.
3. **Brown, D. C.,** The application of cryogenic fluids to the freezing of foods, *Adv. Cryogr. Eng.,* 12, 11–22, 1967.
4. **Fennema, O. R.,** An over-all view of low temperature food preservation, *Cryobiology,* 3, 197–213, 1966. (143 ref.)
5. **Fennema, O. R. and Powrie, W. D.,** Fundamentals of low-temperature food preservation, *Adv. Food Res.,* 13, 219–347, 1964. (379 ref.)
6. **Fennema, O. R., Powrie, W. D., and Marth, E. H., Eds.,** *Low-Temperature Preservation of Foods and Living Matter,* Dekker, New York, 1973, 598 pp.
7. **Giranta, M. E.,** Consideration on the use of cold in the preservation of foods and its comparison with other preservation procedures, *Rev. Frio,* 4(4), 219-230, 1959.
8. **Gortner, W. A., Erdman, F. S., and Masterman, N. K.,** *Principles of Food Freezing,* John Wiley & Sons, New York, 1948.
9. **Hawthorn, J., Ed.,** *Low Temperature Biology of Foodstuffs,* Pergamon Press, Oxford, 1968, 458 pp.
10. **Hayes, K. M., Kane, J. J., Maclinn, W. A., and Collins, S. C.,** Liquified gases – new techniques in food preservation, ASHRAE J., 3(9), 78–83, 1961.
11. **Joslyn, M. A. and Diehl, H. C.,** Physiological aspects of low temperature preservation of plant products, *Annu. Plant Physiol.,* 3, 149–170, 1952.
12. **Kuprianoff, J.,** Fundamentals and practical aspects of the freezing of foodstuffs, in *Aspects Théoriques et Industriels de la Lyophilisation,* Rey, L., Ed., Hermann, Paris, 1964.
13. **McCoy, D. C.,** Refrigeration in food processing, in *Food Processing Operations,* Vol. 1, Joslyn, M. A., and Heid, J. L., Eds., Avi Publishing, Westport, Conn., 1963, 362–400.
14. **Miller, A. T., El-Bisi, H. M., and Sawyer, F. M.,** Microbiological and Public Health Aspects of Prepared Frozen Foods, A Review, University of Massachusetts Bull., 548, 1964.
15. **Nickerson, J. T. R. and Karel, M.,** Preservation of food by freezing, in *Food Processing Operations,* Joslyn, M. A. and Heid, J. L., Eds., Avi Publishing, Westport, Conn., 1964, 259–294.
16. **Pennington, M. E. and Tressler, D. K.,** Food preservation by temperature control, in *The Chemistry and Technology of Food and Food Products,* Vol. 3, Interscience, New York, 1951, 1822–1857.
17. **Pentzer, W. T., Ed.,** *Progress in Refrigeration Science and Technology,* Avi Publishing, Westport, Conn., 1973. (4 vols.)
18. **Rogers, J. L.,** *The Commerce and Technology of Quick Frozen Foods: Processing, Packing, Distribution,* Food Trade Review, Westport, Conn., 1958, 340 pp.
19. **Slavin, J. W.,** Methods for freezing used in the fisheries. A review, *Fish. Gaz. Annu. Rev.,* 75, 176–186, 1958.
20. **Tressler, D. K.,** Cryogenic freezing, in *The Freezing Preservation of Foods,* Tressler, D. K., Van Arsdel, W. B., and Copley, M. J., Eds., Avi Publishing, Westport, Conn., 1968, 153–175.

21. **Tressler, D. K., Van Arsdel, W. B., and Copley, M. J.,** *The Freezing Preservation of Foods,* 4th ed., Avi Publishing, Westport, Conn., 1968. (4 vols.)
22. **Van Arsdel, W. B.,** The time-temperature tolerance of frozen foods, *Food Technol.,* 11, 28–33, 1957.
23. **Weil, B. H. and Sterne, F.,** Literature Search on Preservation of Foods by Freezing, Georgia School of Technology, Atlanta, Ga., 1948.
24. **Woolrich, W. R.,** *Handbook of Refrigerating Engineering,* 4th ed., Avi Publishing, Westport, Conn., 1965, 1966. (2 vols.)
25. **Woolrich, W. R. and Hallowell, E. R.,** *Cold and Freezer Storage Manual,* Avi Publishing, Westport, Conn., 1970, 338 pp.

3. Food Dehydration

1. **Dedert, W. G. and Moore, J. G.,** New trends in evaporation, *Ind. Chem. Eng.,* 55, 57–62, 1963.
2. **Dombrowski, N. and Munday, G.,** Spray drying in *Biochemical and Biological Engineering Science,* Vol. 2, Blakebrough, N., Ed., Academic Press, London, 1968, 209–320.
3. **Feinberg, B.,** Concentration by evaporation, in *Food Processing Operations,* Vol. 3, Joslyn, M. A. and Heid, J. L., Eds., Avi Publishing, Westport, Conn., 1964, 309–334.
4. **Forrest, J. C.,** Drying processes, in *Biochemical and Biological Engineering Science,* Vol. 2, Blakebrough, N., Ed., Academic Press, London, 1968, 97–135.
5. **Greensmith, M.,** *Practical Dehydration,* Avi Publishing, Westport, Conn., 1971, 170 pp.
6. **Hall, C. W. and Hedrick, T. I.,** *Drying Milk and Milk Products,* Avi Publishing, Westport, Conn., 1966.
7. **Hall, C. W. and Hedrick, T. I.,** *Drying of Milk and Milk Products,* 2nd ed., Avi Publishing, Westport, Conn., 1971, 338 pp.
8. **Hunziker, O. F.,** *Condensed Milk and Milk Powder,* Hunziker, La Grange, Ill., 1946.
9. **Joslyn, M. A.,** Food processing by drying and dehydration, in *Food Processing Operations,* Vol. 2, Joslyn, M. A. and Heid, J. L., Eds., Avi Publishing, Westport, Conn., 1963, 545–584.
10. **King, C. J.,** Recent developments in food dehydration technology, in *Proc. SOS/70. 3rd International Congress Food Science and Technology,* Institute of Food Technologists, Chicago, 1971, 565–574. (30 ref.)
11. **Kraybill, H. R.,** Dehydration of meat, *Ind. Eng. Chem.,* 35, 46–51, 1943.
12. **Lapple, W. C. and Clark, W. E.,** Drying methods and equipment, *Chem. Eng.,* 62(10), 191–210, 1955.
13. **Master, K.,** *Spray Drying. An Introduction to Principles, Operational Practices and Applications,* Leonard Hill, London, 1972, 688 pp.
14. **Milleville, H. P.,** High-quality concentrates, *Food Process,* 23(10), 59–62, 1962.
15. **Mrak, E. M. and Mackinney, G.,** The dehydration of foods, in *The Chemistry and Technology of Food and Food Products,* Vol. 3, Jacobs, M. B., Ed., Interscience, New York, 1951, 1773–1821.
16. **Noyes, R.,** *Dehydration Process for Convenience Foods,* Food Processing Review, No. 2, Noyes Development Corp., Park Ridge, N.J., 1969.
17. **Seltzer, E.,** Progress in food dehydration, 1939–1964, *Food Technol.,* 18(9), 117–120, 1964.
18. **Seltzer, E. and Settelmeyen, J. T.,** Spray drying of foods, *Adv. Food Res.,* 6, 399–520, 1949.
19. Society of Chemical Industry, *Fundamental Aspects of the Dehydration of Foodstuffs,* Macmillan, New York, 1958.
20. **Torrey, M.,** *Dehydration of Fruits and Vegetables,* Noyes Data Corp., Park Ridge, N.J., 1974.
21. **Tressler, D. K.,** New developments in the dehydration of fruits and vegetables, *Food Technol.,* 10, 119–124, 1956.
22. **Van Arsdel, W. B.,** *Food Dehydration. Principles,* Vol. 1, Avi Publishing, Westport, Conn., 1963, 196 pp.
23. **Van Arsdel, W. B. and Copley, M. J.,** *Food Dehydration. Processes and Products,* Vol. 2, Avi Publishing, Westport, Conn., 1964, 732 pp.
24. **Van Arsdel, W. B., Copley, M. J., and Morgan, A. I., Jr., Eds.,** *Food Dehydration,* Avi Publishing, Westport, Conn., 1973, 344 pp; Vol. 2, 529 pp.
25. **von Loesecke, H.,** *Drying and Dehydration of Foods,* 2nd ed., Reinhold, New York, 1955.

4. Freeze-Drying

1. **Burke, R. F. and Decareau, R. V.,** Recent advances in the freeze-drying of food products, *Adv. Food Res.* 13, 1–88, 1964. (155 ref.)
2. **Copson, D. A.,** *Freeze Drying Electronic Ovens and Other Applications,* Avi Publishing, Westport, Conn., 1962.
3. **Corridon, G. A.,** Freeze-Drying of Foods. A List of Selected References, U.S. Dep. Agric., National Agriculture Library, List No. 77, 1964.
4. **Cotson, S. and Smith, D. B., Eds.,** *Freeze-drying of Foodstuffs,* Columbine Press, Manchester, 1963, 295 pp.
5. **Fisher, F. R., Ed.,** Freeze-drying of Foods, Conference sponsored jointly by QM Food and Container Institute and National Academy of Sciences, Chicago, Ill., Apr. 12–14, 1961, NAS, Washington, D.C., 1962, 237 pp.
6. **Flosdorf, E. W.,** *Freeze-drying,* Reinhold, New York, 1949.
7. **Ginnette, L. F. and Kaufman, V. F.,** Freeze-drying of foods, in *Freezing Preservation of Foods,* Vol. 4, Tressler, D. K., Van Arsdel, W. B., and Copley, M. J., Avi Publishing, Westport, Conn., 377–403.

8. **Goldblith, S. A.**, Freeze-dehydration of foods, in *Aspects Théoriques et Industriels de la Lyophilisation,* Rey, L., Ed., Hermann, Paris, 1964, 555–572.
9. **Greaves, R. I. N.**, Recent advances in freeze-drying, *J. Pharm. Pharmacol.*, 14, 621–640, 1962. (23 ref.)
10. **Hanson, W. S. F., Ed.**, The Accelerated Freeze-drying (AFD) Method of Food Preservation, HMSO, London, 1961, 169 pp.
11. **Harper, J. C. and Tappel**, A. L., Freeze-drying of food products, *Adv. Food Res.*, 7, 171–234, 1957.
12. **Howard, L. B. and Campbell, H.**, Dehydrofreezing – A new way to preserve foods, *Food Ind.*, 18, 674–675, 1946.
13. **Jackson, S., Rickter, S. L., and Chichester, C. O.**, Freeze drying of fruit, *Food Technol.*, 11, 468–470, 1957.
14. **Joslyn, M. A.**, Concentration by freezing, in *Fruit and Vegetable Juice Processing Technology,* Tressler, D. K. and Joslyn, M. A., Eds., Avi Publishing, Westport, Conn., 1961.
15. **Karel, M. and Goldblith, S. A.**, Processing aspects of freeze dehydration, in *Food Processing Operations,* Vol. 3, Joslyn, M. A. and Heid, M. L., Eds., Avi Publishing, Westport, Conn., 1964, 369–393.
16. **Karel, M.**, Physical and chemical considerations in freeze-dehydrated foods, in *Exploration in Future Food-Processing Techniques,* Goldblith, S. A., Ed., M.I.T. Press, Cambridge, Mass., 1963, 54.
17. **King, C. J.**, Freeze-drying of foodstuffs, *CRC Crit. Rev. Food Technol.*, 1, 379–451, 1970. (232 ref.)
18. **King, C. J.**, *Freeze-drying of Foods,* Chemical Rubber Co., Cleveland, Ohio, 1971, 86 pp.
19. **Lazar, M. E.**, Dehydrofreezing of fruits and vegetables, in *Freezing Preservation of Foods,* Vol. 4, Tressler, D. K., Van Arsdel, W. B. and Copley, M. J., Eds., Avi Publishing, Westport, Conn., 1968, 347–376.
20. **Nair, J. H.**, Freeze-dry systems changing, *Food Eng.*, 34(6), 41–43, 1962.
21. **Noyes, R.**, Freeze Drying of Foods and Biologics, Food Processing Rev. No. 1, Noyes Development Corp., Park Ridge, N.J., 1968.
22. **Peterson, E.**, Industrial accelerated lyophilization of foods, *Rev. Frio,* 6(2), 53–60, 1961.
23. **Rey, L., Ed.**, *Advances in Freeze-Drying,* Hermann, Paris, 1966.
24. **Silverman, G. J., and Goldblith, S. A.**, The microbiology of freeze-dried foods, *Adv. Appl. Microbiol.*, 7, 305–334, 1965. (110 ref.)
25. Symposium. Preparation and preservation of foodstuffs by freezing and freeze-drying, in *Aspects Théoriques et Industriels de la Lyophilisation,* Rey, L., Ed., Hermann, Paris, 1964, 495–601. (7 papers)

5. Heat Preservation

1. **Ball, C. O.**, Thermal Process Time for Canned Foods, Bull. National Research Council 7, 1923.
2. **Blakebrough, N.**, Preservation of biological materials especially by heat treatment, in *Biochemical and Biological Engineering Science,* Vol. 2, Academic Press, London, 1968, 29–63.
3. **Goldblith, S. A.**, Thermal processing of foods. A review, *World Rev. Nutr. Diet.* 13, 165–193, 1971. (27 ref.)
4. **Goldblith, S. A., Joslyn, M. A., and Nickerson, J. T.**, *An Introduction to the Thermal Processing of Foods,* Avi Publishing, Westport, Conn., 1961, 1128 pp.
5. **Hall, C. W. and Trout, G. M.**, *Milk Pasteurization,* Avi Publishing, Westport, Conn., 1968, 234 pp.
6. **Joslyn, M. A.**, Steam in food processing, in *Food Processing Operations,* Joslyn, M. A. and Heid, J. L., Eds., Avi Publishing, Westport, Conn., 1963, 340–361.
7. **Merrill, D. G.**, Heating rates of foods in glass and other containers, *Ind. Eng. Chem.*, 40, 2263–2269, 1948.
8. National Canners Association, Processes for Low Acid Canned Foods in Glass Containers, National Canners Association Bull. 30-L, 1955.
9. National Canners Association, Processes for Low Acid Canned Foods in Metal Containers, 9th ed., National Canners Association Bull. 26-L, 1962.
10. **Peeples, M. L.**, Forced convection heat transfer characteristics of fluid milk products. A review, *J. Dairy Sci.*, 45, 293–302, 1962.
11. **Pflug, I. J., Bock, J. H. and Long, F. E.**, Sterilization of food in flexible packages, *Food Technol.*, 17, 1167–1172, 1963.
12. **Pflug, I. J. and Esselen, W. B.**, Food processing by heat sterilization, in *Food Processing Operations,* Vol. 2, Joslyn, M. A. and Heid, J. L., Eds., Avi Publishing, Westport, Conn., 1963, 410–479.
13. **Wilkinson, G. R. and Baker, L. C.**, Modern trends in steam sterilization, *Prog. Ind. Microbiol.*, 5, 231–282, 1964. (125 ref.)

6. Food Irradiation

1. **Bailey, P. J. and Baines, B. D.**, Irradiation: Present and future applications, *Nucl. Eng.*, 9, 213–216, 1964.
2. **Bellamy, W. D.**, Preservation of foods and drugs by ionizing radiations, *Adv. Appl. Microbiol.*, 1, 49–73, 1959.
3. **Bloch, J. M.**, Use of radiations in sterilizing food products and pharmaceutical preparations, *Bull. Soc. Lorraine Sci.*, 2, 44–61, 1962. (15 ref.)
4. **Bridges, B. A.**, Microbiological aspects of radiation sterilization, *Prog. Ind. Microbiol.*, 5, 283–326, 1964.

5. **Copson, D. A.,** *Microwave Heating,* Avi Publishing, Westport, Conn., 1962, 350 pp.
6. **Desrosier, N. W. and Rosenstock, H. M.,** *Radiation Technology in Food, Agriculture, and Biology,* Avi Publishing, Westport, Conn., 1960, 418 pp.
7. **Fowler, E. E., Shea, K. G., and Dietz, G. R.,** United States Atomic Energy Commission Radiation Processing of Foods Program, International Symposium on Food Irradiation, Karlsruhe, Germany, 1966.
8. **Ginzburg, A. S.,** *Application of Infra-Red Radiation in Food Processing,* Chemical Rubber Co., Cleveland, Ohio, 1969, 412 pp.
9. **Goldblith, S. A.,** Radiation processing of foods and drugs, in *Food Processing Operations,* Joslyn, M. A. and Heid, J. L., Eds., Avi Publishing, Westport, Conn., 1963, 401–446.
10. **Hannan, R. S.,** *Research on the Science and Technology of Food Preservation by Ionizing Radiation,* Chemical Publishing, New York, 1956.
11. International Atomic Energy Agency, Food Irradiation, Proc. International Symposium, IAEA, Vienna, 1966.
12. International Atomic Energy Agency, Food Irradiation, Proc. Joint IAEA/FAO/BARC Seminar, Bhabha Atomic Research Centre, Bombay, 1969.
13. International Atomic Energy Agency, Preservation of Fruit and Vegetables by Radiation, IAEA, Vienna, 1968, 152 pp.
14. International Atomic Energy Agency, Preservation of Fish by Irradiation, IAEA, Vienna, 1970, 163 pp.
15. International Atomic Energy Agency, Radiation Preservation of Food, Proc. Symposium, IAEA, Vienna, 1973, 774 pp.
16. **Jeppson, M. R.,** Consider microwaves, *Food Eng.,* 36(11), 49–52, 1964.
17. **Josephson, E. S. and Frankfort, J. H.,** *Radiation Preservation of Foods, Advances in Chemistry,* Ser. No. 65, American Chemical Society, Washington, D. C., 1967, 184 pp. (15 papers)
18. **Kraybill, H. F. and Brunton, D. C.,** Commercialization technology and economics in radiation processing, *J. Agric. Food Chem.,* 8, 349–356, 1960.
19. **Ley, F. J.,** Recent developments in food irradiation, *Atom,* 130, 224–227, 1967.
20. **Ley, F. J. and Rogers, F.,** Processing with ionizing radiations, *Biochemical and Biological Engineering Science,* Vol. 2, Blakebrough, N., Ed., Academic Press, New York, 1968, 321–364.
21. Low Temperature Research Station, Preservation of food with ionizing radiations. A symposium, *Chem. Ind.,* March 3, 1954, 390–395.
22. **Massey, L. M., Jr.,** Food irradiation, *Agric. Sci. Rev.,* 5(2), 29–36, 1967. (23 ref.)
23. **Niven, C. F., Jr.,** Microbiological aspects of radiation preservation of food, *Annu. Rev. Microbiol.* 12, 507–524, 1958.
24. **Porretta, A.,** Sterilization of foods by ionizing radiation, *Ind. Conserve* (Parma), 36, 207–210, 1961. (15 ref.)
25. Radiation Preservation of Foods, Proc. International Conf., Boston, Mass., Sept. 27–30, 1964, National Academy of Sciences Publ. 1273, Washington, D. C., 1965, 424 pp.
26. **Samoila, A.,** Use of ionizing radiation to preserve food products. General considerations, *Ind. Alimen. Produse Animale* 10, 67–71, 1962.
27. **Schultz, H. W. and Lee, J. S.,** Food preservation by irradiation. Present status. *Food. Technol.,* 20 136–141, 1966.
28. **Sreenivasan, A.,** Food irradiation: Progress, problems and prospects, in *Proc. SOS/70. 3rd International Congress Food Science and Technology,* Institute of Food Technologists, Chicago, 1971, pp. 595–603.
29. **Urbain, W. M.,** Progress and problems in radiation processing of food, *J. Agric. Food Chem.,* 8, 340–342, 1960.
30. **Urrows, G. M.,** Food Preservation by Irradiation, U. S. Atomic Energy Commission, Washington, D. C., 1964.
31. U. S. Army Quartermaster Corps, Radiation Preservation of Food, Res. Dev. Ser. No. 1, U. S. Gov. Print. Off., Washington, D. C., 1957, 461 pp.
32. **Young, F. R., Bamell, H. C., Barnes, J. M., Boyland, E., and Frazer, A. C.,** Report of the Working Party on Irradiation of Food, HMSO, London, 1964.

7. Fermentation. Brining. Smoking

a. Fermentation

1. **Fabian, F. W.,** Food preservation by use of microorganisms, in *The Chemistry and Technology of Food and Food Products,* Jacobs, M. B., Ed., Interscience, New York, 1951, Vol. 3, 1888–1935.
2. **Pederson, C. S.,** Processing by fermentation, in *Food Processing Operations,* Vol. 2, Joslyn, M. A. and Heid, J. L., Eds., Avi Publishing, Westport, Conn., 1963, 480–512.
3. **Underkofler, L. A. and Hickey, R. J., Eds.,** *Industrial Fermentations,* Chemical Publishing, New York, 1954, 565 pp.
4. **White, J.,** *Yeast Technology,* John Wiley & Sons, New York, 1954.

b. Brining

1. **Binsted, R., Devey, J. D., and Dakin, J. C.,** *Pickle and Sauce Making,* 3rd ed., Avi Publishing, Westport, Conn., 1971.
2. **Harris, F.,** *Pickling and Preserving,* Abeland Press, New York, 1951.

3. **Joslyn, M. A. and Timmons, A.,** Salt – use in food processing, in *Food Processing Operations,* Vol. 3, Joslyn, M. A. and Heid, J. L., Eds., Avi Publishing, Westport, Conn. 1964, 50–77.
4. **Pederson, C. S.,** Sauerkraut, *Adv. Food Res.,* 10, 223–291, 1960.

c. Smoking

1. **Brissey, G. E. and Goeser, P. A.,** Aging, curing and smoking of meats, in *Food Processing Operations,* Vol. 1, Joslyn, M. A. and Heid, J. L., Eds., Avi Publishing, Westport, Conn., 1963, 600–634.
2. Meat Curing and Sausage Making, Morton Salt Co., Chicago, Ill., 1946.
3. **Tilgner, I. D. J.,** New knowledge with regard to smoke-curing processes, *Die Fleischwirtschaft,* 10, 649–660, 751–762, 1958.

8. Chemical Preservation

1. **Bauernfeind, J. C. and Pinkert, D. M.,** Food processing with added ascorbic acid, *Adv. Food Res.,* 18, 219–315, 1970. (406 ref.)
2. **Chichester, D. F. and Tanner, F. W., Jr.,** Antimicrobial food additives, in *Handbook of Food Additives,* Furia, T. E., Ed., Chemical Rubber Co., Cleveland, Ohio, 1968, 137–207.
3. **Ienistea, C.,** Use of antibiotics for the preservation of foods, *Igiena* (Bucharest), 10, 57–62, 1961. (28 ref.)
4. **Ingram, M.,** Technical aspects of the commercial use of antimicrobial chemicals as food preservatives, *Chem. Ind.* (Lond.) 552–557, 1959.
5. **Ingram, M., Ottaway, F. J. H., and Coppock, J. B. M.,** The preservative action of acid substances in food, *Chem. Ind.* (Lond.) 1154–1163, 1956.
6. **Jacobs, M. B.,** Chemical preservatives, in *The Chemistry and Technology of Food and Food Products,* Vol. 3, Interscience, New York, 1951, 1936–1991.
7. **Nickerson, J. T. R.,** Preservatives and antioxidants, in *Food Processing Operations,* Vol. 2, Joslyn, M. A. and Heid, J. L., Eds., Avi Publishing, Westport, Conn., 1963, 218–247.
8. **Piragibe, L.,** Sorbic acid as a preservative. A review, *Arq. Bromatol.* (Rio de J.), 7, 109–120, 1950. (32 ref.)
9. **Smith, W. L., Jr.,** Chemical treatment to reduce postharvest spoilage of fruits and vegetables, *Bot. Rev.,* 28, 411–445, 1962. (175 ref.)
10. **Swisher, H. E. and Swisher, L. H.,** Use of acids in food processing, in *Food Processing Operations,* Vol. 2, Joslyn, M. A. and Heid, J. L., Eds., Avi Publishing, Westport, Conn., 1963, 139–173.
11. **Vaughn, R. H. and Stewart, G. F.,** Antibiotics as food preservatives, *JAMA,* 174, 1308–1310, 1960.

9. Food Packaging

a. General

1. **Ball, C. O.,** Food packaging, in *Handbook of Food and Agriculture,* Blanck, F. C., Ed., Reinhold, New York, 1955, 691–732.
2. **Brody, A. L.,** Flexible packaging of foods, *CRC Crit. Rev. Food Technol.,* 1, 71–155, 1968. (109 ref.)
3. **Brody, A. L.,** *Flexible Packaging of Foods,* Chemical Rubber Co., Cleveland, Ohio, 1971, 103 pp.
4. **Browne, C. W.,** Packaging, in *The Chemistry and Technology of Food and Food Products,* Vol. 3, Jacobs, M. B., Ed., Interscience, New York, 1951, 1092–2017.
5. **Cavett, J. J.,** The effects of new forms of packaging on the microbiology and storage life of meats, poultry, and fish, *Prog. Ind. Microbiol.,* 7, 77–123, 1968. (164 ref.)
6. **Sacharow, S. and Griffin, R. C., Jr.,** *Food Packaging. A Guide for the Supplier, Processor and Distributor,* Avi Publishing, Westport, Conn., 1970, 412 pp.

b. Canning

1. A Complete Course in Canning, 7th ed., Canning Trade, Baltimore, MD, 1946.
2. American Can Co., *Sterilization of Canned Foods. Operation Procedures. Practical Aspects of New Methods,* American Can Co., Maywood, Ill., 1950, 42 pp.
3. American Can Co., *Sterilization of Canned Foods. Theoretical Considerations in the Sterilization of Canned Foods,* American Can Co., Maywood, Ill., 1950, 244 pp.
4. American Can Co., *The Canned Food Reference Manual,* 3rd ed., American Can Co., New York, 1949, 638 pp.
5. **Ball, C. O.,** Advancement in sterilization methods for canned foods, *Food Res.,* 3, 13–52, 1938.
6. **Baumgartner, J. G.,** *Canned Foods, An Introduction to Their Microbiology,* 3rd ed., Churchill, London, 1949, 278 pp.
7. **Bitting, A. W.,** *Appertizing or the Art of Canning. Its History and Development,* Trade Pressroom, San Francisco, Calif., 1937, 852 pp.
8. **Brody, A. L.,** Food canning in rigid and flexible packages, *CRC Crit. Rev. Food Technol.,* 2, 187–243, 1971. (143 ref.)
9. **Coonen, N. H. and Mason, S. I.,** Recent advances in rigid metal containers, in *Proc. SOS/70. 3rd International Congress Food Science and Technology,* Institute of Food Technologists, Chicago, 1971, 589–594.

10. **Herbert, D. A.**, Impact of plastics on food packaging and preservation, *Proc. SOS/70. 3rd International Congress Food Science and Technology,* Institute of Food Technologists, Chicago, 1971, 585–588.
11. **Hersom, A. C. and Hulland, E. D.**, *Canned Foods. An Introduction to their Microbiology,* Chemical Publishing Co., New York, 1964, 291 pp.
12. **Jones, O. and Jones, T. W.**, *Canning Practice and Control,* Chapman & Hall, London, 1937.
13. **Kohman, E. F.**, Vitamins in Canned Food, Bull. 19-L, National Canners Association, Washington, D.C., 1937.
14. **Lang, O. W.**, *Thermal Processes for Canned Marine Products,* University of California Press, Berkeley, 1935.
15. **Lock, A.**, *Practical Canning,* 3rd ed., Avi Publishing, Westport, Conn., 1969, 428 pp.
16. National Canners Association Staff, *Laboratory Manual for Food Canners and Processors,* 3rd ed., Avi Publishing, Westport, Conn., 1968. (2 vols.)
17. **Rogers, J. L.**, *A Course in Canning,* 4th ed., Food Trade Press, London, 1966, 112 pp.
18. The Safety of Canned Foods, Royal Society of Health, London, 1966. 163 pp.
19. **Williams, O. B.**, Preservation of food in hermetically sealed containers, in *The Chemistry and Technology of Food and Food Products,* Jacobs, M. B., Ed., Vol. 3, 2nd ed., Interscience, New York, 1951, 1858–1887.

10. Food Storage

1. **Anderson, J. A. and Alcock, A. W., Eds.**, *Storage of Cereal Grains and their Products,* American Association of Cereal Chemists, St. Paul, Minn., 1954, 515 pp.
2. **Ashby, B. H.**, Protecting Perishable Foods during Transport by Motortruck, Agric. Handb. No. 105, USDA, ARS, U.S. Gov. Print. Off., Washington, D.C., 1970, 141 pp.
3. Food and Agriculture Organization, Preservation of Grains in Storage, FAO Agric. Stud. No. 2, FAO, Rome, 1952, 174 pp.
4. Food and Agriculture Organization, Bigger Crops – and Better Storage. The Role of Storage in World Food Supplies, World Food Problems No. 9, FAO, Rome, 1969, 49 pp.
5. **Genel, M. R.**, Successful methods for reducing losses during storage in tropical environments, in *Some Issues Emerging from Recent Breakthroughs in Food Production,* Turk, K. L., Ed., N.Y. State College of Agriculture, Cornell University, Ithaca, 1971, 319–325.
6. **Hall, D. W.**, Handling and Storage of Food Grains in Tropical and Subtropical Areas, FAO Agric. Dev. Papers No. 90, FAO, Rome, 1970, 359 pp.
7. **Hearne, J. F.**, Long-term storage of foods, *Food Technol.,* 18(3), 60–65, 1964.
8. **Kirk, L. E., Ling, L., and Oxley, T. A.**, Storing and Drying Grain in Canada, in the United States, in the United Kingdom, FAO Agric. Stud. No. 6, FAO, Washington, D.C., 1948, 41 pp.
9. **Majumder, S. K.**, Protecting food from deterioration during storage, handling, and distribution in technologically less developed countries of the world, *Proc. SOS/70. 3rd International Congress Food Science and Technology,* Institute of Food Technologists, Chicago, 1971, 518–531. (30 ref.)
10. **Oxley, T. A.**, *The Scientific Principles of Grain Storage,* Vorthem Publishing, Liverpool, 1948, 103 pp.
11. **Pentzer, W. T.**, Storage of agricultural raw products, in *Handbook of Food and Agriculture,* Blanck, F. C., Ed., Reinhold, New York, 1955, 305–329.
12. **Ryall, A. L.**, Protecting the quality of fruits and vegetables after harvest, in *Food Quality,* Irving, G. W., Jr. and Hoover, S. R., Eds., American Association for the Advancement of Science, Washington, D.C., 1965, 47–56.
13. **Thieme, J. G.**, Handling and storage of agricultural products, in Summary Proc. on Agriculture of the United Nations Conference on the Application of Science and Technology for the Benefit of the Less Developed Areas, FAO, Washington, D.C., 1963, 281–287.

VIII. UNCONVENTIONAL SOURCES OF FOOD

A. General

1. **Irving, G. W.,** Food from nontraditional sources – a challange to agricultural science, Agric. Board, Agric. Res. Serv., National Academy of Sciences, Washington, D. C., 1969.
2. **Johnson, W.,** Current innovation in world food production, *J. Law Econ. Dev.*, (1), 190–207, 1968.
3. **Pirie, N. W.,** Future sources of food supply: Scientific problems, *J. R. Stat. Soc.*, 125 A, 399–417, 1962.
4. **Pirie, N. W.,** Orthodox and unorthodox methods of meeting world food needs, *Sci. Am.*, 216, 27–34, 1967.
5. **Pirie, N. W.,** *Food Resources, Conventional and Novel,* Penguin Books, Baltimore, Md., 1969, 208 pp.
6. **Pirie, N. W.,** Production and use of unconventional sources of food, in *Man, Food, and Nutrition: Strategies and Technological Measures for Alleviating the World Food Problem,* Rechcigl, M., Jr., Ed., CRC Press, Cleveland, Ohio, 1973, 189–202.
7. Symposium. Unusual foods for human consumption, *Proc. Nutr. Soc.*, 12, 32–82, 1953.

B. New Plant and Animal Sources of Food

1. Plant Sources

a. Aquatic Plants

1. **Boyd, C. E.,** Fresh-water plants: A potential source of protein, *Econ. Bot.*, 22, 359–368, 1968.
2. **Boyd, C. E.,** A bibliography of interest in the utilization of vascular aquatic plants, *Econ. Bot.*, 26, 74–84, 1972. (233 ref.)
3. **Chapman, V. J.,** *Seaweeds and Their Uses,* Pitman, New York, 1952, 240 pp.
4. **Little, E. C.,** Handbook of Utilization of Aquatic Plants, FAO, Rome, 1968.

2. Animal Sources

a. Insects

1. **Bodenheimer, F. S.,** *Insects as Human Food,* Junk, The Hague, 1951, 352 pp.
2. **Meyer-Rochow, V. B.,** Edible insects in three different ethnic groups of Papua and New Guinea, *Am. J. Clin. Nutr.*, 26, 673–677, 1973.
3. **Meyer-Rochow, V. B.,** Insekten als Nahrungs-mittel, *Selecta,* in press.

b. Rodents

1. **den Hartog, A. P. and de Vos, A.,** The use of rodents as food in tropical Africa, *FAO Nutr. Newsletter,* 11(2), 1–14, 1973.

c. Fish

1. **Reader, D. E.,** The Utilization of Sharks, HMSO, London, 1952, 284 pp.

C. Novel Protein Sources

1. **Altschul, A. M.,** Using unusual protein supplies, IAD Newsletter, International Agriculture Development Service, U.S.Dep.Agric., February 1968.
2. **Bender, A. E., Kihlberg, R., Löfqvist, B., and Bar Munck, L., Eds.,** *Evaluation of Novel Protein Products,* Pergamon Press, Oxford, 1970, 390 pp.
3. **Bieber, H., Ed.,** *Engineering of Unconventional Protein Production,* Symp. Ser. 93, Vol. 65, American Institute of Chemical Engineers, New York, 1969, 127 pp.
4. International Symposium on Novel Protein Sources for Human Nutrition, Amsterdam, 1968, Stichting tot Wetenschappelijke Voorlickting of Voedingsgebied, Haag, 1968.
5. **Mauron, J.,** Technology of protein synthesis and protein-rich foods, in *Nutrition and Technology of Foods for Growing Humans,* Somogyi, J. C., Ed., *Bibl. Nutr. Dieta* No. 18, S. Karger, Basel, 1973, 24–44.
6. **Milner, M.,** Status of development and use of some unconventional proteins, in *Protein-Enriched Cereal Foods for World Needs,* Milner, M., Ed., American Association of Cereal Chemists, St., Paul, Minn., 1969, 97–104.
7. **Parisher, E. R.,** Proteins of aquatic origin as human food, *Food Technol.*, 25, 1162–1174, 1971.
8. **Parisher, E. R.,** Proteins of aquatic origin as foods for human consumption, *Adv. Food Res.*, 20, 189–216, 1973.
9. **Pirie, N. W.,** The merits of food proteins from novel sources, *Sci. Prog.*, 54, 401–412, 1966.
10. **Pirie, N. W.,** Complementary ways of meeting the world's protein needs in *Proteins as Human Food,* Lawrie, R. A., Ed., Avi Publishing, Westport, Conn., 1970, 46–61.
11. **Spensley, P. C., Halliday, D., and Orr, E.,** The Prospects for "Non-Conventional" Protein Resources, Tropical Products Institute, London, 1970.
12. **Spensley, P. C., Halliday, D., and Orr, E.,** The prospects for non-conventional protein resources, *Trop. Sci.*, 14, 203–233, 1972. (52 ref.)
13. Symposium. New sources of food protein, *Proc. Nutr. Soc.*, 28, 76–109, 1969. (6 papers)
14. Symposium. New sources of protein, in *Proc. SOS/70, 3rd International Congress Food Science and Technology,* Institute of Food Technologists, Washington, D. C., 1971, 234–309. (12 papers)

15. **Worgan, J. T.**, World supplies of proteins from unconventional sources, in *Proteins in Human Nutrition*, Porter, J. W. G. and Rolls, B. A., Eds., Academic Press, London, 1973, 47–74.

D. Novel Animal Proteins

1. Livestock Proteins

1. **Anderson, P. E. and Jul, M.**, Animal husbandry proteins, in *Evaluation of Novel Protein Products*, Bender, A. E., Kihlberg, R., Löfqvist, B., and Munck, L., Eds., Pergamon Press, Oxford, 1970, 77–83.

2. Fish Proteins

1. **Borgström, F.**, Fish protein, in *Evaluation of Novel Protein Products*, Bender, A. E., Kihlberg, R., Löfqvist, B. and Munck, L., Eds., Pergamon Press, Oxford, 1970, 55–76.
2. Conference on Fish Protein Concentrate, Ottawa, Oct. 24–25, 1967.
3. **Finch, R.**, Fish protein for human foods, *CRC Crit. Rev. Food Technol.*, 1, 519–580, 1970. (288 ref.)
4. Fish Protein Concentrate. A Comprehensive Bibliography, U.S. Library of Congress, Washington, D.C., 1970, 77 pp.
5. Food and Agriculture Organization, Report of an Expert Panel on Fish Meal and Fish Flour for Human Consumption, Washington, D.C., Sept. 28 and 29, 1961, FAO, Rome, 1962.
6. **Hadjimarkos, D. M.**, Fish flour and fluoride, *Science*, 152, 1567–1568, 1966.
7. **Hallgren, B., Sjöberg, L-B., and Stelleman, J.**, New uses for fish proteins, in *Proteins in Human Nutrition*, Porter, J. W. G. and Rolls, B. A., Eds., Academic Press, London, 1973, 369–381.
8. **Halliday, D. and Disney, J. G.**, Fish Protein Concentrate: a Review, Tropical Products Institute, Publ. G58, London, 1971, 18 pp.
9. **Liston, J. and Pigott, G. M.**, Fish protein concentrate, in Proc. SOS/70. 3rd International Congress Food Science and Technology, Institute of Food Technologists, Washington, D.C., 1971, 285–289. (24 ref.)
10. **Lovern, J. A.**, General outlook for edible fish protein concentrates, *Advances in Chemistry*, Ser. No. 57, 37–51, 1966. (25 ref.)
11. **Lovern, J. A.**, Problems in the development of fish protein concentrates, *Proc. Nutr. Soc.* 28, 81–85, 1969.
12. **Meade, T. L.**, Recent advances in the fish by-products industry, *CRC Crit. Rev. Food Technol.*, 2, 1–19, 1971. (37 ref.)
13. **Olden, J. H.**, Fish flour for human consumption, *Commer. Fish. Rev.*, 22(1), 12–18, 1960.
14. Protein Advisory Group of the United Nations System, The Potential of Fish Protein Concentrate for Developing Countries, PAG Statement No. 16, August 19, 1971.
15. **Randall, J.**, FPC – New hope for undernourished, *Bioscience*, 17, 257–258, 1967.
16. **Stillings, B. R. and Knobl, G. M., Jr.**, Fish protein concentrate: A new source of dietary protein, *J. Am. Oil Chem. Soc.*, 48, 412–414, 1971. (7 ref.)
17. **Tannenbaum, S., Stillings, B. R., and Scrimshaw, N. S., Eds.**, *Economics, Marketing and Technology of Fish Protein Concentrate*, M.I.T. Press, Cambridge, Mass., 1974, 500 pp.
18. U.S. Department of Interior, Fish-Protein Concentrate. Lifeline for the Future, Fish and Wildlife Bureau of Commercial Fisheries, Sept. 1962.
19. U.S. Department of Interior, Marine Protein Concentrate, U.S. Bureau Commercial Fish Leaflet 584, April 1966, 27 pp.

E. Novel Plant Proteins

1. General

1. **Altschul, A., Ed.**, *Processed Plant Protein Foodstuffs*, Academic Press, New York, 1958, 955 pp. (33 papers)
2. **Bird, K. M.**, Plant proteins: Progress and problems, *Food Technol.*, 28(3), 31–39, 1974.
3. **Dean, R. F. A.**, Plant Proteins in Child Feeding, Special Rep. Ser. No. 279, Medical Research Council, London, 1953.
4. **Dean, R. F. A.**, Use of processed plant proteins as human food, in *Processed Plant Protein Foodstuffs*, Altschul, A. M., Ed., Academic Press, New York, 1958, 205–247. (99 ref.)
5. **Narayana Rao, M. and Swaminathan, M.**, Plant proteins in the amelioration of protein deficiency states, *World Rev. Nutr. Diet.*, 11, 106–141, 1969. (135 ref.)
6. **Parpia, H. A. B.**, Novel routes to plant protein, *Sci. J.*, 4(5), 66–71, 1968.
7. **Parpia, H. A. B. and Subramanian, N.**, Plant proteins in India, in *World Protein Resources, Advances in Chemistry*, Ser. No. 57, American Chemical Society, Washington, D.C., 1966, 112–132. (16 ref.)
8. **Pirie, N. W.**, Progress in biochemical engineering broadens our choice of crop plants, *Econ. Bot.*, 15, 302–310, 1961.
9. **Pirie, N. W.**, Use of plant protein concentrates as human food, *Chem. Ind.*, June 29, 1968, 864–866.
10. Symposium. Integrated research in economic botany. VII. Protein for food, *Econ. Bot.*, 22, 3–79, 1968. (10 papers)
11. **Teply, L. J. and Gyorgy, P.**, Vegetable proteins in infant feeding, *J. Pediatr.*, 61, 925–933, 1962.
12. Textured vegetable proteins, *PAG Bull.*, 2(1), 22–26, 1970.

2. Seed Proteins

1. **Altschul, A. M.,** Seed protein and world food problems, *Econ. Bot.*, 16, 2–13, 1962. (46 ref.)
2. **Altschul, A. M.,** Seed proteins, in *Symposium on Foods: Proteins and Their Reactions*, Schultz, H. W. and Anglemier, A. F., Eds., Avi Publishing, Westport, Conn., 1964, 295–313.
3. **Altschul, A. M.,** Research accomplishments with seed proteins for human food, *Proc. 14th Annu. Meeting of the Agricultural Research Institute, October 18–19, 1965*, National Academy of Sciences, Washington, D.C., 1965, 41–50.
4. **Altschul, A. M.,** Edible seed protein concentrates: Their role in control of malnutrition, *Isr. J. Med. Sci.*, 1(3), 471–479, 1965.
5. **Altschul, A. M.,** Food proteins: New sources from seeds, *Science*, 158, 221–226, 1967.
6. **Altschul, A. M., Talluto, K. F., and Sharar, B. A., Eds.,** Proc. Seed Protein Conference, New Orleans, January 21–23, 1963, USDA, ARS, Southern Utilization Research and Development Div., New Orleans, 1963, 292 pp.
7. **Altschul, A. M., Yatsu, L. Y., Ory, R. L., and Engleman, E. M.,** Seed proteins, *Annu. Rev. Plant Physiol.*, 17, 113–136, 1966.
8. **Dechary, J. M. and Altschul, A. M.,** Major seed proteins and the concept of aleurins, in *World Protein Resources*, in *Advances in Chemistry*, Ser. No. 57, American Chemical Society, Washington, D.C., 1966, 148–158.
9. **Inglett, G. E., Ed.,** *Symposium: Seed Proteins*, Avi Publishing, Westport, Conn., 1972, 320 pp. (23 papers)
10. **Milner, M.,** General outlook for seed protein concentrates, in *World Protein Resources*, in *Advances in Chemistry*, Ser. No. 57, American Chemical Society, Washington, D.C., 1966, 52–64. (8 ref.)
11. **Van Etten, C. H., Kwolek, W. F., Peters, J. E., and Barclay, A. S.,** Plant seeds as protein sources for food or feed. Evaluation based on amino acid composition of 379 species, *J. Agric. Food Chem.*, 15, 1077–1089, 1967.

3. Oilseed Proteins

a. General

1. **Altschul, A. M.,** Present status of proteins from oilseeds, in *Progress in Meeting Protein Needs of Infants and Pre-School Children*, National Academy of Sciences, Washington, D.C., 1961, 517–530.
2. **Altschul, A. M.,** Oilseed proteins as related to the world food problem, in *Evaluation of Novel Protein Products*, Bender, A. E., Kihlberg, R., Löfqvist, B., and Munck, L., Eds., Pergamon Press, Oxford, 1970, 41–53.
3. **Johnson, D. W.,** Oilseed proteins. Properties and applications, *Food Prod. Dev.*, 3(8), 78, 80, 84, 87, 1970.
4. **Johnson, D. W.,** Functional properties of oilseed proteins, *J. Am. Oil Chem. Soc.*, 47, 402–407, 1970.
5. **Jones, D. B.,** Nutritive value of soya bean and peanut proteins, *Fed. Proc.*, 3, 116–120, 1944.
6. **Krishnaswamy, P. R.,** Limitations and scope of oilseed proteins, in *Proc. SOS/70. 3rd International Congress Food Science and Technology*, Institute of Food Technologists, Washington, D.C., 1971, 280–284.
7. **Orr, E. and Adair, D.,** *The Production of Protein Foods and Concentrates from Oilseeds*, Tropical Products Institute Report 6–31, Lond., June, 1967, 156 pp.
8. **Rice, E. E.,** Nutritive value of oilseed proteins, *J. Am. Oil Chem. Soc.*, 48, 408–413, 1971. (16 ref.)
9. **Rosen, G. D.,** Factors governing the use of soya-bean and other oilseed proteins, in *Proteins in Human Nutrition*, Porter, J. W. G. and Rolls, B. A., Eds., Academic Press, London, 1973, 383–395.
10. **Smith, R. L.,** Lipid-protein isolates, in *World Protein Resources*, in *Advances in Chemistry*, Ser. No. 57, American Chemical Society, Washington, D.C., 1966, 133–147. (17 ref.)
11. Symposium: Oilseed processors challenged by world protein needs, *J. Am. Chem. Soc.*, 48, 400–419, 1971. (5 papers)
12. Symposium: Oilseed processors challenged by world protein needs, *J. Am. Chem. Chem. Soc.*, 48, 473–494, 1971. (6 papers)
13. Symposium: Oilseed processors challenged by world protein needs, *J. Am. Chem. Soc.*, 48, 599–628, 1971. (6 papers)
14. Symposium: Cruciferous oilseeds, *J. Am. Oil Chem. Soc.*,, 48, 718–739, 1971.
15. **Vix, H. L. E. and Decossas, K. M.,** Processing oilseeds into edible products: Solvent extraction, *Chem. Eng. Prog. Symp. Ser.*, 65(93), 49–56, 1969.
16. **Wilcke, H. L., Martinez, W. H., and Calvert, F. E.,** Oilseed meals and flours for food use, *Soybean Dig.*, 28(8), 18, 20, 1969.
17. **Woodham, A. A.,** Oilseed protein: Present and future, *World Rev. Nutr. Diet.*, 10, 44–76, 1969. (143 ref.)

b. Coconut

1. CFTRI, Report on use of coconut preparations as a protein supplement in child feeding: Prospects, PAG–WHO/UNICEF Meeting, Geneva, August 1963.
2. **Curtin, L. V.,** Coconut oil meal, in *Processed Plant Protein Foodstuffs*, Altschul, A. M., Ed., Academic Press, New York, 1958, 645–676. (104 ref.)
3. **Strength, D. R.,** Preparation, characterization, and evaluation of coconut protein, in *Proc. SOS/70. 3rd International Congress Food Science and Technology*, Institute of Food Technologists, Chicago, 1971, 272–279.

c. Cottonseed

1. **Altschul, A. M., Lyman, C. M., and Thurber, F. H.,** Cottonseed meal, in *Processed Plant Protein Foodstuffs,* Altschul, A. M., Ed., Academic Press, New York, 1958, 469–534. (207 ref.)
2. **Harper, G. A.,** Research to improve cottonseed meal, *Oil Mill Gaz.*, 69, 10–14, 1964.
3. **Harper, G. A. and Smith, K. J.,** Status of cottonseed protein, *Econ. Bot.*, 22, 63–72, 1968. (39 ref.)
4. **Krishnamoorthi, V.,** Cottonseed protein from degosypolized and glandless seed, *Food Technol.*, 19, 1085–1090, 1965.
5. **Martinez, W. H., Berardi, L. C., and Goldblatt, L. A.,** Potential of cottonseed: products, composition, and use, in *Proc. SOS/70. 3rd International Congress Food Science and Technology,* Washington, D.C., Institute of Food Technologists, Chicago, 1971, 248–261.
6. **Milner, M.,** Cottonseed proteins: Problems and opportunities for world protein deficits, *Food Technol.*, 18(10), 1539–1540, 1964.
7. **Milner, M.,** Cottonseed protein concentrates, *PAG Bull.*, 5, 68–73, 1965.
8. **Phelps, R. A.,** Cottonseed meal for poultry: From research to practical application, *World's Poult. Sci. J.*, 22, 86–112, 1966.
9. **Phelps, R. A.,** Cottonseed meal for nonruminants, Proc. Semi-Annual Mtg. AFMA Nutr. Council, 1962, 10–14.
10. **Phelps, R. A.,** Increasing the amount and performance of cottonseed concentrates in nonruminants rations, Proc. Conf. Inactivation of Gossypol with Mineral Salts, National Cottonseed Products Association, Memphis, Tenn., 1966, 5–10.
11. Proc. 3rd Conf. Processing as Related to Nutritive Value of Cottonseed Meal, 1953, U.S. Dep. Agric., Southern Regional Research Laboratory, New Orleans, La., 1953.
12. Proc. 4th Conf. Processing as Related to Nutritive Value of Cottonseed Meal, 1957, U.S. Dep. Agric., Southern Regional Research Laboratory, New Orleans, La. 1957.
13. Proc. Conf. on Cottonseed Protein Concentrates, New Orleans, La., Jan. 15–17, 1964, USDA, ARS–72–38, April 1965, 243 pp.
14. **Robinson, W. L.,** Cottonseed Meal for Pigs, Ohio Agric. Exp. Sta. Bull. 534, 1934, 44 pp.
15. **Smith, K. J.,** Cottonseed protein: Status and potential, in *Engineering of Unconventional Protein Production,* Bieber, H., Ed., American Institute of Chemical Engineers, New York, 1969, *Chem. Eng. Prog. Symp.*, Ser. No. 93, Vol. 65, 30–35.
16. **Smith, K. J.,** Nutrient composition of cottonseed meal, *Feedstuffs,* 42(16), 19–20, 1970.
17. **Swan, T. H. and Altschul, A. M., Eds.,** Proc. Conf. on Cottonseed Protein for Animals and Man, New Orleans, La., Nov. 14–16, 1960, USDA, ARS–72–24, Jan. 1962, 77 pp.
18. **Vix, H. L. E., Gardner, H. K., Jr., Lambou, M. G., and Rollins, M. L.,** Ultrastructure related to cottonseed and peanut processing products, in *Symposium: Seed Proteins,* Inglett, G. E., Ed., Avi Publishing, Westport, Conn., 1972, 212–230.

d. Linseed

1. **Peterson, S. W.,** Linseed oil meal, in *Processed Plant Protein Foodstuffs,* Altschul, A. M., Ed., Academic Press, New York, 1958, 593–617. (72 ref.)

e. Palm Kernel

1. **Collingwood, J. G.,** Palm kernel meal, in *Processed Plant Protein Foodstuffs,* Altschul, A. M., Ed., Academic Press, New York, 1958, 677–701. (41 ref.)

f. Peanut

1. **Arthur, J. C., Jr.,** Peanut protein for industrial utilization: A literature survey, *J. South. Res.*, 1, 6–14, 1949.
2. **Arthur, J. C., Jr.,** Peanut protein: Isolation, composition, and properties, *Adv. Protein Chem.*, 8, 393–414, 1953.
3. **Bhatia, D. S., Kalbag, S. S., and Subrahmanyan, V.,** Technology of peanut protein isolate, Proc. 1st International Congress Food Science and Technology, London, Sept. 18–21, 1962.
4. **Bhatia, D. S. and Parpia, H. A. B.,** Peanut protein isolate production and properties, International Symposium Oilseed Protein Foods, Japan, May 11–15, 1964.
5. **Burnett, R. S.,** Manufacture and use of peanut protein, *Chem. Eng. News,* 24, 478–480, 1946.
6. Food and Agriculture Organization, Technology of Production of Edible Flours and Protein Products from Groundnuts, Agric. Serv. Bull. No. 10, FAO, Rome, 1971, 88 pp.
7. **Milner, M.,** Peanuts as a protein resource in international feeding programmes, *Food Technol.*, 16(7), 46–53, 1962.
8. **Pominski, J., Gordon, W. O., McCourtney, E. J., Vix, H. L. E., and Gastrock, E. A.,** Pilot plant production of peanut protein, *Ind. Eng. Chem.*, 44, 925–928, 1952.
9. **Rosen, G. D.,** Groundnuts (peanuts) and groundnut meal, in *Processed Plant Protein Foodstuffs,* Altschul, A. M., Ed., Academic Press, New York, 1958, 419–468. (361 ref.)
10. **St. Angelo, A. J. and Mann, G. E.,** Peanut proteins, in *Peanuts – Culture and Uses,* American Peanut Research and Education Association, Stone Printing, Roanoke, Va., 1972, 559–592.

g. Rapeseed

1. **Bell, J. M.,** The nutritional value of rapeseed oilmeal: A review, *Can. J. Agric. Sci.,* 35, 242–251, 1955.
2. **Bowland, J. P., Clandinin, D. R., and Wetter, L. R., Eds.,** Rapeseed Meal for Livestock and Poultry – A review, Publ. 1257, Can. Dep. Agric., Ottawa, 1965.
3. **Christian, B. C.,** Rapeseed, mustard-seed, and poppy-seed meals, in *Processed Plant Protein Foodstuffs,* Altschul, A. M., Ed., Academic Press, New York, 1958, 577–592. (65 ref.)
4. **Rocquelin, G., Sergiel, J. P., Martin, B., Leclerc, J., and Cluzan, R.,** The nutritive value of refined rapeseed oils: A review, *J. Am. Oil Chem. Soc.,* 48, 728–732, 1971. (66 ref.)
5. **Rutkowski, A.,** The feed value of rapeseed meal, *J. Am. Oil Chem. Soc.,* 48, 863–868, 1971. (53 ref.)
6. **Sims, R. P. A. and Nunes, A. C.,** Rapeseed and sunflower protein, in *Proc. SOS/70. 3rd International Congress Food Science and Technology,* Institute of Food Technologists, Chicago, 1971, 262–271.
7. **Tape, N. W., Sabry, Z. I., and Eapen, K. E.,** Production of rapeseed flour for human consumption, *Can. Inst. Technol. J.,* 3, 78–81, 1970.

h. Safflower

1. **Kohler, G. O.,** Safflower, a potential source of protein for human food, in *World Protein Resources,* in *Advances in Chemistry,* Ser. No. 57, American Chemical Society, Washington, D.C., 1966, 243–253. (33 ref.)

i. Sesame

1. **Caldwell, R. W.,** Sesame meal, in *Processed Plant Protein Foodstuffs,* Altschul, A. M., Ed., Academic Press, New York, 1958, 535–556. (42 ref.)

j. Soybean

1. **Berk, Z.,** Soy protein concentrates and isolates, in *Proc. SOS/70. 3rd International Congress Food Science and Technology,* Institute of Food Technologists, Chicago, 1971, 242–244. (9 ref.)
2. **Circle, S. J. and Johnson, D. W.,** Edible isolated soybean protein, in *Processed Plant Protein Foodstuffs,* Altschul, A. M., Ed., Academic Press, New York, 1958, 399–418. (93 ref.)
3. **Circle, S. J. and Smith, A. K.,** Processing soy flours, protein concentrates and protein isolates, in *Soybeans: Chemistry and Technology,* Vol. 1, *Proteins,* Smith, A. K. and Circle, S. J., Eds., Avi Publishing, Westport, Conn., 1972, 294–338.
4. **Cravens, W. W. and Sipos, E.,** Soybean oil meal, in *Processed Plant Protein Foodstuffs,* Altschul, A. M., Ed., Academic Press, New York, 1958, 353–397. (198 ref.)
5. **Dimpler, R. J.,** Oilseed protein sources and potentials: Soybeans, in *Engineering of Unconventional Protein Production,* Bieber, H., Ed., American Institute of Chemical Engineers, *Chem. Eng. Prog. Symp.,* Ser. No. 93, Vol. 65, New York, 1969, 22–29.
6. **Hartman, W. E.,** Textured soy proteins, *Proc. SOS/70. 3rd International Congress Food Science and Technology,* Institute of Food Technologists, Chicago, 1971, 245–247.
7. **Koury, S. D. and Hodges, R. E.,** Soybean protein for human diet, *J. Am. Diet. Assoc.,* 52, 480–484, 1968.
8. **Liener, I. E.,** Nutritional value of food protein products, in *Soybeans: Chemistry and Technology.* Vol. 1, *Proteins,* Smith, A. K. and Circle, S. J., Eds., Avi Publishing, Westport, Conn., 1972, 203–277.
9. **Meyer, E. W.,** Soya protein isolates for food, in *Proteins as Human Food,* Lawrie, R. A., Ed., Avi Publishing, Westport, Conn., 1970, 346–362.
10. **Meyer, E. W.,** Soybean flours and grits, in *Proc. SOS/70. 3rd International Congress Food Science and Technology,* Institute of Food Technologists, Chicago, 1971, 235–241.
11. **Mustakas, G. C., Griffin, E. L., Jr., and Sohns, V. E.,** Full-fat soybean flours by continuous extrusion cooking, in *World Protein Resources,* in *Advances in Chemistry,* Ser. No. 57, American Chemical Society, Washington, D.C., 1966, 101–11.
12. Nutrition Symposium. Nutritional properties of soybean protein, *Fed. Proc.,* 24, 1488–1500, 1965.
13. **Payne, D. S. and Stuart, L. S.,** Soybean protein in human nutrition, *Adv. Protein Chem.,* 1, 187–208, 1944.
14. **Pomeranz, Y.,** Soy flour in breadmaking – a review of its chemical composition, nutritional value and functional properties, *Baker's Dig.,* 40(60), 44–48, 78, 1966.
15. Proc. Conference on Soybean Products for Protein in Human Foods, Peoria, Ill., Sept. 13–15, 1961, USDA, ARS, Washington, D.C., 1962.
16. **Rakosky, J.,** Soy proteins – their preparation and use in comminuted meat products, *Meat Hyg.,* U.S. Dep. Agric. Consumer Marketing Serv., 8(6), 1–11, 1967.
17. **Rakosky, J., Jr.,** Soy products for the meat industry, *J. Agric. Food Chem,,* 18, 1005–1009, 1970.
18. **Rock, H., Sipos, E. F., and Meyer, E. W.,** Soy protein – its role in processed meat production, *Meat,* 32, 52–56, 1966.
19. **Smith, A. K. and Circle, S. J., Eds.,** *Soybeans: Chemistry and Technology,* Vol. 1, *Proteins,* Avi Publishing, Westport, Conn., 1972, 470 pp.
20. **Smith, A. K. and Wolf, W. J.,** Food uses and properties of soybean protein. 1. Food uses, *Food Technol.,* 15, 4–6, 8, 10, 1961.

21. **Wolf, W. J.,** Chemical and physical properties of soybean protein, *Baker's Dig.*, October 1969, 30–37.
22. **Wolf, W. J.,** Soybean proteins: Their functional, chemical, and physical properties, *J. Agric. Food Chem.*, 18, 969–976, 1970.
23. **Wolf, W. J.,** Soybean ultrastructure and its relationship to processing, in *Symposium: Seed Proteins*, Avi Publishing, Westport, Conn., 1972, 231–241.
24. **Wolf, W. J. and Smith, A. K.,** Food uses and properties of soybean protein. 2. Physical and chemical properties of soybean protein, *Food Technol.*, 15(5), 12, 13, 16, 18, 21, 23, 26, 28, 31, 33, 1961.

k. Sunflower

1. **Clandinin, D. R.,** Sunflower seed oil meal, in *Processed Plant Protein Foodstuffs*, Altschul, A. M., Ed., Academic Press, New York, 1958, 557–575. (56 ref.)
2. **Smith, K. J.,** A review of the nutritional value of sunflower meal, *Feedstuffs*, 30(23), 20–24, 1968.

l. Other Plant Proteins

1. **Boruff, C. S. and van Lanen, J. M.,** Fermentation feedstuffs, in *Processed Plant Protein Foodstuffs*, Altschul, A. M., Ed., Academic Press, New York, 1958, 739–759. (61 ref.)
2. **Kneeland, J. A.,** Minor oilseed and tree nut meals, in *Processed Plant Protein Foodstuffs*, Altschul, A. M., Ed., Academic Press, New York, 1958, 619–644. (57 ref.)
3. **Seeley, R. D.,** Milling Feeds, in *Processed Plant Protein Foodstuffs*, Altschul, A. M., Ed., Academic Press, New York, 1958, 761–787. (66 ref.)

4. Leaf Protein

1. **Akeson, W. R. and Stahmann, M. A.,** Leaf protein concentrates: A comparison of protein production per acre of forage with that from seed and animal crops, *Econ. Bot.*, 20, 244–250, 1966.
2. **Chayen, I. H., Smith, R. H., Tristram, G. R., Thirkill, D., and Webb, T.,** The isolation of leaf components. I., *J. Sci. Food Agric.*, 12(7), 505–512, 1961.
3. **Davys, M. N. G. and Pirie, N. W.,** Protein from leaves by bulk extraction, *Engineering*, 190, 274–275, 1960.
4. **Kohler, G. O. and Bickoff, E. M.,** Leaf protein, in *Proc. SOS/70. 3rd International Congress Food Science and Technology*, Institute of Food Technologists, Chicago, 1971, 290–295.
5. **Oelshlegel, F. J., Jr., Schroeder, J. R., and Stahmann, M. A.,** Potential for proteins from alfalfa and waste green plant material, *J. Agric. Food Chem.*, 17, 791–795, 1969.
6. **Pirie, N. W.,** Leaf proteins, *Annu. Rev. Plant Physiol.*, 10, 33–52, 1959. (151 ref.)
7. **Pirie, N. W.,** Leaf protein as a human food, *Science*, 152, 1701–1705, 1966.
8. **Pirie, N. W.,** Leaf protein in research in the international biological program, *Agric. Sci. Rev.*, 5(4), 17–21, 1967.
9. **Pirie, N. W.,** The present position of research on the use of leaf protein as a human food, *Plant Foods Hum. Nutr.*, 1, 237–246, 1969.
10. **Pirie, N. W.,** The production and use of leaf protein, *Proc. Nutr. Soc.*, 28, 85–91, 1969.
11. **Pirie, N. W.,** Leaf proteins, in *Evaluation of Novel Protein Products*, Bender, A. E., Kihlberg, R., Löfqvist, B., and Munk, L., Eds., Pergamon Press, Oxford, 1970, 87–91.
12. **Pirie, N. W., Ed.,** *Leaf Protein. Its Agronomy, Preparation, Quality and Use*, IBP Handb. No. 20, Blackwell, Oxford, 1971, 192 pp. (16 papers)
13. **Stahmann, M. A.,** The potential for protein production from green plants, *Econ. Bot.*, 22, 73–83, 1968. (7 ref.)
14. **Thompson, C. R.,** Alfalfa and other leaf meals, in *Processed Plant Protein Foodstuffs*, Altschul, A. M., Ed., Academic Press, New York, 1958, 703–737. (57 ref.)
15. **Tilley, J. M. A. and Raymond, W. F.,** The extraction and utilization of leaf protein, *Herbage Abstr.*, 27, 235–245, 1957.
16. **Woodham, A. A.,** The nutritive value of leaf protein concentrates, *Proc. Nutr. Soc.*, 24, XIV–XXV, 1965.

F. Microbial Food. Single Cell Protein

1. General

1. **Bhattacharje, J. K.,** Microorganisms as potential sources of food, *Adv. Appl. Microbiol.*, 13, 139–161, 1970.
2. **Bunker, H. J.,** Microbial food, in *Biochemistry of Industrial Microorganisms*, Rainbow, C. and Rose, A. H., Eds., Academic Press, 1963, 34–67.
3. **Bunker, H. J.,** Microbial food, in *Global Impacts of Applied Microbiology*, Starr, M. P., Ed., Almqvist & Wiksell, Stockholm, 1964, 234–240.
4. **Coty, V. F. and Leavitt, R. I.,** Microbial protein from hydrocarbons, *Dev. Ind. Microbiol.*, 12, 61–71, 1971.
5. **Dabbah, R.,** Protein from microorganisms, *Food Technol.*, 24, 659–666, 1970.
6. **Dawson, R. C.,** Potential for increasing food production through microbiology, *Bacteriol. Rev.*, 251–266, 1965. (169 ref.)
7. **Enebo, L.,** Single-cell protein, in *Evaluation of Novel Protein Products*, Bender, A. E., Kihlberg, R., Löfqvist, B., and Munck, L., Eds., Pergamon Press, Oxford, 1970, 93–103. (8 ref.)

8. **Ghose, T. K.,** Foods of the future, *Process Biochem.*, 4(12), 43–46, 1969.
9. **Gray, W. D.,** Microbial protein for the space age, *Dev. Ind. Microbiol.*, 3, 63-71, 1962.
10. **Humphrey, A. E.,** Engineering of single cell protein. State of the art, in *Engineering of Unconventional Protein Production,* Bieber, H., Ed., American Institute of Chemical Engineers, New York, 1969, 60–65.
11. **Humphrey, A. E.,** Microbial protein from petroleum, *Process Biochem.*, 5(6), 19–22, 1970. (45 ref.)
12. **Johnson, M. J.,** Growth of microbial cells on hydrocarbons, *Science,* 155, 1515–1519, 1967.
13. **Kihlberg, R.,** The microbe as a source of food, *Annu. Rev. Microbiol.*, 26, 427–466, 1972. (403 ref.)
14. **Lipinsky, E. S., Kinne, I. L., and Litchfield, J. H.,** Technical-economic comparison of microbial, animal, and plant proteins for food and food uses, *Dev. Ind. Microbiol.*, 10, 112–124, 1969.
15. **Lipinsky, E. S. and Litchfield, J. H.,** Algae, bacteria, and yeasts as food or feed, *CRC Crit. Rev. Food Technol.*, 1(1), 581–618, 1970. (135 ref.)
16. **Lipinsky, E. S. and Litchfield, J. H.,** Single-cell protein in perspective, *Food Technol.*, 28(5), 16–24, 40, 1974.
17. **Mateles, R. I. and Tannenbaum, S. R., Eds.,** *Single-cell Protein,* M.I.T. Press, Cambridge, Mass., 1968, 480 pp.
18. **Mateles, R. I. and Tannenbaum, S. R.,** Single cell protein, *Econ. Bot.*, 22, 42–50, 1967.
19. Panel discussion: Microbial production of amino acids, in *Recent Progress in Microbiology,* Gibbons, N. E., Ed., 8th International Congress of Microbiology, University of Toronto Press, Toronto, 1963, 313–352. (6 papers)
20. Problems of nutrition and the contribution of microorganisms to their solution, in *Biotechnology and Bioengineering Symposium,* Gaden, E. L., Jr., Ed., 2nd International Conference on Global Impacts of Applied Microbiology, Addis Ababa, Ethiopia, Nov. 6–11, 1967, Interscience, New York, 1969, 31–124. (10 papers)
21. *Proc. 2nd International Symposium on Single Cell Protein,* M.I.T. Press, Cambridge, Mass., 1973.
22. **Ribbons, D. W.,** Potential of microbiological protein, *Chem. Ind.*, 26, 867–870, 1968.
23. **Silverman, M. P., Gordon, J. N., and Wender, I.,** Microbial synthesis of food from coal-derived material, in *World Protein Resources,* in *Advances in Chemistry,* Ser. No. 57, American Chemical Society, Washington, D.C., 1966, 269–279. (19 ref.)
24. **Snyder, H. E.,** Microbial sources of protein, *Adv. Food Res.*, 18, 85–140, 1970. (112 ref.)
25. **Spicer, A.,** Single-cell protein, in *Proc. SOS/70. 3rd International Congress Food Science and Technology,* Institute of Food Technologists, Chicago, 1971, 296–298.
26. **Stokes, J. L.,** Microbial proteins, in *Processed Plant Protein Foodstuffs,* Altschul, A. M., Ed., Academic Press, New York, 1958, 789–804. (56 ref.)
27. Symposium on microorganisms as potential food sources, *Dev. Ind. Microbiol.*, 7, 201–228, 1966 (6 papers)
28. **Tannenbaum, S. R.,** Single-cell protein – food of the protein, *Food Technol.*, 25, 962–967, 1971.
29. **Tannenbaum, S. R. and Mateles, R. I.,** Single cell protein, *Sci. J.*, 4 (May), 87–92, 1966.
30. **Tannenbaum, S. R., and Mateles, R. I.,** Single cell protein, *Econ. Bot.*, 22, 42–50, 1968.
31. **Wang, D. I. C.,** Proteins from petroleum, *Chem. Eng.*, 75(18), 99–108, 1968.
32. **Wilkinson, J. F.,** Hydrocarbons as a source of single-cell protein, in *Microbes and Biological Productivity,* 21st Symp. Soc. Gen. Microbiol., Cambridge University Press, Cambridge, 1971, 15–46.
33. **Woodbine, M.,** Microbial fat: Microorganisms as potential fat producers, *Prog. Ind. Microbiol.*, 1, 179–245, 1959. (332 ref.)

2. Bacteria

1. **Ogur, M.,** The evolution of bacteria as a potential source of food, *Dev. Ind. Microbiol.*, 7, 216–220, 1966.
2. **Tannenbaum, S. R., Mateles, R. I., and Capco, G. R.,** Processing of bacteria for production of protein concentrates, in *World Protein Resources,* in *Advances in Chemistry,* Ser. No. 57, American Chemical Society, Washington, D.C., 1966, 254–260. (11 ref.)

3. Fungi

1. **Falanghe, H.,** Mushroom mycelium as another potential source of protein, *Food Technol.*, 21, 157–159, 1967.
2. **Gray, W. D.,** Fungi as a nutrient source, Biologistics for Space Systems Symp., May 1962, Tech. Doc. Rep. AMRL–TDR–62–116, Oct. 1962, Air Force Systems Command, Wright-Patterson Air Force Base, Ohio, 1962, 356–381.
3. **Gray, W. D.,** Fungal protein for food and feeds. I. Introduction, *Econ. Bot.*, 20, 89–93, 1966.
4. **Gray, W. D.,** Fungi and world protein supply, in *World Protein Resources,* in *Advances in Chemistry,* Ser. 57, 261–268, 1966. (13 ref.)
5. **Gray, W. D.,** The use of fungi as food and in food processing, *CRC Crit. Rev. Food Technol.*, 1, 255–329, 1970. (348 ref.)
6. **Gray, W. D. and Abou-El-Seoud, M.,** *Fungi imperfecti* as source of protein, *Dev. Ind. Microbiol.*, 7, 221–225, 1966.
7. **Gray, W. D., Och, F. F., and Abou-El-Seoud, M.,** *Fungi imperfecti* as a potential source of protein, *Dev. Ind. Microbiol.*, 5, 384–389, 1964.
8. **Litchfield J. H.,** The mass cultivation of *Morchella* species in submerged cultures and their potential uses as sources of protein, in *Global Impact of Applied Microbiology,* Starr, M. P., Ed., Almqvist & Wiksell, Stockholm, 1964, 327–337.

9. **Litchfield, J. H.,** The production of fungi, in *Single-Cell Protein,* Mateles, R. I. and Tannenbaum, S. R., Eds., M.I.T. Press, Cambridge, Mass., 1968, 309–329.
10. **Spicer, A.,** Protein production by microfungi, *Trop. Sci.,* 13, 239–250, 1971.
11. **Thatcher, F. S.,** Foods and feeds from fungi, *Annu. Rev. Microbiol.,* 8, 449–472, 1954.

4. Yeast

1. **Bressani, R.,** The use of yeast in human foods, in *Single-Cell Protein,* Mateles, R. I. and Tannenbaum, S. R., Eds., M.I.T. Press, Cambridge, Mass., 1968, 90–121.
2. **Lindegren, C. C.,** Food yeast, *Dev. Ind. Microbiol.,* 7, 216–220, 1966.
3. **Peppler, H. J.,** Food yeasts, in *The Yeasts,* Vol. 3, Rose, A. H. and Harrison, J. S., Eds., Academic Press, New York, 1970, 421–462.
4. **Pyke, M.,** The technology of yeast, in *The Chemistry and Biology of Yeasts,* Cook, A. H., Ed., Academic Press, New York, 1958, 535–586.
5. **Reed, G. and Peppler, H. J.,** *Yeast Technology,* Avi Publishing, Westport, Conn., 1973.
6. **Shacklady, C. A.,** Hydrocarbon-grown yeasts in nutrition, in *Proteins as Human Food,* Lawrie, R. A., Ed., Avi Publishing, Westport, Conn., 1970, 317–327.
7. **Shacklady, C. A.,** Yeasts grown on hydrocarbons as a new source of protein, *World Rev. Nutr. Diet.,* 14, 154–179, 1972. (10 ref.)

5. Algae

1. **Black, W. A. P.,** The algae, in *Processed Plant Protein Foodstuffs,* Altschul, A. M., Ed., Academic Press, New York, 1958, 805–827. (93 ref.)
2. **Burlew, J. S., Ed.,** *Algal Culture: From Laboratory to Pilot Plant,* Carnegie Institute of Washington, Publ. 600, Washington, D.C., 1953, 357 pp.
3. **Casey, R. P. and Lubits, J. A.,** Algae as food for space travel, *Food Technol.,* 17(11), 48–56, 1963.
4. **Enebo, L.,** Growth of algae for protein: State of the art, in *Engineering of Unconventional Protein Production,* Bieber, H., Ed., American Institute of Chemical Engineers, New York, 1969, in *Chem. Eng. Prog. Symp.,* Ser. No. 93, Vol. 65, 80–86, 1969.
5. Evaluation of algae as a food for human diets, *Nutr. Rev.,* 21, 7–8, 1963.
6. **Golueke, C. G. and Oswald, W. J.,** Harvesting and processing sewage-grown planktonic algae, *Water Pollut. Contr. Fed. J.,* 37, 471–498, 1965.
7. **Golueke, C. G., Oswald, W. J., Gee, H. K., and Cook, B. B.,** Production of low cost algal protein, in Proc. Symposium on Algology, Indian Council of Agricultural Research, New Delhi, 1960, 174–184.
8. **Gordon, J. F.,** Algal proteins and human diet, in *Proteins as Human Food,* Lawrie, R. A., Ed., Avi Publishing, Westport, Conn., 1970, 328–345.
9. **Jackson, D. F.,** *Algae and Man,* Plenum Press, New York, 1965, 434 pp.
10. **Krauss, R. W.,** Mass culture of algae for food and other organic compounds, *Am. J. Bot.,* 49, 425–435, 1962.
11. **Lavery, J. and Tischer, R. G.,** Food from Algae. Review of the Literature, Quartermaster Food and Container Inst. for the Armed Forces Library Bull. No. 1, Chicago, 1958.
12. **McDowell, M. E. and Leveille, G. A.,** Feeding experiments with algae, *Fed. Proc.,* 22, 1431–1438, 1963.
13. **McDowell, M. E. and Leveille, G. A.,** Algae systems, in Conference on Nutrition in Space and Related Waste Problems, NASA SP-70, 1964, 317–322.
14. **Matthern, R. O.,** The Potential of Algae for Food, T.P. 73 U.S. Army Laboratories, Natick, Mass., 1965.
15. **Milner, H. W.,** Algae as food, *Sci. Am.,* 189, 31–35, 1953.
16. **Oswald, W. J. and Golueke, C. G.,** Large-scale production of algae, in *Single-Cell Protein,* Mateles, R. I. and Tannenbaum, S. R., Eds., M.I.T. Press, Cambridge, Mass., 1968, 271–305.
17. **Mitsuda, H., Tonomura, B., and Yasumoto, K.,** Algal and other microbiol isolated proteins, in *Proc. SOS/70. 3rd International Congress Food Science and Technology,* Institute of Food Technologists, Chicago, 1971, 299–304. (7 ref.)
18. Proteins from microalgae and microfungi, *PAG Bull.,* 2(1), 19–22, 1972.
19. **Rose, M.,** Plankton as food, *Rev. Brasil Quim.* (Sao Paulo), 46, 548–568, 1958.
20. **Schmid, W. E. and Verduin, J.,** Mass-culture-produced algae as a food source, *Dev. Ind. Microbiol.,* 7, 210–215, 1966.
21. **Tamiya, H.,** Mass culture of algae, *Annu. Rev. Plant Physiol.,* 8, 309–334, 1957.
22. **Verduin, J. and Schmid, W.,** Evaluation of algal culture as a source of food supply, *Dev. Ind. Microbiol.,* 7, 205–209, 1966.
23. **Vincent, W. A.,** Algae and lithotrophic bacteria as food sources, in *Microbes and Biological Productivity,* 21st Symp. Soc. Gen. Microbiol., Cambridge University Press, Cambridge, Mass., 1971, 47–76.

G. Synthetic Foods

1. Ahead: Synthetics for the starving, *Chem. Week,* Aug. 1964, 79, 80, 86.
2. **Belikov, V. M., Nesmeyanov, A. N., Golovnia, R. V., and Rogoshin, S. V.,** Contribution of organic chemistry to the synthesis of new foods and food components, in *Proc. SOS/70. 3rd International Congress Food Science and Technology,* Institute of Food Technologists, Chicago, 1971, 647–653.
3. **Bender, A. E.,** Synthetic and substitute foods, *Dairy Ind.,* 35, 144–146, 1970.
4. **Champagnat, A.,** Protein from petroleum, *Sci. Am.,* 213, 13–70, 1965.
5. **Champagnat, A., Vernet, C., Lainé, B., and Filosa, J.,** Biosynthesis of protein-vitamin concentrates from petroleum, *Nature,* 197, 13, 14, 1963.
6. Ethyl Corporation, *Food for America's Future,* McGraw-Hill, New York, 1960.
7. **Fox, S. W.,** The outlook for synthetic foods, *Food Technol.,* 17, 388–392, 1963.
8. **Fox, S. W.,** Prospects for chemical synthesis of proteinaceous foodstuffs, Conference Closed Life Support Systems, Ames Research Center, 1966.
9. **Gounelle de Pontanel, H., Ed.,** *Protein from Hydrocarbons,* Academic Press, New York, 1971.
10. **Katchalski, E. and Sela, M.,** Synthesis and chemical properties of poly-α-amino acids, *Adv. Protein Chem.,* 13, 244–492, 1958.
11. **McPherson, A. T.,** Chemistry, food and civilization, *J. Wash. Acad. Sci.,* 50, 1–14, 1960.
12. **McPherson, A. T.,** The synthesis of food, *Food Technol.,* 16(11), 34–39, 1962.
13. **McPherson, A. T.,** Chemical and biochemical production of food for man and animal, *J. Anim. Sci.,* 25, 575–581, 1966.
14. **McPherson, A. T.,** Production of lysine and methionine, in *World Protein Resources,* in *Advances in Chemistry,* Ser. No. 57, 65–74, 1966. (26 ref.)
15. **McPherson, A. T.,** Food synthesis as a supplement to agriculture, *Proc. 7th International Congress of Nutrition,* Vol. 4, Pergamon Press, Oxford, 1967, 920–924.
16. **McPherson, A. T.,** Synthetic foods: Their present and potential contribution to the world food supply, *Indian J. Nutr. Diet.,* 9, 285–308, 1972. (44 ref.)
17. **Mauron, J.,** Technology of protein synthesis and protein-rich foods, *Bibl. Nutr. Dieta,* 18, 24–44, 1973. (28 ref.)
18. **Pyke, M.,** Synthetic foods, *Sci. J.,* 4(5), 93–98, 1968.
19. **Pyke, M.,** *Synthetic Food,* Murray, London, 1971, 158 pp.
20. **Rosenzweig, M. and Ushio, S.,** Protein from methanol, *Chem. Eng.,* 81(1), 62–63, 1974.
21. Symposium on synthetics and substitutes for the food industry, *J. Agric. Food Chem.,* 18, 961–1014, 1970. (10 papers)
22. **Winitz, M., Graff, J., Gallagher, N., Narkin, A., and Seedman, D.,** Evaluation of chemical diets as nutrition for man-in-space, *Nature,* 205, 741–743, 1965.

H. Food From Food Wastes

1. **Ben-Gera, I. and Kramer, A.,** The utilization of food industries wastes, *Adv. Food Res.,* 17, 77–152, 1969. (377 ref.)
2. **Bolley, D. S. and Holmes, R. L.,** Inedible oilseed meals, in *Processed Plant Protein Foodstuffs,* Atlschul, A. M., Ed., Academic Press, New York, 1958, 829–857. (127 ref.)
3. **Inglett, G. E., Ed.,** *Symposium: Processing Agricultural and Municipal Wastes,* Avi Publishing, Westport, Conn., 1973, 221 pp.
4. **Kelley, E. G.,** Plant residues and pomaces, in *Processed Plant Protein Foodstuffs,* Altschul, A. M., Ed., Academic Press, New York, 1958, 859–875. (31 ref.)

IX. IMPROVING NUTRITIONAL QUALITY OF FOOD

A. Food Quality In General

1. **Herschdoerfer, S. M., Ed.,** *Quality Control in the Food Industry,* Academic Press, New York, 1967–1972. (3 vols.)
2. **Irving, G. W., Jr., and Hoover, S. R., Eds.,** Food Quality. Effect of Production Practices and Processing, American Association for the Advancement of Science, Washington, D.C., 1965, 298 pp. (23 papers)
3. **Kramer, A.,** Food quality and quality control, in *Handbook of Food and Agriculture,* Blanck, F. C., Ed., Reinhold, New York, 1955, 733–750.
4. **Kramer, A.,** Quality control: Objectives, organization, methods, equipment, in *Food Processing Operations,* Vol. 1, Joslyn, M. A. and Heid, J. L., Eds., Avi Publishing, Westport, Conn., 1963, 93–129.
5. **Kramer, A. and Twigg, B. E.,** *Quality Control for the Food Industry,* 3rd ed., Avi Publishing, Westport, Conn., 1970–1973. (2 vols.)

B. Increasing Nutritive Value of Plants

1. Nutritive Value of Plants

1. **Goodall, D. W. and Gregory, F. G.,** Chemical composition of plants as an index of their nutritional status, Imp. Bur. Hort. and Plant Crops Tech. Comm. No. 17, Kent, East Malling, 1947, 167 pp.
2. **Macy, P. F.,** The critical nutrient composition of plants, *Plant Physiol.,* 11, 749, 1935.
3. **Schuphan, W.,** *Nutritional Values in Crops and Plants,* Faber & Faber, London, 1965, 280 pp.
4. **White, P. L. and Selvey, N., Eds.,** *Nutritional Qualities of Fresh Fruits and Vegetables,* Futura Publishing, Mount Kisco, N.Y., 1974.

2. Measures to Increase Nutritive Value of Plants

a. General

1. **Harris, R. S.,** The effects of agricultural practices on the composition of foods. A. Foods of plants origin, in *Nutritional Evaluation of Food Processing,* Harris, R. S. and von Loesecke, H., Eds., Avi Publishing, Westport, Conn., 1960, 5–24.
2. **Hopkins, H. T.,** Improving nutrient composition of plants through genetics and agronomic practices, in *Man, Food and Nutrition. Strategies and Technological Measures for Alleviating the World Food Problem,* Rechcigl, M., Jr., Ed., CRC Press, Cleveland, 1973, 203–214. (71 ref.)
3. **Schuphan, W.,** Quality production of vegetables according to the viewpoint of nutritional physiology, *Ind. Obst-Gemuseverwertung,* 52(5), 145–148, 1967.
4. **Schuphan, W.,** Control of plant proteins. The influence of genetics and ecology on plant foods, in *Proteins as Human Food,* Lawrie, R. A., Ed., Avi Publishing, Westport, Conn., 1970, 245–265.
5. U.S. Department of Agriculture, Factors Affecting the Nutritive Value of Foods, U.S. Dep. Agric. Misc. Publ. 664, U.S. Gov. Print. Off., Washington, D.C., 1948.
6. **Wallace, H. A.,** Development of improved crops, in *Human Nutrition, Historic and Scientific,* Monogr. 3, International Universities Press, New York, 1959, 143–170.
7. Ways of improving nutritive value of vegetable production, *Proc. 8th International Congress of Nutrition,* Prague, 1969, Excerpta Medica, Amsterdam, 1970, 722–741. (6 papers)

b. Environmental Measures

1. **Allaway, W. H.,** Agronomic controls over the environmental cycling of trace elements, *Adv. Agron.,* 20, 235–274, 1968.
2. **Armstrong, R. W.,** Environmental factors involved in studying the relationship between soil elements and disease, *Am. J. Public Health,* 54, 1536–1544, 1964. (18 ref.)
3. **Beeson, K. C.,** The effect of mineral supply on the mineral concentration and nutritional quality of plants, *Bot. Rev.,* 12, 424–455, 1946.
4. **Beeson, K. C.,** The soil factor in human nutritional problems, *Nutr. Rev.,* 7, 353, 1949.
5. **Beeson, K. C.,** Effects of fertilizers on nutrition quality of crops and health of animals and man, *Plant Food J.,* 5,(4), 6–11, 1951.
6. **Beeson, K. C.,** The effect of fertilizers on the nutritional quality of crops, Symposium on Nutrition of Plants, Animals, and Man, Michigan State University Centennial, East Lansing, 1955, 45.
7. **Boswell, V. R.,** Effects of insecticides on flavor and quality of food products, *J. Econ. Entomol.,* 48(5), 495–499, 1955.
8. **Bould, C.,** Soil and plant nutrient content in relation to crop yield, *Proc. Nutr. Soc.,* 24, 21, 1965.
9. **Carpenter, K. J.,** Nutritional considerations in attempt to change chemical composition of crops, *Proc. Nutr. Soc.,* 29, 3–12, 1970.
10. **Hemphill, D. D.,** Availability of trace elements to plants with respect to soil-plant interaction, *Ann. N. Y. Acad. Sci.,* 199, 46–60, 1972. (55 ref.)

11. **Hodgson, J. F.,** Chemistry of trace elements in soils with reference to trace element concentration in plants, Proc. Univ. Missouri 3rd Annu. Conf. Trace Sub. Environ. Health, University of Missouri, Columbia, Mo., 1969, 45–48.
12. **Hopkins, H. T., Stevenson, E. H., and Harris, P. L.,** Soil factors and food composition, *Am. J. Clin. Nutr.*, 18, 390–395, 1966. (28 ref.)
13. **Horvath, D. J.,** An overview of soil/plant/animal relationships with respect to utilization of trace elements, *Ann. N.Y. Acad. Sci.*, 199, 82–94, 1972. (58 ref.)
14. **Maynard, L. A.,** Soils and health, *JAMA*, 143, 807–812, 1950.
15. **Maynard, L. A.,** Effects of fertilizers on the nutritional value of foods, JAMA, 161, 1478–1480, 1956.
16. **Norman, A. G.,** Influence of environmental factors on plant composition, in Proc. Centennial Symposium on Nutrition of Plants, Animals, Man, Michigan State University, East Lansing, 1955, 14–19.
17. **Singhi, B., Wu, M. T., and Salunkhe, D. K.,** The use of chemicals to increase leaf and seed proteins of certain crops, *Plant Foods Hum. Nutr.*, 2, 171–184, 1972.
18. **Specht, A. W.,** Evidence for and some implications of variation-control mechanisms in plant composition, *Soil Sci.* 89, 83–91, 1960.
19. **Speirs, M. et al.,** Influence of environment on the chemical composition of plants, Southern Cooperative Ser. Bull. 42, 1955, Puerto Rico Agric. Exp. Sta. Rio Piedras, P. R., 1955.
20. Symposium: The effects of soils, fertilizers and environment on the yield and nutrient content of plants, *Proc. Nutr. Soc.*, 24, 1–48, 1965. (7 papers)
21. U.S. Department of Agriculture, *The Effect of Soils and Fertilizers on the Nutritional Quality of Plants,* USDA, ARS, Agric. Inf. Bull. 299, Washington, D.C., 1965, 24 pp.
22. U.S. Department of Agriculture, Influence of Environment on the Chemical Composition of Plants, Southern Cooperative Ser. Bull. 36, 1954.

c. Genetics

(1) General

1. Symposium: Plant breeding and nutrition, *Proc. Nutr. Soc.*, 29, 1–80, 1970. (12 papers)
2. **Swaminathan, M. S.,** Genetic upgrading of nutritional quality in food plants, Proc. 1st Asian Congress of Nutrition, Hyderabad 1971, Tulpule, P. G. and Jaya Rao, K. S., Eds., Nutrition Society of India, Hyderabad, 1972, 69–82. (26 ref.)

(2) Plant Proteins

1. **Dumanović, J. and Ehrenberg, L.,** Modern breeding methods for improving protein quality and quantity in plants, in Isotope Studies on the Nitrogen Chain, Proc. Symp., Vienna, 1967, IAEA, Vienna, 1968, 325–336.
2. **House, L. R.,** Breeding for protein quality and quantity in foods, *Int. Symp. on Protein Foods and Concentrates,* Central Food Technological Research Institute, Mysore, 1967.
3. Improving Plant Protein by Nuclear Techniques, Proc. Symposium on Plant Resources: Their Implications through the Application of Nuclear Techniques jointly organized by the IAEA and FAO, Vienna, June 8–12, 1970, International Atomic Energy Commission, Vienna, 1970, 460 pp.
4. **Johnson, V. A., Schmidt, J. W. and Mattern, P. J.,** Cereal breeding for better protein impact, *Econ. Bot.*, 22, 16–25, 1968. (19 ref.)
5. **Mattern, P. J.,** Protein improvement by breeding, in *Protein-Enriched Cereal Foods for World Needs,* Milner, M., Ed., American Association of Cereal Chemists, St. Paul, Minn., 1969, 234–244.
6. **Nelson, E. O.,** Genetic modifications of protein quality in plants, *Adv. Agron.*, 21, 171–194, 1969.
New Approaches to Breeding for Improved Plant Protein, Proc. Panel Meeting on New Approaches to Breeding for
7. Improved Plant Protein, Röstånga, Sweden, June 17–21, 1968, IAEA, Vienna, 1969, 195 pp. (18 papers)
8. **Swaminathan, M. S., Naik, M. S., Kaul, A. K., and Austin, A.,** The choice of strategy for the genetic upgrading of protein properties in cereals, millets and pulses, *Plant Foods Hum. Nutr.*, 2, 119–131, 1972.
9. Symposium on quality improvement of plant protein, *J. Agric. Food Chem.*, 22, 547–566, 1974.
10. **Whitehouse, R. N. H.,** The potential of cereal grain crops for protein production, in *The Biological Efficiency of Protein Production,* Jones, J. G. W., Ed., Cambridge University Press, London, 1973, 83–99.

3. Nutritional Improvement of Specific Crops

a. Cereals

1. **Brown, W. L.,** Performance and quality in cereals as influenced by genotype and environment, in Food Quality, Irving, G. W., Jr. and Hoover, S. R., Eds., American Association for the Advancement of Science, Washington, D.C., 1965, 83–98.
2. **Munck, L.,** New cereal varieties with high protein quality, in *Evaluation of Navel Protein Products,* Bender, A. E., Kihlberg, R., Löfqvist, B., and Munch, L., Eds., Pergamon Press, Oxford, 1970, 85, 86.

(1) Barley

1. **Munck, L.**, Barley seed proteins, in *Symposium: Seed Proteins*, Inglett, G. E., Ed., Avi Publishing, Westport, Conn., 1972, 144–164.

(2) Corn

(a) General

1. **Hageman, R. H., Leng, E. R., and Dudley, J. W.**, A biochemical approach to corn breeding, *Adv. Agron.*, 19, 45–86, 1967.
2. **Hamilton, T. S., Hamilton, B. C., Johnston, B. C., and Mitchell, H. H.**, The dependence of the physical and chemical composition of the corn kernel on soil fertility and cropping system, *Cereal Chem.*, 28(3), 163–176, 1951.
3. Proceedings of the Conference on Nutritional Improvement of Corn, INCAP, Guatemala, 1972.

(b) Protein Quality

1. **Frost, H. C. and Robinson, D.**, High lysine corn – what lies ahead, *J. Am. Oil Chem. Soc.*, 48, 407–411, 1971. (38 ref.)
2. **Mertz, E. T.**, High lysine corn, *Agric Sci. Rev.*, 6(3) 1–6, 1968 (24 ref.)
3. **Mertz, E. T.**, High-lysine corn, in *Proc. SOS/70. 3rd International Congress Food Science and Technology*, Institute of Food Technologists, Chicago, 1971, 305–309.
4. **Mertz, E. T.**, Recent improvements in corn proteins, in *Symposium: Seed Proteins*, Inglett, G. E., Ed., Avi Publishing, Westport, Conn., 1972, 136–143.
5. **Mertz, E. T., Bates, L. S., and Nelson, E. R.**, Mutant gene that changes protein composition and increases lysine content of maize endosperm, *Science*, 145, 279, 1964.
6. **Mertz, E. T. and Nelson, O. E., Eds.**, Proc. High Lysine Corn Conference, Corn Industries Foundation, Washington, D.C., 1966.
7. **Mertz, E. T., Nelson, O. E., Bates, L. S., and Veron, O. A.**, Better protein quality in maize, in *World Protein Resources*, in *Advances in Chemistry*, Ser. No. 57, 228–242, 1966.
8. **Nelson, O. E.**, Opaque-2, floury-2 and high protein maize, Proc. High Lysine Corn Conference, Corn Industries Foundation, Washington, D.C., 1966, 156–160.

(3) Oats

1. **Clark, W. L. and Potter, G. C.**, The composition and nutritional properties of protein in selected oat varieties, in *Symposium: Seed Proteins*, Inglett, G. E., Ed., Avi Publishing, Westport, Conn., 1972, 165–175.

(4) Rice

1. **Juliano, B. O.**, Studies on protein quality and quantity of rice, in *Symposium: Seed Proteins*, Inglett, G. E., Ed., Avi Publishing Westport, Conn., 1972, 114–125.
2. **Juliano, B. O., Ignacio, C. C., Panganiban, V. M., and Perez, C. M.**, Screening for high protein-rice varieties, *Cereal Sci. Today*, 13, 299–301, 313, 1968.

(5) Sorghum

1. **Pickett, R. C. and Oswalt, D. L.**, Sorghum nutritional quality improvement, in *Sorghum in Seventies*, Ganga Prasada Rao, N., and House, L. R., Eds., Oxford and I. B. H. Publishing, New Delhi, 1972, 445–464.

(6) Wheat

1. **Hehn, E. R. and Barmore, M. A.**, Breeding wheat for quality, *Adv. Agron.*, 17, 85–114, 1965.
2. **Johnson, V. A., Mattern, P. J., and Schmidt, J. W.**, The breeding of wheat and maize with improved nutritional value, *Proc. Nutr. Soc.*, 29, 20–31, 1970.
3. **Johnson, V. A., Mattern, P. J., and Schmidt, J. W.**, Genetic studies of wheat protein, in *Symposium: Seed Proteins*, Inglett, G. E., Ed., Avi Publishing, Westport, Conn., 1972, 126–135.
4. **Johnson, V. A., Mattern, P. J., Whited, D. A., and Schmidt, J. W.**, Breeding for high protein content and quality in wheat, in *Symposium: New Approaches to Breeding for Improved Plant Protein*, Röstånga, Sweden, June 17–21, 1968.
5. **Johnson, V. A., Whited, D. A., Mattern, P. J., and Schmidt, J. W.**, Nutritional improvement of wheat by breeding, *Proc. 3rd International Wheat Genetics Symposium*, Australian Academy of Sciences, Canberra, 1968, 457–461.
6. **Sullivan, B.**, Wheat protein research. Fifty years of progress, *Cereal Sci. Today*, 10, 338–361, 1965.

b. Pulses and Oil Seeds

1. **Howell, R. W., Brin, C. A., and Rinne, R. W.**, The plant geneticists' contribution toward changing lipid and amino acid composition of soybeans, *J. Am. Oil Chem. Soc.*, 49, 30–32, 1972.
2. **Knowless, P. F.**, The plant geneticists' contribution toward changing lipid and amino acid composition of safflower, *J. Am. Oil Chem. Soc.*, 49, 27–29, 1972.

3. **Miravalle, R. J.,** The plant geneticists' contribution toward changing the lipids and amino acid composition of cottonseed, *J. Am. Oil Chem. Soc.*, 49, 24–26, 1972.
4. Nutritional Improvement of Food Legumes by Breeding Proc. Symposium: Protein Advisory Group of the United Nations System, New York, 1973, 389 pp.
5. Symposium: The plant geneticist's contribution toward changing the lipid and amino acid composition of oil seeds, *J. Am. Oil Chem. Soc.*, 49, 15–37, 1972. (7 papers.)

c. Roots and Tubers

1. **Stevenson, F. J., Akeley, R. V., and Cunningham, C. E.,** Potato quality as related to heredity and environment, in *Food Quality*, Irving, G. and Hoover, S. R., Eds., American Association for the Advancement of Science, Washington, D.C., 1965, 19–34.

d. Fruits and Vegetables

1. **Lee, T. C. and Chichester, C. O.,** The influence of harvest time on nutritional value, in *Nutritional Qualities of Fresh Fruits and Vegetables*, White, P. L. and Selvey, N., Eds., Futura Publishing, Mount Kisco, N.Y., 1974, 111–120, (39 ref.)
2. **Salamon, M.,** Importance of agronomic practices on nutritional values, in *Nutritional Qualities of Fresh Fruits and Vegetables*, White, P. L. and Selvey, N., Eds., Futura Publishing, Mount Kisco, N.Y., 1974, 81–86. (20 ref.)
3. **Stevens, M. A.,** Varietal influence on nutritional value, in *Nutritional Qualities of Fresh Fruits and Vegetables*, White, P. L. and Selvey, N., Eds., Futura Publishing, Mount Kisco, N.Y., 1974, 87–110.

C. Improving Nutrient Composition of Animals

1. General

1. **Bowman, J. C.,** Possibilities for changing by genetic means the biological efficiency of protein production by whole animals, in *The Biological Efficiency of Protein Production*, Jones, J. G. W., Ed., Cambridge University Press, London, 1973, 173–182.
2. **Byerly, T. C.,** The effect of agricultural practices on the composition of foods. B. Foods of animal origin, in *Nutritional Evaluation of Food Processing*, Harris, R. S. and Von Loesecke, H. Eds., Avi Publishing, Westport, Conn., 1960, 25–57.
3. CICRA, Proc. Conf. Composition of Animal Products, Dublin, 1967, An Foras Taluntais, Dublin, Ireland, 1968.
4. **Lodge, G. A.,** Improving the nutrient composition of meat-producing animals, in *Man, Food, and Nutrition. Strategies and Technological Measures for Alleviating the World Food Problem*, Rechcigl, M. Jr., Ed., CRC Press, Cleveland, 1973, 215–230. (157 ref.)

2. Carcass Composition

1. **Braude, R.,** Variations in the composition of pig meat, CICRA Proc. Conf. Composition of Animal Products, An Foras Taluntais, Dublin, Ireland, 1968, 21–55.
2. **Byerly, T. C.,** Meat quality: Genetic and environmental factors in development and performance, in *Food Quality*, Irving, G. W., Jr. and Hoover, S. R., Eds., American Association for the Advancement of Science, 1965, 251–266.
3. **Hedrick, H. B.,** Bovine Growth and Composition, Mo. Agric Exp. Sta. Res. Bull. No. 928, 1968.
4. **Lodge, G. A.,** Quantitative and qualitative control of proteins in meat animals, in *Proteins as Human Food*, Lawrie, R. A., Ed., Avi Publishing, Westport, Conn., 1970, 141–166.
5. **Pearson, A. M.,** The influence of nutrition and management upon meat quality, in *Food Quality*, Irving, G. W., Jr. and Hoover, S. R., Eds., American Association for the Advancement of Science, Washington, D.C., 1965, 267–286.
6. **Tribe, D. E., Ed.,** Carcass Composition and Appraisal of Meat Animals, Commonwealth Scientific and Industrial Research Organization, Australia, 1964.

3. Milk

1. **Bayley, N. D. and Plowman, R. D.,** Milk quality as influenced by the genetics of milk composition, in *Food Quality*, Irving, G. W., Jr. and Hoover, S. R., Eds., American Association for the Advancement of Science, Washington, D.C., 1965, 143–152.
2. **Kirchgessner, M., Friesecke, H. and Koch, G.,** *Nutrition and the Composition of Milk*, Lippincott, Philadelphia, 1967, 273 pp.
3. **Kirchmeier, O.,** Influence of nutrition on the composition of milk proteins, *Proc. 8th International Congress of Nutrition*, Prague, 1969, Experta Medica, Amsterdam, 1970, 750–754.
4. **Roberts, W. M.,** Protection of milk quality through management practices, in *Food Quality*, Irving, G. W., Jr. and Hoover, S. R., Eds., American Association for the Advancement of Science, Washington, D.C., 1965, 153–163.

4. Eggs

1. **Baker, C. M. A.,** The genetic basis of egg quality, *Br. Poult. Sci.*, 1, 3–16, 1960.

2. **Coppock, J. B. M. and Daniels, N. W. R.**, Influence of diet and husbandry on the nutritional value of the hen's egg, *J. Sci. Food Agric.*, 13, 459–469, 1962.
3. **Everson, G. J. and Souders, H. J.**, Composition and nutritive importance of eggs, *J. Am. Diet. Assoc.*, 33, 1244–1254, 1957.
4. **Linneweaver, H.**, Quality feeds yield good end products – improved poultry and egg products, Int. Animal Feed Symp., U.S. Dep. Agric., Foreign Agric. Serv., Washington, D.C., 1959, 121–129.
5. **Stadelman, W. J.**, Genetic and environmental factors influencing eggs and poultry meat, in *Food Quality*, Irving, G. W., Jr. and Hoover, S. R., Eds., American Association for the Advancement of Science, Washington, D.C., 1965, 193–212.

D. Increasing Availability of Nutrients

1. **Krehl, W. A. and Barboriak, J. J.**, Factors affecting utilization of food nutrients by the normal organism, in *Nutritional Evaluation of Food Processing*, Harris, R. S. and von Loesecke, H., Eds., Avi Publishing, Westport, Conn., 1960, 521–541.
2. **Liener, I. E. and Kakade, M. L.**, The increased availability of nutrients from plant foodstuffs through processing, in *Man, Food, and Nutrition. Strategies and Technological Measures for Alleviating the World Food Problem*, Rechcigl, M., Jr., Ed., CRC Press, Cleveland, Ohio, 231–241.

E. Food Fortification and Enrichment

1. Process of Fortification and Enrichment

a. General

1. Symposium, American Medical Association. Council on Foods and Nutrition, Improvement of nutritive quality of foods, *JAMA*, 205(12), 868–869, 1968.
2. **Andrews, J.**, Enrichment: Remembrance of things past, *Cereal Sci. Today*, 11, 258,1966.
3. **Borenstein, B.**, Rationale and technology of food fortification with vitamins, minerals, and amino acids, *CRC Crit. Rev. Food Technol.*, 2(2) 171–186, 1971. (48 ref.)
4. Enrichment: The quiet miracle, *Cereal Sci. Today*, 11, 221–282, 1966. (9 papers)
5. **Forman, M. J.**, Enrichment and food for peace, *Cereal Sci. Today*, 11(6), 231, 232, 1966.
6. **Friend, B.**, Enrichment and fortification of foods 1957–1961, National Food Situation, 106, USDA, ERS, Nov. 1963.
7. **Harris, R. S.**, Attitudes and approaches to supplementation of foods with nutrients, *J. Agric. Food. Chem.*, 16, 149–152, 1968.
8. Joint FAO/WHO Expert Committee on Nutrition, Eighth Report, Food Fortification: Protein-Calorie Malnutrition, WHO Rep. Ser. No. 477, Geneva, 1971.
9. **Kübler, W.**, Enrichment and fortification of foods for infants and children, *Bibl. Nutr. Dieta*, 18, 224–245, 1973. (82 ref.)
10. **Lachance, P. A.**, Nutrification, *Food Technol.*, 24, 724, 1970.
11. **Levinson, F. J.**, Food fortification in low-income countries. A new approach to an old standby, *Am. J. Public Health*, 62, 715–718, 1972.
12. **Menden, E. and Cremer, H. D.**, The problem of improving nutritive value with special reference to enrichment of food, *Food Manuf.*, 33(7), 293–298, 1958; 33(8), 330–334, 340, 1958; 33(9), 379–382, 1958; 33(10), 417, 418, 1958; 33(11), 463–467, 470, 1958; 34(2), 65–69, 80, 1959; 34(3), 121–124, 126, 1958.
13. **Mitsuda, H.**, New approaches in amino acid and vitamin enrichment in Japan, in *Protein-Enriched Cereal Foods for World Needs*, Milner, M., Ed., American Association of Cereal Chemists, St. Paul, Minn., 1969, 208–219.
14. **Mitsuda, H.**, Approaches to amino acid and vitamin enrichment programmes in Japan, in *Proc. First Asian Congress of Nutrition*, Tulpule, P. G. and Jaya Rao, K. S., Eds., Nutrition Society of India, Hyderabad, 1972, 457–471.
15. **Nelson, E. M.**, The philosophy of food fortification, *J. Am. Diet, Assoc.*, 30, 984–986, 1954.
16. Pan American Health Organization, Guidelines for Food Fortification in Latin America and the Caribbean, Report of a Technical Group Meeting, PAHO, Pan American Sanitary Bureau, Washington, D.C., 1972.
17. **Rubin, S. H.**, Nutrition research and enrichment, *Cereal Sci. Today*, 11, 234–239, 1966.
18. **Sebrell, W. H.**, Enrichment, a public health approach to better nutrition, *Public Health Rep.*, 68(8), 741–746, 1953.
19. **Sebrell, W. H.**, Enrichment: Good gift of yesterday, *Cereal Sci. Today*, 11, 228–230, 1966.
20. **Sebrell, W. H.**, Fortification of foods with synthetic nutrients, *Technol. Rev.*, 72, 32–35, 1970.
21. **Sebrell, W. H., Jr.**, The concept of the fortification of foods with synthetic nutrients, in *Amino Acid Fortification of Protein Foods*, Scrimshaw, N. S. and Altschul, A. M., Eds., M.I.T. Press, Cambridge, Mass., 1971, 63–76.
22. **Sebrell, W. H., Jr.**, Chemical aspects of updating diet quality, *J. Agric. Food Chem.*, 20, 518–522, 1972.
23. **Somogyi, J. C.**, Some problems of the improvement of the nutritive quality of foods, *Proc. 8th International Congress of Nutrition*, Prague, 1969 Experta Medica, Amsterdam, 1970, 773–777.
24. Symposium on enrichment and fortification of food with nutrients, *J. Agric. Food Chem.*, 16, 149–189, 1966. (8 papers)

25. **Teply, L. J.,** Food fortification, in *Man, Food, and Nutrition. Strategies and Technological Measures for Alleviating the World Food Problem,* Rechcigl, M., Jr., CRC Press, Cleveland, Ohio, 1973, 243–249.
26. **Walker, A. R.,** Problems in nutritional supplementation and enrichment, *Am. J. Clin. Nutr.,* 12, 157–160, 1963.
27. **Williams, R. R.,** Food enrichment, progress and controversial issues, *J. Agric. Food Chem.,* 2, 770–774, 1954.
28. **Williams, R. R.,** Fortification and restoration of processed foods, *Ind. Eng. Chem.,* 33(6), 718–720, 1941.

b. Amino Acids

1. **Altschul, A. M.,** Amino acid fortification, *Proc. SOS/70. 3rd International Congress Food Science and Technology, 1970,* Institute of Food Technologists, Chicago, 1971, 208–212.
2. **Altschul, A. M. and Rosenfield, D.,** Protein supplementation. Satisfying man's food needs, *Prog. Unilever Q.,* 54(305), 26–84, 1970.
3. Amino acid fortification of foods, PAG Statement No. 9, August 24, 1970, 20 pp.
4. **Beigler, M. A.,** Practical problems of amino acid fortification, in *Protein-Enriched Cereal Foods for World Needs,* Milner, M., Ed., American Association of Cereal Chemists, St. Paul, Minn., 1969, 200–207.
5. **Block, R. J.,** Methods of increasing the nutritional value of foods. B. Addition of amino acids, in *Nutritional Evaluation of Food Processing,* Harris, R. S., and von Loesecke, H., Eds., Avi Publishing, Westport, Conn., 1960, 508–520.
6. **Bressani, R., Elías, L. G., and Gómez Brenes, R. A.,** Improvement of protein quality by amino acid and protein supplementation, in *Protein and Amino Acid Functions,* Bigwood, E. J., Ed., *International Encyclopaedia of Food and Nutrition,* Vol. 11, Pergamon Press, Oxford, 1972, 475–540.
7. **Cole, W. H., Ed.,** *Some Aspects of Amino Acid Supplementation,* Rutgers University Press, New Brunswick, N.J., 1956, 85 pp.
8. **Flodin, N. W.,** Economic aspects of amino acid supplementation, *Ann. N.Y. Acad. Sci.,* 69, 1061–1063, 1958.
9. **Hegsted, D. M.,** Amino acid fortification and the protein problem, *Am. J. Clin. Nutr.,* 21, 688–692, 1968. (16 ref.)
10. **Howe, E. E.,** Amino acid supplementation, *Borden's Rev. Nutr. Res.,* 19, 19–34, 1958. (83 ref.)
11. **Howe, E. E.,** Summary of progress on the use of purified amino acids in foods, in *Progress in Meeting Protein Needs of Infants and Preschool Children,* National Academy of Sciences, Washington, D.C., 1961, 495–507. (52 ref.)
12. **Howe, E. E., Gilfillan, E. W., and Milner, M.,** Amino acid supplementation of protein concentrates as related to the world food supply, *Am. J. Clin. Nutr.,* 16, 321–326, 1965.
13. **Howe, E. E., Jansen, G. R., and Gilfillan, E. W.,** Amino acid supplementation of cereal grains as related to the world food supply, *Am. J. Clin. Nutr.,* 16, 315–320, 1965.
14. **Jansen, G. R.,** Amino acid supplementation and the world food problem, in *Evaluation of Novel Protein Products,* Bender, A. E., Kihlberg, P., Lofqvist, B., and Munck, L., Eds., Pergamon Press, Oxford, 1970, 105–114.
15. **Kato, J. and Muramatsu, N.,** Amino acid supplementation of grain, *J. Am. Oil Chem. Soc.,* 48, 415–419, 1971. (28 ref.)
16. **McGandy, R. B.,** Fortification of cereals with amino acids, *Econ. Bot.,* 22, 26–28, 1968.
17. **Ottenheym, H. H. and Jenneskens, P. J.,** Synthetic amino acids and their use in fortifying foods, *J. Agric. Food Chem.,* 18, 1010–1014, 1970.
18. **Rosenberg, H. R.,** Amino acid supplementation of foods and feeds, in *Protein and Amino Acid Nutrition,* Albanese, A. A., Ed., Academic Press, New York, 1959, 381–417.
19. **Rosenberg, H. R.,** Supplementation of foods with amino acids, *J. Agric. Food Chem.,* 7, 316–321, 1959. (13 ref.)
20. **Rosenberg, H. R. and Eckert, R. E.,** Multiple amino acid supplementation of proteins, in *Progress in Meeting Protein Needs of Infants and Preschool Children,* National Academy Sciences, Washington, D.C., 1961, 451–467. (19 ref.)
21. **Rosenfield, D.,** Enrichment of plant proteins, in *Proteins as Human Food,* Lawrie, R. A., Ed., Avi Publishing, Westport, Conn., 1970, 266–279.
22. **Scrimshaw, N. S. and Altschul, A. M., Eds.,** *Amino Acid Fortification of Protein Foods,* M.I.T. Press, Cambridge, Mass., 1971, 664 pp.
23. **Senti, F. R., and Pence, J. W.,** Technological Aspects of Adding Amino Acids to Foods, USDA, ARS, 1970.
24. **Teply, L. J.,** Amino acid fortification, in *Proc. 3rd Western Hemisphere Nutrition Congress,* Futura Publishing, Mount Kisco, N.Y., 1972, 37–39.
25. **Waddel, J.,** Supplementation of plant proteins with amino acids, in *Processed Plant Protein Foodstuffs,* Academic Press, New York, 1958, 307–351. (122 ref.)
26. **Zucker, H.,** The value of amino acid supplementation, in *Proteins in Human Nutrition,* Porter, J. W. G. and Rolls, B. A., Eds., Academic Press, London, 1973, 469–479.

c. Vitamins and Minerals

(1) General

1. **Aylward, F. and Morton, I. D.,** Vitamin fortification of foods, in *Proc. SOS/70. 3rd International Congress of Food Science and Technology,* Institute of Food Technologists, Chicago, 1971, 192–199. (28 ref.)
2. **Bauernfeind, J. C.,** Vitamin fortification and nutrified foods, in *Proc. SOS/70. 3rd International Congress Food Science and Technology,* Institute of Food Technologists, Chicago, 1971, 217–232. (65 ref.)

3. **Francois, A. C. and Lhuissier, M.**, L'adjouction do vitamines aux alimentes de l'homme et des animaux, *Ann. Nutr. Aliment.*, 24, B477–B551, 1970. (143 ref.)
4. **Gage, J. W.**, *The Cost of Fortifying Foods with Vitamins*, Hoffman, La Roche, 1970.
5. **Harris, R. S.**, Supplementation of foods with vitamins, *J. Agric. Food Chem.*, 7, 88–102, 1959. (51 ref.)
6. **Harris, R. S.**, Mineral fortification of foods and food products, in *Proc. SOS/70. 3rd International Congress of Food Science and Technology*, Institute of Food Technologists, Chicago, 1971, 200–207. (39 ref.)
7. **Parman, G. K.**, Vitaminization of food outside the USA, *Ann. N.Y. Acad. Sci.*, 98, 607–613, 1942.
8. **Parman, G. K. and Salinard, G. J.**, Vitamins as ingredients in food processing, in *Food Processing Operations*, Vol. 2, Joslyn, M. A. and Heid, M. L., Eds., Avi Publishing, Westport, Conn., 1963, 192–217.
9. **Rubin, S. H. and Cort, W. M.**, Aspects of vitamin and mineral enrichment, *Protein-Enriched Cereal Foods for World Needs*, Milner, M., Ed., American Association of Cereal Chemists, St. Paul, Minn., 1969, 220–233.
10. **Siemers, G. F.**, Methods of increasing the nutritive value of foods. A. Enrichment with vitamins and minerals, in *Nutritional Evaluation of Food Processing*, Harris, R. S. and von Loesecke, H., Eds., Avi Publishing, Westport, Conn., 1960, 492–508.

(2) Specific Nutrients

1. **Finch, C. A. and Monsen, E. R.**, Iron nutrition and the fortification of food with iron, *JAMA*, 219, 1462–1465, 1972.
2. **Kuhajek, E. J. and Fiedelman, H. W.**, Nutritional iodine in processed foods, *Food Technol.*, 27(1), 52–53, 1973.
3. **Martiñez-Torres, C. and Layrisse, M.**, Interest for the study of dietary absorption and iron fortification, *World Rev. Nutr. Diet*, 19, 51–70, 1974.
4. **Monsen, E. R.**, The need for iron fortification, *J. Nutr. Educ.*, 2, 152–155, 1971. (14 ref.)

2. Enriched Foods

a. Grain Products

1. **Brooke, C.**, Enrichment and fortification of cereals and cereal products with vitamins and minerals, *J. Agric. Food Chem.*, 16, 163–167, 1968.
2. **Conway, H. F. and Anderson, R. A.**, Protein fortified extruded food products, *Cereal Sci. Today*, 18(4), 94–97, 1973.
3. Council of Foods and Nutrition, Iron in enriched wheat flour, farina, bread, buns, and rolls, *JAMA*, 220, 855–859, 1972. (25 ref.)
4. **Goldsmith, G. A.**, Iron enrichment of bread and flour, *Am. J. Clin. Nutr.*, 26, 131–132, 1973.
5. **Graham, R. P., Morgan, A. I., Jr., Hart, M. R., and Pence, J. W.**, Mechanics of fortifying cereal grains and products, *Cereal Sci. Today*, 13, 224–227, 253, 1968.
6. **Hulse, J. H.**, The protein enrichment of bread and baked products, in *New Protein Foods*, Vol. 1A Altschul, A.M., Ed., Academic Press, New York, 1974, 155–229.
7. **Jansen, G. R.**, The amino acid fortification of cereals, in *New Protein Foods*, Vol. 1A, Altschul, A.M., Ed., Academic Press, New York, 1974, 39–120.
8. **Milner, M., Ed.**, *Protein-Enriched Cereal Foods for World Needs*, American Association of Cereal Chemists, St. Paul, Minn., 1969, 343 pp.
9. National Research Council, Enrichment of Flour and Bread, A History of Movement, Bull. No. 110, National Academy of Sciences, Washington, D.C., 1944.
10. National Research Council, Committee on Cereals, *Cereal Enrichment in Perspective*, National Academy of Sciences, Washington, D.C., 1958.
11. National Research Council, *Proposed Fortification Policy for Cereal Grain Products*, National Academy of Sciences, Washington, D.C., 1974.
12. **Parman, G. K.**, Fortification of cereals and cereal products with proteins and amino acids, *J. Agric. Food Chem.*, 16, 168–171, 1968.
13. **Pomeranz, Y.**, Protein-enriched bread, *CRC Crit. Rev. Food Technol.*, 1(3), 453–478, 1970. (37 ref.)
14. **Rosenfield, D. and Berntson, B. L.**, Economics and technology of cereal fortification, in *Symposium: Seed Proteins*, Inglett, G. E., Ed., Avi Publishing, Westport, Conn., 1972, 5–18.
15. **Sebrell, W. H.**, 20 years of enrichment, Twentieth Anniversary of Bread Enrichment Meeting, New York City, Feb. 24, 1961.
16. **Senti, F. R.**, Impact of grain product fortification on the nutrient content of the U.S. diet, *Cereal Sci. Today*, 16, 92–102, 1971.
17. **Senti, F. R., Copley, M. J., and Pence, J. W.**, Protein-fortified grain products for world use, *Cereal Sci. Today*, 12, 426–430, 441, 1969.
18. United Kingdom, Ministry of Health, Iron in Flour, Reports on Public Health and Medical Subjects No. 117, London, 1968.

b. Dairy Products

1. **Bauernfeind, J. C. and Allen, L. E.,** Vitamin A and D enrichment of nonfat dry milk, *J. Dairy Sci.*, 46, 245–254, 1963. (15 ref.)
2. **Coulter, S. T. and Thomas, E. L.,** Enrichment and fortification of dairy products and margarine, *J. Agric. Food Chem.*, 16, 158–162, 1968.

c. Water

1. **Arnold, F. A., Jr.,** The rise of fluoride compounds for the prevention of dental caries, *Int. Dent. J.*, 7, 54–72, 1957.
2. **Browner, F.,** Fluoridation or obsession? *Am. J. Clin. Nutr.*, 22, 1346–1348, 1969.
3. **Dean H. T., Arnold, F. A., Jr., Jay, P., and Knutson, J. W.,** Studies on mass control of dental caries through fluoridation of the public water health supply, *Public Health Rep.*, 65, 1403–1408, 1950.

d. Salt

1. **Hunnikin, C.,** The iodization of crude salt for the prophylaxis of goiter, *Food Technol.*, 18, 1544–1547, 1964.
2. **Levinson, F. J. and Berg, A.,** . . . with a grain of fortified salt, *Food Technol.*, 23(9), 1174–1176, 1969.
3. **Lowenstein, F. W.,** Iodized salt in the prevention of endemic goiter: A world-wide survey of present programs, *Am. J. Public Health*, 57, 1815–1823, 1967. (12 ref.)

e. Sugar

1. **Navia, J. M.,** Enrichment of sugar and sugar products, *J. Agric. Food Chem.*, 16, 172–176, 1968.

f. Tea

1. **Brooke, C. L. and Cort, W. M.,** Vitamin A fortification of tea, *Food Technol.*, 26(6), 50–52, 58, 1972.

g. Fruit Products

1. **Bunnell, R. H.,** Enrichment of fruit products and fruit juices, *J. Agric. Food Chem.*, 16, 177–183, 1968.

h. Special Dietary Products

1. **Filer, L. J., Jr.,** Enrichment of special dietary products, *J. Agric. Food Chem.*, 16, 184–189, 1968.

F. Formulation of New Nutritious Food Products

1. General

1. **Altschul, A. M. and Hornstein, I.,** Foods of the future, *J. Agric. Food Chem.*, 20, 532–537, 1972.
2. **Bressani, R. and Elias, L. G.,** Development of new highly nutritious food products, in *Man, Food and Nutrition. Strategies and Technological Measures for Alleviating the World Food Problem*, Rechcigl, M., Jr., Eds., CRC Press, Cleveland, Ohio, 1973, 251–274. (87 ref.)
3. **Clausi, A. S.,** Improving the nutritional quality of food, *Food Technol.*, 27(6), 36–40, 1973.
4. **Glicksman, M.,** Fabricated foods, *CRC Crit. Rev. Food Technol.*, 2(1), 21–43, 1971. (102 ref.)
5. **Harrington, J.,** Unique products to meet nutritional needs, in *Proc. Western Hemisphere Nutrition Congress*, American Medical Association, Chicago, 1966, 76–78.
6. **Jelliffe, D. B.,** Commerciogenic malnutrition? Time for a dialogue, *Food Technol.*, 25(2), 55, 56, 1971.
7. **Lachaence, P. A.,** Nutrification: A concept of assuring nutritional quality by primary intervention in feeding systems, *J. Agric. Food Chem.*, 20, 522–525, 1972.
8. **Parpia, H. A. B.,** New development of infant and weaning foods, in Proc. 1st Asian Congress of Nutrition, Tulpule, P. G. and Jaya Rao, K. S., Eds., Nutrition Society of India, Hyderabad, 1972, 472–489. (25 ref.)
9. **Popkin, B. M. and Latham, M. C.,** The limitations and dangers of commerciogenic nutritious foods, *Am. J. Clin. Nutr.*, 26, 1015–1023, 1973. (22 ref.)
10. **Sarett, H. P.,** Nutritional value of commercially produced foods for infants, *Bibl. Nutr. Dieta.*, No. 18, S. Karger, Basel, 1973, 246–266. (43 ref.)

2. Low-Cost Foods

1. **Altschul, A. M.,** Low-cost foods: Fortified cereals and protein beverages, in *Protein-Enriched Cereal Foods for World Needs*, Milner, M., Ed., American Association of Cereal Chemists, St. Paul, Minn., 1969, 82–96.
2. **Elder, A. and Weisberg, S. M.,** High nutrition-low cost foods. Their impact in developing countries, *Proc. SOS/70. 3rd International Congress of Food Science and Technology*, Institute of Food Technologists, Chicago, 1971.

3. Textured Foods

1. **Arenson, S. W.,** Imitation dairy products. Their formulation, processing, quality control, *Food Eng.*, 41(4), 76–79, 1969.
2. **Gutcho, M.,** *Textured Foods and Allied Products*, Noyes Data Corp., Park Ridge, N. J., 1973, 315 pp.

3. **Horan, F. E.,** Meat analogs, in *New Protein Foods,* Vol. IA, Altschul, A. M., Ed., Academic Press, New York, 1974, 366–413.
4. Textured vegetable proteins, *PAG Bull.,* 2(1), 22–26, 1970.

4. High Protein Foods

1. **Altschul, A. M., Ed.,** *New Protein Foods,* Academic Press, New York, 1974.
2. **Cordaro, J. B.,** An Inquiry into the Agency for International Development's Commercial Studies High Protein Food Program. Master's Thesis, Cornell University, Ithaca, N.Y., 1972.
3. **Cordaro, J. B. and Call, D. L.,** Nutritional protection of vulnerable groups through protein-rich mixtures: A critical review, *Proc. 9th International Congress of Nutrition,* Mexico City, 1972, in press.
4. **De Muelenaere, H. J. H.,** Development, production, and marketing of high-protein foods, in *Protein-Enriched Cereal Foods for World Needs,* Milner, M., Ed., American Association of Cereal Chemists, St. Paul, Minn., 1969, 266–277.
5. Harvard University Graduate School of Business Administration, The Protein Paradox, Malnutrition, Protein-Rich Foods and the Role of Business, Management Rep., Boston, 1965.
6. **Hundley, J. M.,** Enrichment of foods with protein, *Ann. N.Y. Acad. Sci.,* 69, 1042–1060, 1958. (14 ref.)
7. **Kapsiotis, G. D.,** History and status of specific-protein-rich foods. FAO/WHO/UNICEF protein food program and products, in *Protein-Enriched Cereal Foods for World Needs,* Milner, M., Ed., American Association of Cereal Chemists, St. Paul, Minn., 1969, 255–265.

New Protein Foods, Report of a Conference, Univ. of Reading, July 1971, UK Agricultural Dev. Assoc., 1971, 79 pp.

9. **Noyes, R.,** *Protein Food Supplements 1969,* Noyes Development Corp., Park Ridge, N.J., 1969, 412 pp.
10. **Orr, E.,** The Use of Protein-Rich Foods for the Relief of Malnutrition in Developing Countries: An Analysis of Experience, Tropical Products Institute, London, 1972, 71 pp.
11. Protein Fortification of Foods, Proc. Seminar at Jadarpur University, Calcutta, Feb. 15–16, 1969, Association of Food Technologists, Eastern Regional Branch, Jadarpur University, Calcutta, 1969.
12. Symposium, Protein enriched foods for less developed countries, *Cereal Sci. Today,* 12, 415–462, 1967. (9 papers)

5. Plant Protein Mixtures and Foods

a. General

1. **Aiyar, A. S. and Sreenivasan, A.,** Protein-rich food from plant sources, in Proc. 1st Asian Congress of Nutrition, Hyderabad 1971, Tulpule, P. G. and Jaya Rao, K.S., Eds., Nutrition Society of India, Hyderabad, 1972, 427–456.
2. **Bressani, R.,** Development of food products. Formulated vegetable mixtures, in *Proc. Western Hemisphere Nutrition Congress – 1965,* American Medical Association, Chicago, 1966, 86–90.
3. **Bressani, R.,** Application of food science and technology to the exploitation of new protein sources, in *Proc. 3rd Western Hemisphere Nutrition Congress,* Futura Publishing Co., Mount Kisco, N.Y., 1972, 368–377.
4. **Bressani, R. and Béhar, M.,** The use of plant protein foods in preventing malnutrition, *Proc. 6th International Congress of Nutrition,* E. S. Livingstone, Edinburgh, 1964, 181–206.
5. **Bressani, R. and Elias, L. G.,** Processed vegetable protein mixtures for human consumption, *Adv. Food Res.,* 16, 126–228, 1968.
6. **Chandrasekhara, M. R., Shurpalekar, S. R., Aswathanarayana, S., and Parpia, H. A. B.,** Foods based on vegetable proteins for infant nutrition, *Proc. 7th International Congress of Nutrition,* Vol. 3, Kuhnau, J., Ed., Friedr. Vieweg & Sohn, Braunschweig, Pergamon Press, Oxford, 165–172, 1967.
7. **Dean, R. F. A.,** Plant Proteins in Child Feeding, Med. Res. Counc. Spec. Rep. Ser., 279, 1953.
8. **Hartman, W. E.,** Vegetarian protein foods, *Food Technol.,* 20, 39–40, 1966.
9. Indian Council of Medical Research, Milk Substitutes of Vegetable Origin, Spec. Rep. Ser. 31, New Delhi, 1955.
10. **Jelliffe, D. B. and Jelliffe, E. F. P.,** The role of legumes and dark green leafy vegetables in domestic multimixes for the transitional, *PAG Bull.,* 3(2), 40–45, 1973.
11. **Jelliffe, E. F. P.,** A new look at weaning multimixes in the Commonwealth Caribbean, *J. Trop. Pediatr.,* 17, 135–150, 1971.
12. **Narayana Rao, M. and Swaminathan, M.,** Processed protein foods of vegetable origin, *Annu. Rev. Food Technol.* (India), 1, 73–99, 1960.
13. **Parpia, H. A. B.,** Protein foods of India based on cereals, legumes, and oilseed meals, in *Protein-Enriched Cereal Foods for World Needs,* Milner, M., Ed., American Association of Cereal Chemists, St. Paul, Minn., 1969, 129–139.
14. **Parpia, H. A. B. and Subramanian, N.,** Plant protein foods in India, in *World Protein Resources,* in *Advances in Chemistry,* Ser. No. 57, American Chemical Society, Washington, D.C., 1966, 112–132. (16 ref.)
15. **Parpia, H. A. B., Swaminathan, M. S., and Subrahmanyan, V.,** Plan for the manufacture of Indian multipurpose food, *Food Sci.,* 6, 96–99, 1957.
16. **Rajalakshmi, R., Sail, S. S., Ramachandran, K., Chandrasekaran, K. N., and Sulbulakshmi, G.,** Formulation and evaluation of meals based on plant proteins for pre-school children, *Plant Foods Hum. Nutr.,* 2, 151–165, 1972.
17. **Schaffer, G. E.,** New protein foods from plant sources. A system for economic evaluation, *Econ. Bot.,* 22, 29–36, 1968.

18. **Scrimshaw, N. S. and Bressani, R.,** Vegetable protein mixtures for human consumption, *Fed. Proc.*, 20(1), III, Suppl. 7, 80–88, 1961.
19. **Shaw, R. L.,** Incaparina: A low cost vegetable mixture and its commercial application, *Plant Foods Hum. Nutr.*, 1, 99–107, 1969.
20. **Shaw, R. L.,** Incaparina in Central America, in *Protein-Enriched Cereal Foods for World Needs,* American Association of Cereal Chemists, St. Paul, Minn., 1969, 320–339.
21. **Shaw, R. L.,** Incaparina: The market development of a protein food, *Trop. Sci.*, 14, 347–371, 1972.
22. Symposium. Vegetable mixtures as protein-sources for young children, in *Proc. 7th International Congress of Nutrition,* Vol. 3, Kuhnau, J., Ed., Friedr. Vieweg & Sohn, Braunschweig, Pergamon Press, Oxford, 1967, 149–219.

b. Cereal-Based Foods

(1) General

1. **Agren, G., Hofvander, Y., Selinius, R., and Vahlquist, B.,** Faffa: A supplementary cereal-based weaning food in Ethiopia, in *Protein-Enriched Cereal Foods for World Needs,* Milner, M., Ed., American Association of Cereal Chemists, St. Paul, Minn., 1969, 278–287.
2. **Bacigalupo, A.,** Protein-rich cereal foods in Peru, in *Protein-Enriched Cereal Foods for World Needs,* Milner, M., Ed., American Association of Cereal Chemists, St. Paul, Minn., 1969, 288–304.
3. **Milner, M., Ed.,** *Protein-Enriched Cereal Foods for World Needs,* American Association of Cereal Chemists, St. Paul, Minn., 1969, 343 pp.
4. **Senti, F. R.,** Formulated cereal foods in the U.S. Food for Peace Program, in *Protein-Enriched Cereal Foods,* Milner, M., Ed., American Association of Cereal Chemists, St. Paul, Minn., 1969, 246–254.
5. **Senti, F. R., Copley, M. J., and Pence, J. W.,** Protein-fortified grain products for world use, *Cereal Sci. Today,* 12, 426–430, 1969.

(2) Rice Products

1. **Lynn, L.,** Edible rice bran foods, in *Protein-Enriched Cereal Foods for World Needs,* Milner, M., Ed., American Association of Cereal Chemists, St. Paul, Minn., 1969, 154–172.

(3) Wheat Products

1. **Bass, E. J.,** Wheat protein concentrates and related food products, in *Protein-Enriched Cereal Food for World Needs,* Milner, M., Ed., American Association of Cereal Chemists, St. Paul, Minn., 1969, 117–128. (9 ref.)
2. **Haley, W. L. and Pence, J. W.,** Bulgor, an ancient wheat food, *Cereal Sci. Today,* 5, 203–207, 214, 1960.
3. **Overley, V. A. and Greenwood, M. L.,** Developing wafers and biscuits of varying protein content, *J. Am. Diet. Assoc.,* 45, 342–346, 1964.
4. **Pence, J. W.,** New wheat foods, *Nutr. Rev.,* 26, 291–294, 1968.
5. **Pence, J. W.,** History and status of specific protein-rich foods, Bulgur, WURLD wheat, and wheat protein beverages, in *Protein-Enriched Cereal Foods for World Needs,* Milner, M., Ed., American Association of Cereal Chemists, St. Paul, Minn., 1969, 105–116. (34 ref.)
6. **Shepherd, A. D.,** Beverage products from wheat, Proc. 4th National Conference on Wheat Utilization Research, Boise, Idaho, Nov. 3–5 1965, USDA, ARS-74-35, April 1966.
7. **Sullivan, B.,** Wheat-based products for world use, *Cereal Sci. Today,* 12, 446–448, 462, 1967.

c. Oilseed Protein Foods

(1) General

1. **Abbott, J. C.** Protein-rich foods from oilseeds, *Food. Technol.,* 19, 929–933, 1965.
2. **Anson, M. L.,** Potential uses of isolated oilseed protein in foodstuffs, in *Processed Plant Protein Foodstuffs,* Altschul, A. M., Ed., Academic Press, New York, 1958, 277–289.
3. **Anson, M. L.,** Oilseed proteins in foods, *Arch. Biochem. Biophys.,* Suppl. 1, 68–77, 1962. (23 ref.)
4. **Bacigalupo, A.,** Oilseed protein foods, *Proc. Western Hemisphere Nutrition Congress,* American Medical Association, Chicago, 1966, 90–94.
5. **Bressani, R.,** Formulation and testing of weaning and supplementary foods containing oilseed proteins, in *Protein-Enriched Cereal Foods for World Needs,* Milner, M., Ed., American Association of Cereal Chemists, St. Paul, Minn., 1969, 49–66.
6. **Bressani, R. and Elias, L. G.,** Legume foods, in *New Protein Foods,* Vol. IA, Altschul, A. M., Ed., Academic Press, New York, 1974, 230–297.
7. Proc. Conf. on Protein-Rich Foods from Oilseeds, New Orleans, May 15–16, 1968, USDA, ARS-72-71, 1969.
8. **Swaminathan, M. and Parpia, H. A. B.,** Milk substitutes based on oilseeds and nuts, *World Rev. Nutr. Diet.,* 8, 184–206, 1968.
9. **Walker, D. B., Horan, F. E., and Burke, R. E.,** Engineered foods – the place for oilseed proteins, *Food Technol.,* 25, 813–818, 1971.

(2) Cottonseed Protein Foods

1. **Black, H. C.,** Edible cottonseed oil products, in *Cottonseed and Cottonseed Products, Their Chemistry and Chemical Technology*, Bailey, A. E., Ed., John Wiley & Sons, New York, 1948, 732–762.
2. **Bressani, R.,** The use of cottonseed protein in human foods, *Food Technol.*, 19, 1655–1662, 1965.
3. **Bressani, R., Elías, L. G., and Braham, E.,** Cottonseed protein in human foods, *World Protein Resources*, Advances in Chemistry Ser. No. 57, American Chemical Society, Washington, D.C., 1966, 75–100. (78 ref.)
4. **Martinez, W. H., Berardi, L. C., and Goldblatt, L. A.,** Cottonseed protein products – composition and functionality, *J. Agric. Food Chem.*, 18, 961–968, 1970.

(3) Peanut Protein Foods

1. **Parpia, H. A. B., Swaminathan, M., and Bhatia, D. S.,** Peanut protein foods for protein foods for protein-poor countries, *J. Food. Sci. Technol.*, 2, 17–33, 1965.

(4.) Soybean Protein Foods

1. **Burnett, R. S.,** Soybean protein food products, in *Soybeans and Soybean Products*, Vol. 2., Markley, K. S., Ed., John Wiley & Sons, New York, 1951, 949–1002.
2. **Circle, S. J. and Smith, A. K.,** Functional properties of commercial edible soybean protein products, in *Symposium: Seed Proteins*, Inglett, G. E., Ed., Avi Publishing, Westport, Conn., 1972, 242–254.
3. **Eley, C. P.,** Food uses of soy protein, in Marketing and Transportation Situation, USDA, ERS 388, MTS-170, 27–30, 1968.
4. **Hesseltine, C. W. and Wang, H. L.,** Fermented soybean food products, in *Soybeans: Chemistry and Technology*, Vol. 1, *Proteins*, Smith, A. K. and Circle S. J., Eds., Avi Publishing, Westport, Conn., 1972, 389–419.
5. Proc. Int. Conf. on Soybean Protein Foods, Peoria, Ill., Oct. 17–19, 1966, USDA, ARS-71–35, Washington, D.C., May 1967.
6. **Rakosky, J.,** Soy products for the meat industry, *J. Agric. Food Chem.*, 18, 1005–1009. 1970.
7. **Smith, A. K. and Circle, S. J.,** Protein products as food ingredients, in *Soybeans: Chemistry and Technology*, Vol. 1, *Proteins*, Smith, A. K. and Circle, S. J., Eds., Avi Publishing, Westport, Conn., 1972, 339–388.
8. **Turro, E. J. and Sipos, E.,** Soy protein products in commercial cake formulations, *Baker's Dig.*, 44(1) 58–64, 1970.
9. **Wood, J. C.,** Soy flour in food products. Ingredient survey, *Food Manuf.*, 42, 11–15, 1967.

X. FOOD HAZARDS AND FOOD SAFETY

A. General References

1. Food Hazards and Hygiene

1. **Ayres, J. C., Kraft, A. A., Snyder, H. E., and Walker, H. W., Eds.,** *Chemical and Biological Hazards in Food,* Iowa State University Press, Ames, 1962, 383 pp. (20 papers)
2. **Baltzer, J.,** Food handling and nutrition. Some hygienic aspects, *Proc. 7th International Congress of Nutrition,* Vol. 4, Hamburg, 1966, Pergamon Press, Oxford, 1967, 420–426.
3. **Birch, G. G., Green, L. F., and Plaskett, L. G., Eds.,** *Health and Food,* John Wiley & Sons, New York, 1972, 224 pp.
4. **Brandley, P. J., Migaki, G., and Taylor, K. E.,** *Meat Hygiene,* 3rd ed., Lea & Febiger, Philadelphia, 1966, 789 pp.
5. **Bressani, R.,** Safe and nutritious food supply, in *The Safety of Foods,* Graham, H. D., Ed., Avi Publishing, Westport, Conn., 1968, 245–261.
6. **Christie, A. B. and Christie, M. C.,** *Food Hygiene and Food Hazards,* Faber & Faber, London, 1971.
7. **Collins, C. P.,** *Food Hygiene,* Pitman Medical, London, 1963.
8. Food and Agriculture Organization, Meat Hygiene, FAO Agric. Stud. No. 34, FAO, Rome, 1957, 527 pp.
9. **Genigeorgis, G. A. and Riemann, H.,** Food safety and food poisoning, in *Food, Nutrition and Health, A Multidisciplinary Treatise Addressed to the Major Nutrition Problems from a World Wide Perspective,* Rechcigl, M., Jr., Ed., *World Rev. Nutr. Dieta,* Vol. 16, S. Karger, Basel, 1973, 363–397.
10. **Golberg, L.,** Chemical and biochemical implications of human and animal exposure to toxic substances in food, *Pure Appl. Chem.,* 21, 309–330, 1970. (123 ref.)
11. **Graham, H. D., Ed.,** *The Safety of Foods,* Avi Publishing, Westport, Conn., 1968, 367 pp. (40 papers)
12. **Hafen, B. Q., Ed.,** *Man, Health and Environment,* Burgess, Minneapolis, Minn., 1972, 270 pp.
13. **Hamilton, A. and Hardy, H. L.,** *Industrial Toxicology,* 3rd ed., Publishing Sciences Group, Acton, Mass., 1974, 575 pp.
14. **Horwitz, A.,** Food and protection of health, *Fed. Proc.,* 20(1), III, Suppl. 7, 398–402, 1961.
15. **Roe, F. J. C., Ed.,** *Metabolic Aspects of Food Safety,* Academic Press, New York, 1971, 612 pp. (20 papers)
16. **Sapeika, N.,** *Food Pharmacology,* Charles C Thomas, Springfield, Ill., 1969.
17. Symposium: Hygienic aspects of nutrition, *Proc. 8th International Congress of Nutrition,* Prague, 1969, Excerpta Medica, Amsterdam, 1970, 611–649. (11 papers)
18. Symposium. Problems in food protection and toxicology, in *Proc. Western Hemisphere Nutrition Congress, Chicago, Nov. 8–11, 1965,* Am. Med. Assoc., Chicago, 1966, 115–150.
19. World Health Organization, Milk Hygiene, WHO Monogr. Ser. No. 48, WHO, Geneva, 1962.
20. **White, P. and Robbins, D., Eds.,** *Environmental Quality and Food Supply,* Futura Publishing, Mount Kisco, N.Y., 1974, 272 pp.
21. World Health Organization, Food, in *Health Hazards of the Human Environment,* WHO, Geneva, 1972, 72–93.

2. Food-Borne Diseases

1. **Angelotti, R.,** Prevention of food-borne infection, in *The Safety of Foods,* Graham, H. D., Ed., Avi Publishing, Westport, Conn., 1968, 194–203.
2. **Brandley, P. J., Migaki, G., and Taylor, K. E.,** Food-borne illness, in *Meat Hygiene,* 3rd ed., Lea & Febiger, Philadelphia, 1966, 324–353.
3. **Bryan, F. L.,** Diseases Transmitted by Foods. A Classification and Summary, Center for Disease Control, Atlanta, 1971.
4. **Dack, G. M.,** *Food Poisoning,* 3rd ed., University of Chicago Press, Chicago, 1956.
5. **Dewberry, E.,** *Food Poisoning,* 4th ed., Hill Books, London, 1959.
6. **Dolman, C. E.,** Epidemiology of meat-borne diseases, in Meat Hygiene, WHO Mongr. Ser. No. 33, WHO, Geneva, 1957, 11–108.
7. **Galton, M. M. and Arustrin, P.,** *Poultry Diseases in Public Health. Review for Epidemiologists,* U.S. Public Health Service, Publ. No. 767, U.S. Gov. Print. Off., Washington, D.C., 1960.
8. **Gleason, M. N., Gesselin, R. E., Hodge, H. C. and Smith, R. P.,** *Clinical Toxicology of Commercial Products. Acute Poisoning,* 3rd ed., Williams & Wilkins, Baltimore, MD, 1969.
9. **Hobbs, B. C.,** *Food Poisoning and Food Hygiene,* 2nd ed., Arnold, London, 1968, 252 pp.
10. **Hull, T. G., Ed.,** *Diseases Transmitted From Animals To Man,* 5th ed., Charles C Thomas, Springfield, Ill., 1963.
11. **Kaplan, M. M., Abdussalan, M., and Bijlenga, G.,** Diseases transmitted through milk, in Milk Hygiene, WHO Monogr. Ser. No. 48, WHO, Geneva, 1962, 11–74.
12. National Research Council. Food Protection Committee, *An Evaluation of Public Health Hazards from Microbiological Contamination of Foods,* National Academy of Sciences, Washington, D.C., 1964.
13. **O'Leary, S. B.,** Poisoning in man from eating plants, *Arch. Environ. Health,* 9, 216–242, 1964. (29 ref.)
14. **Riemann, H., Ed.,** *Food-borne Infections and Intoxications,* Academic Press, New York, 1968.
15. **Sadler, W. W.,** Food-borne diseases of animal origin, in *The Safety of Foods,* Graham, H. D., Ed., Avi Publishing, Westport, Conn., 1968, 204–216.

16. **Steele, J. H.,** Animal Diseases and Human Health, FHHC Basic Study No. 3, FAO, Rome, 1962, 50 pp.
17. **Tanner, F. W. and Tanner, L. P.,** *Food-borne Infections and Intoxications,* 2nd ed., Garrard Press, Champaign, Ill., 1953.
18. **Waldbott, G. L.,** *Health Effects of Environmental Pollutants,* C. V. Mosby, St. Louis, 1973, 448 pp.

3. Food Spoilage and Deterioration

a. General

1. **Ayres, J. C.,** Reducing food spoilage and deterioration, in *Man, Food, and Nutrition. Strategies and Technological Measures for Alleviating the World Food Problem,* Rechcigl, M., Jr., Ed., CRC Press, Cleveland, 1973, 177–188.
2. **Goresline, H. E.,** Food spoilage and deterioration, in *Handbook of Food and Agriculture,* Blanck, F. C., Ed., Reinhold, New York, 1955, 389–410.
3. **Halverson, H. O.,** Food spoilage and food poisoning, in *The Chemistry and Technology of Food and Food Products,* 2nd ed., Vol. 1, Jacobs, M. B., Ed., Interscience, New York, 1951, 410–450.

b. Microbiological Aspects

1. **Brandley, P. J., Migaki, G., and Taylor, K. E.,** Deteriorative changes in meats, in *Meat Hygiene,* 3rd ed., Lea & Febiger, Philadelphia, 1966, 301–323.
2. **Chichester, C. O. and Graham, H. D., Eds.,** *Microbial Safety of Fishery Products,* Academic Press, New York, 1973, 334 pp.
3. **Eddy, B. P.,** Microbial spoilage, in *Biochemistry of Industrial Micro-Organisms,* Rainbow, C. and Rose, A. H., Eds., Academic Press, London, 1963, 489–507.
4. **Hobbs, B. C. and Christian, J. H. B., Eds.,** *The Microbiological Safety of Foods,* Academic Press, New York, 1974,
5. **Ingram, M.,** Yeasts in food spoilage, in *The Chemistry and Biology in Yeasts,* Cook, A. H., Ed., Academic Press, New York, 1958, 603–633.
6. **Mossel, D. A. A. and Ingram, M.,** The physiology of the microbial spoilage of foods, *J. Appl. Bacteriol.,* 18, 233–268, 1955.
7. **Mossel, D. A. A.,** Microbiological spoilage of proteinaceous foods, in *Proteins as Human Food,* Lawrie, R. A., Ed., Avi. Publishing, Westport, Conn., 1970, 89–106.
8. **Ordal, Z. J.,** The prevention and control of microbial food spoilage, *Dev. Ind. Microbial.,* 10, 101–111, 1969.
9. **Ordal, Z. J.,** Sources of food spoilage microorganisms, in *The Safety of Foods,* Graham, H. D., Ed., Avi Publishing, Westport, Conn., 1968, 183–193.
10. **Scott, W. J.,** Water relations of food spoilage microorganisms, *Adv. Food Res.,* 7, 83–127, 1957.
11. **Starr, M. P.,** Microbes as spoilers, in *Global Impact of Applied Microbiology,* Almqvist & Wiksell, Stockholm, 1964, 166–172.
12. **Walker, H. W. and Ayres, J. C.,** Yeasts as spoilage organisms, in *The Yeasts,* Vol. 3, Rose, A. H. and Harrison, J. S., Eds., Academic Press, New York, 1970, 463–527.

c. Chemical Aspects

1. **Aylward, F. and Haisman, D. R.,** Oxidation systems in fruit and vegetables – Their relation to the quality of preserved products, *Adv. Food Res.,* 17, 1–76, 1969. (349 ref.)
2. **Finot, P. A.,** Non-enzymic browning, in *Proteins in Human Nutrition,* Porter, J. W. G. and Rolls, B. A., Eds. Academic Press, London, 1973, 501–514.
3. **Holman, R. T.,** Autoxidation of fats and related substances, in *Progress in Chemistry of Fats and Lipids,* Vol. 2, Holman, R. T., Lundberg, W. O., and Malkin, T., Eds., Pergamon Press, New York, 1954, 51–98.
4. **Mathew, A. G. and Parpia, H. A. B.,** Food browning as a polyphenol reaction, *Adv. Food Res.,* 19, 75–145, 1971.
5. Reactions and interactions of carbohydrates, in *Symposium on Foods: Carbohydrates and Their Roles,* Schultz, H. W., Cain, R. F., and Wrolstad, R. W., Eds., Avi Publishing, Westport, Conn., 1969, 205–297. (4 papers)
6. **Reynolds, T. M.,** Chemistry of non-enzymic browning. I. The reaction between aldoses and amines, *Adv. Food Res.,* 12, 1–52, 1963.
7. **Reynolds, T. M.,** Chemistry of non-enzymic browning. II. *Adv. Food Res.,* 14, 167–283, 1965. (113 ref.)
8. **Riemenschneider, R. W.,** Oxidative rancidity and antioxidants, in *Handbook of Food and Agriculture,* Blanck, C. F., Ed., Reinhold, New York, 1955, 237–278. (339 ref.)
9. **Schultz, H. W., Ed.,** *Symposium on Foods: Lipids and Their Oxidation,* Avi Publishing, Westport, Conn., 1962, 442 pp.
10. **Schultz, H. W., Ed.,** *Symposium on Foods: Proteins and Their Reactions,* Avi Publishing, Westport, Conn., 1964, 472 pp.
11. **Schultz, H. W., Ed.,** *Symposium on Foods: The Chemistry and Physiology of Flavors,* Avi Publishing, Westport Conn., 1967, 552 pp.

4. Food Microbiology

1. **Eddy, B. P., Ed.,** The Microbiology of Fish and Meat Curing Brines, HMSO, London, 1958, 336 pp.

2. **Fjerdingstad, E.,** Microbial criteria of environment qualities, *Annu. Rev. Microbiol.,* 25, 563–582, 1971. (582 ref.)
3. **Ford, B. J.,** *Microbiology and Food,* Northwood, London, 1970, 212 pp.
4. **Frazier, W. C.,** *Food Microbiology,* 2nd ed., McGraw-Hill, New York, 1967, 537 pp.
5. **Hammer, B. W. and Babel, F. J.,** *Dairy Bacteriology,* 4th ed., John Wiley & Sons, New York, 1957.
6. **Jay, J. M.,** *Modern Food Microbiology,* Van Nostrand Reinhold, New York, 1970, 328 pp.
7. **Jensen, L. B.,** *Microbiology of Meats,* Garrard Press, Champaign, Ill., 1954.
8. **Nickerson, J. T. R. and Sinskey, A. J.,** *Microbiology of Foods and Food Processing,* American Elsevier, New York, 1972, 306 pp.
9. **Pederson, C. S.,** *Microbiology of Food Fermentations,* Avi Publishing, Westport, Conn., 1971, 284 pp.
10. **Prescott, S. C. and Dunn, M. G.,** *Industrial Microbiology,* 2nd ed., McGraw-Hill, New York, 1949, 541 pp.
11. **Rangaswami, G.,** *Agricultural Microbiology,* Asia Publishing, New York, 1966, 413 pp.
12. **Slanetz, L. W., Chichester, C. O., Gaufin, A. R., and Ordal, Z. J., Eds.,** *Microbial Quality of Food,* Academic Press, New York, 1963, 274 pp.
13. **Stumbo, C. R.,** *Thermobacteriology in Food Processing,* Academic Press, New York, 1973, 329 pp.
14. Symposium. The industrial aspects of food microbiology, *Dev. Ind. Microbiol.,* 10, 39–76, 1969. (7 papers)
15. **Tanner, F. W.,** *The Microbiology of Foods,* Garrard Press, Champaign, Ill., 1944, 1196 pp.
16. **Thatcher, F. S.,** Microbiological standards for foods, *Food Technol.,* 12, 117–122, 1958.
17. **Weiser, H. H.,** *Practical Food Microbiology and Technology,* Avi Publishing, Westport, Conn., 1962, 345 pp.
18. **Weiser, H. H., Mountney, G. J., and Gould, W. A.,** *Practical Food Microbiology and Technology,* Avi Publishing, Westport, Conn., 1971, 360 pp.

5. Food Sanitation

1. Association of Food Industry Sanitarians, *Sanitation for the Food Preservation Industries,* McGraw-Hill, New York, 1952.
2. **Benarde, M. A.,** *Disinfection,* Dekker, New York, 1970, 466 pp.
3. **Graham-Rack, B. and Binsted, R.,** *Hygiene in Food Manufacturing and Handling,* Food Trade Review, Westport, Conn., 1964, 148 pp.
4. Grain and Cereal Products Sanitation Conference, University of Minnesota, Institute of Agriculture, Feb. 16–17, 1967.
5. **Guthrie, R. K.,** *Food Sanitation,* Avi Publishing, Westport, Conn., 1972, 248 pp.
6. **Longreé, K.,** *Quality Food Sanitation,* Interscience, New York, 1967, 397 pp.
7. **Longreé, K.,** *Sanitary Techniques in Food Service,* John Wiley & Sons, New York, 1971, 225 pp.
8. **Majumder, S. K.,** Grain Sanitation, Academy of Pest Control Sciences, Mysore, India, 1969.
9. **Parker, M. E. and Litcheld, J. H.,** *Food Plant Sanitation,* Reinhold, New York, 1962, 401 pp.
10. **Reddish, G. F.,** *Antiseptics, Disinfectants, Fungicides, and Chemical and Physical Sterilisation,* Kimpton, London, 1954.
11. **Richardson, T. M.,** *Sanitation of Food Service Workers,* Avi Publishing, Westport, Conn., 1969.
12. **Sykes, G.,** *Disinfection and Sterilisation,* 2nd ed., Spon, London, 1965.
13. **Vaughn, R. H.,** Food plant sanitation, in *Food Processing Operations,* Vol. 1, Joslyn, M. A. and Heid, J. L., Eds., Avi Publishing, Westport, Conn., 1963, 223–245.

B. Naturally Occurring Toxicants

1. General

1. **DeVries, C. A., and Kochva, E.,** *Toxins of Animal and Plant Origin,* Gordon & Breach, New York, 1972.
2. **Gontzea, I. and Sutzescu, P.,** *Antinutritive Substances in Foodstuffs and Forages,* S. Karger, Basel, 1968, 184 pp.
3. **Keegan, H. L. and MacFarlane, W. V.,** *Venomous and Poisonous Animals and Noxious Plants of the Pacific Region,* Pergamon Press, New York, 1963.
4. **Lewis, H. B.,** Natural toxicants and nutrition, *Nutr. Rev.,* 2, 97–99, 1944.
5. **Liener, I. E.,** Toxic factors in protein foods, in *Proteins in Human Nutrition,* Porter, J. W. G. and Rolls, B. A., Eds., Academic Press, London, 1973, 481–500.
6. **Liener, I. E., Ed.,** *Toxic Constituents of Animal Feedstuffs,* Academic Press, New York, 1974, 222 pp.
7. **Mickelsen, O. and Yang, M. G.,** Naturally occurring toxicants in foods, *Fed. Proc.,* 25, 104–123, 1966.
8. **Mickelsen, O., Yang, M. G., and Goodhart, R. S.,** Naturally occurring toxic foods, in *Modern Nutrition in Health and Disease,* 5th ed., Goodhart, R. S. and Shils, M. E., Eds., Lea & Febiger, Philadelphia, 1973, 412–433. (208 ref.)
9. National Research Council. Food Protection Committee, *Toxicants Occurring Naturally in Foods,* 2nd ed., National Academy of Sciences, Washington, D.C., 1973, 624 pp. (26 papers)
10. **Philp, J. M.,** Toxicity and naturally occurring chemicals in food, *R. Soc. Health J.,* 90, 237–242, 1970. (71 ref.)
11. **Scott, H. G.,** Poisonous plants and animals, in *Food-borne Infections and Intoxications,* Riemann, H., Ed., Academic Press, New York, 1969, 543–604.
12. **Strong, F. M.,** Naturally occurring toxic factors in plants and animals used as food, *Can. Med. Assoc. J.,* 94, 568–573, 1966.

13. **Strong, F. M.,** Toxicants occurring naturally in foods, *Nutr. Rev.*, 32(8), 225–231, 1974.
14. Symposium on natural food toxicants, *J. Agric. Food Chem.*, 17, 413–538, 1969. (19 papers)

2. Toxins of Animal Origin

1. **Bücherl, W. and Buckley, E. E.,** *Venomous Animals and Their Venoms,* Academic Press, New York, 1971. (3 vols.)
2. **Diniz, C. R. and Corrado, A. P.,** Venoms of insects and arachnids, in *Pharmacology and Toxicology of Naturally Occurring Toxins,* Rašková, H., Ed., International Encyclopedia of Pharmacology and Therapeutics, Pergamon Press, Oxford, 1971, 117–140.
3. **Henriques, S. B. and Henriques, O. B.,** Pharmacology and toxicology of snake venoms, in *Pharmacology and Toxicology of Naturally Occurring Toxins,* Rašková, H., Ed., International Encyclopedia of Pharmacology and Therapeutics, Pergamon Press, Oxford, 1971, 215–368.
4. **Keegan, H. L. and Toshioka, S.,** Some Venomous Animals of the Far East, U.S. Army 406th Medical General Laboratory, Camp Zama, Japan, 1958.
5. **Russell, F. E. and Saunders, P. R., Eds.,** *Animal Toxins. A Collection of Papers,* Proc. 1st International Symposium on Animal Toxins, Pergamon Press, Oxford, 1967, 428 pp. (43 papers)

3. Seafood Toxins

1. **Fuhrman, F. A., Fuhrman, G. J., Dull, D. L., and Mosher, H. S.,** Toxins from eggs of fishes and amphibia, *J. Agric Food Chem.*, 17, 417–424, 1969.
2. **Halstead, B. W.,** *Dangerous Marine Animals,* Cornell Maritime Press, Cambridge, MD, 1959.
3. **Halstead, B. W.,** *Poisonous and Venomous Marine Animals of the World,* U.S. Gov. Print. Off., Washington, D.C. 1965–1968. (3 vols.)
4. **Jensen, E. T., Ed.,** *Conference on Shellfish Toxicology,* U.S. Public Health Service, Washington, D.C., 1957, 126 pp
5. **Russell, F. E.,** *Marine Toxins and Venomous and Poisonous Marine Animals,* Academic Press, London, 1965, 176 pp.
6. **Russell, F. E.,** Marine toxins and venomous and poisonous marine animals, *Adv. Marine Biol.*, 3, 256–384, 1965.
7. **Russell, F. E.,** Pharmacology of toxins of marine organisms, in *Pharmacology and Toxicology of Naturally Occurring Toxins,* Rašková, H., Ed., International Encyclopedia of Pharmacology and Therapeutics, Pergamon Press Oxford, 1971, 3–114.
8. **Russell, F. E.,** Poisonous marine animals, in *The Safety of Foods,* Graham, H. D., Ed., Avi Publishing, Westport Conn., 1968, 68–81.
9. **Schantz, E. J.,** Studies on shellfish poisons, *J. Agric. Food Chem.*, 17, 413–416, 1969.
10. **Scheuer, P. J.,** Toxins from fish and other marine organisms, *Adv. Food Res.*, 18, 141–161, 1970.
11. **Wills, J. H., Jr.,** Seafood toxins, in *Toxicants Occurring Naturally in Foods,* National Research Council, Nationa Academy of Sciences, Washington, D.C., 1966, 147–163. (82 ref.)

4. Toxins of Plant Origin

1. **Brocq-Rousseau, D. and Fabre, R.,** *Les toxines végétales,* Herrmann, Paris, 1947.
2. **Ford, W. W.,** Plant poisons and their antibodies, *Zentralbl. Bakteriol. Parasitenkd. Infektionskr., Hyg. Abt. 1, Orig.* Ref. 58, 129–161, 193–222, 1913.
3. **Garner, R. J.,** Poisonous plants, in *Veterinary Toxicology,* Bailliere, Tindall & Cox, London, 1957, 263–356.
4. **Jaffé, W.,** Plant toxins, in *The Safety of Foods,* Graham, H. D., Ed., Avi Publishing, Westport, Conn., 1968, 60–67
5. **Kingsbury, J. M.,** *Poisonous Plants in the United States and Canada,* Prentice-Hall, Englewood Cliffs, N.J., 1964 626 pp. (1715 ref.)
6. **Lampe, K. F. and Fagerström, R.,** *Plant Toxicity and Dermatitis,* Williams & Wilkins, Baltimore, 1968, 231 pp.
7. **Leopold, A. C. and Ardrey, R.,** Toxic substances in plants and food habits of early man, *Science,* 176, 512–514 1972.
8. **Liener, I. E.,** Toxic factors in edible legumes and their elimination, *Am. J. Clin. Nutr.*, 11, 281–298, 1962.
9. **Liener, I. E.,** Toxic substances associated with seed proteins, in *World Protein Resources,* in *Advances in Chemistry,* Ser. No. 57, American Chemical Society, Washington, D.C., 1966, 178–194. (65 ref.)
10. **Liener, I. E., Ed.,** *Toxic Constituents of Plant Foodstuffs,* Academic Press, New York, 1969, 500 pp. (13 papers)
11. **Liener, I. E.,** Naturally occurring toxicants of horticultural significance, *Hortic. Sci.*, 8(2), 112–116, 1973. (18 ref.,
12. **Long, H. C.,** *Plants Poisonous to Livestock,* Cambridge University Press, Cambridge, 1917, 119 pp.
13. **Muenscher, W. C.,** *Poisonous Plants of the United States,* Macmillan, New York, 1961, 277 pp.
14. **Radeleff, R. D.,** Poisonous plants, in *Veterinary Toxicology,* Lea & Febiger, Philadelphia, 1964, 42–157.
15. **Schuphan, W.,** Food plants and environmental toxicology, *Proc. 8th International Congress of Nutrition,* Prague 1969, Excerpta Medica, Amsterdam, 1970, 722–730.
16. **Watt, J. M. and Breyer-Brandwijk, M. A.,** *The Medical and Poisonous Plants of Southern and Eastern Africa,* Livingstone, Edinburgh, 1932, 1457 pp.

5. Specific Toxins

a. Alkaloids

1. **Keeler, R. F.,** Toxic and teratogenic alkaloids of Western Range plants, *J. Agric. Food Chem.*, 17, 473–482, 1969.

b. Allergens

1. **Bleumink, E.,** Food allergy. The chemical nature of the substances eliciting symptoms, *World Rev. Nutr. Diet.*, 12, 506–570, 1970. (148 ref.)
2. **Cooke, R. A.,** Protein derivatives as factors in allergy, *Ann. Intern. Med.*, 16, 71–80, 1942. (30 ref.)
3. **Frankland, A. W.,** Food allergies, *R. Soc. Health J.*, 90, 243–247, 1970. (22 ref.)
4. **Goldstein, G. B. and Heiner, D. C.,** Clinical and immunological perspectives in food sensitivity. A review, *J. Allergy*, 46, 270–291, 1970. (173 ref.)
5. **Masesmann, H. C., Jr.,** Foods as antigens and allergens, in *Toxicants Occurring Naturally in Foods*, National Research Council, National Academy of Sciences, Washington, D.C., 1966, 72–93. (50 ref.)
6. Nutrition and allergy, *Proc. 7th International Congress of Nutrition*, Vol. 1, Hamburg, 1966, Pergamon Press, Oxford, 1967, 352–369. (4 papers)
7. **Perlman, F.,** Food allergy and vegetable proteins, *Food Technol.*, 20, 1438–1442, 1966.
8. **Perlman, F.,** Allergens, in *Toxic Constituents of Plant Foodstuffs*, Liener, I. E., Ed., Academic Press, New York, 1969, 319–348.
9. **Rinkle, H. J. and Randolph, T. C.,** *Food Allergy*, Charles C Thomas, Springfield, Ill., 1951.
10. **Rowe, A. H.,** Food allergy. Reasons for delayed recognition and control by physicians, *Q. Rev. Allergy Appl. Immunol.*, 8, 391–403, 1954.
11. **Spies, J. R.** Allergens, *J. Agric. Food Chem.*, 22, 30–36, 1974.
12. **Withers, O. R. and Hale, R.,** Food allergy. A review of the literature, *Ann. Allergy*, 14, 384–456, 1956.

c. Amino Acids and Peptides (Toxic)

1. **Fowden, L., Lewis, D., and Tristram, H.,** Toxic amino acids: Their action as antimetabolites, *Adv. Enzymol.*, 29, 89–163, 1967.
2. **Hylin, J. W.,** Toxic peptides and amino acids in foods and feeds, *J. Agric. Food Chem.*, 17, 493–496, 1969.

d. Antienzyme Factors

1. **Ambrose, A. M.,** Naturally occurring antienzymes (inhibitors), in *Toxicants Occurring Naturally in Foods*, National Research Council, National Academy of Sciences, Washington, D.C., 1966, 105–111. (35 ref.)
2. **Crosby, D. G.,** Natural cholinesterase inhibitors in food, in *Toxicants Occurring Naturally in Foods*, National Research Council, National Academy of Sciences, Washington, D.C., 1966, 112–116.

e. Antivitamins

1. **Klosterman, H. J.,** Vitamin B_6 antagonists of natural origin, *J. Agric. Food Chem.*, 22, 13–16, 1974.
2. **Lepkovsky, S.,** Antivitamins in foods, in *Toxicants Occurring Naturally in Foods*, National Research Council, National Academy of Sciences, Washington, D.C., 1966, 98–104. (51 ref.)
3. **Link, K. P.,** The discovery of dicumerol and its sequels, *Circulation*, 19, 97–107, 1959.

f. Carcinogens

1. **Armstrong, R. W., Jr.,** Is there a particular kind of soil or geologic environment that predisposes to cancer? *Ann. N.Y. Acad. Sci.*, 199, 239–248, 1972. (59 ref.)
2. **Basu, T. K., Dickerson, J. W. T., and Williams, D. C.,** Inter-relationships of nutrition and cancer, *Ecol. Food Nutr.*, 2, 193–199, 1973.
3. **Miller, J. A.,** Tumorigenic and carcinogenic natural products, in *Toxicants Occurring Naturally in Foods*, National Research Council, National Academy of Sciences, Washington, D.C., 1966, 24–39. (94 ref.)
4. **Roe, F. J. C. and Lancaster, M. C.,** Natural metallic and other substances as carcinogens, *Br. Med. Bull.*, 20, 127–133, 1964.
5. **Wogan, G. N.,** Naturally occurring carcinogens in foods, *Prog. Exp. Tumor Res.*, 11, 134–162, 1969. (110 ref.)

g. Cyanogens

1. **Auld, S. J. M.,** Cyanogenesis under digestive condtions, *J. Agric. Sci.*, 5, 409–417, 1912–13.
2. **Conn, E. E.,** Cyanogenic glycosides, *J. Agric. Food Chem.*, 17, 519–526, 1969.
3. **Gettler, A. O. and Baine, J. O.,** The toxicology of cyamide, *Am. J. Med. Sci.*, 195, 182–198, 1938.
4. **Liener, I. E.,** Cyanogenetic glycosides, in *Toxicants Occurring Naturally in Foods*, National Research Council, National Academy of Sciences, Washington, D.C., 1966, 58–61. (17 ref.)
5. **Montgomery, R. D.,** Cyanogens, in *Toxic Constituents of Plant Foodstuffs*, Liener, I. E., Ed., Academic Press, New York, 1969, 143–157.

h. Cycasin

1. **Laqueur, G. L.,** Gnotobiotes as useful tools in studies in carcinogenesis with cycasin, a natural glucoside, in *Germ-free Biology,* Mirand, E. A. and Buck, N., Eds., *Adv. Exp. Med. Biol.,* Vol. 3, Plenum Press, New York, 1969 163–167.
2. **Laqueur, G. L. and Spatz, M.,** Toxicology of cycasin, *Cancer Res.,* 28, 2262–2267, 1968.(61 ref.)
3. **Mugera, G. M. and Nderito, P.,** Cycad as hepatic carcinogens in Kenya, in *Cancer in Africa,* Clifford, P., Linsell, C. A. and Timms, G. L., Eds., East Africa Medical Journal, Nairobi, 1968, 323.
4. **Spatz, M.,** Toxic and carcinogenic alkylating agents from cycads, *Ann. N.Y. Acad. Sci.,* 163, 848–859, 1969. (52 ref.)
5. **Whiting, M. G.,** Toxicity of cycads, *Econ. Bot.,* 17, 271–302, 1963.
6. **Whiting, M., Spatz, M., and Matsumoto, H.,** Research progress on cycads, *Econ. Bot.,* 20, 98–102, 1966.
7. **Yang, M. G., Kobayashi, A., and Mickelsen, O.,** Bibliography of cycad research, *Fed. Proc.,* 31, 1543–1546, 1972 (120 ref.)
8. **Yang, M. G. and Mickelsen, O.,** Cycads, in *Toxic Constituents of Plant Foodstuffs,* Liener, I. E., Ed., Academio Press, New York, 1969, 159–167.

i. Essential Nutrients

(1) General

1. **Boyd, E. M.,** *Toxicity of Pure Foods,* CRC Press, Cleveland, Ohio, 1973, 200 pp.
2. **Scrimshaw, N. S.,** Specific nutrients, in *Chemical and Biological Hazards in Foods,* Ayres, J. C., Kraft, A. A. Snyder, H. E., and Walker, H. W., Eds., Iowa State University Press, Ames, 1962, 27–49.

(2) Amino Acids

1. **Benevenga, N. J.,** Toxicities of methionine and other amino acids, *J. Agric. Food Chem.,* 22, 2–9, 1974.
2. **Harper, A. E.,** Excess of indispensable amino acids, in *Toxicants Occurring Naturally in Foods,* National Research Council, National Academy of Sciences, Washington, D.C., 1966, 221–228. (20 ref.)

(3) Minerals

1. **Davis, G. K.,** Toxicity of the essential minerals, in *Toxicants Occurring Naturally in Foods,* National Research Council, National Academy of Sciences, Washington, D.C., 1966, 229–235. (35 ref.)
2. **Mennely, G. R.,** Toxic effects of dietary sodium chloride and the protective effect of potassium, in *Toxicants Occurring Naturally in Foods,* National Research Council, National Academy of Sciences, Washington, D.C.; 1966, 267–279.

(4) Vitamins

1. **Ostwald, R. and Briggs, G. M.,** Toxicity of the vitamins, in *Toxicants Occurring Naturally in Foods,* National Research Council, National Academy of Sciences, Washington, D.C., 1966, 183–220. (186 ref.)

j. Estrogens

1. **Bickoff, E. M., Livingston, A. L., Hendrickson, A. P., and Booth, A. N.,** Relative potencies of several estrogen-like compounds found in forages, *J. Agric. Food Chem.,* 10, 410–414, 1962.
2. **Bradbury, R. B. and White, D. E.,** Estrogens and related substances in plants, *Vitamins Hormones,* 12, 207–233, 1954.
3. **Burroughs, W., Cheng, E. W., Yoder, L., and Story, C. D.,** Estrogenic activity of some naturally occurring isoflavones, *Ann. N.Y. Acad. Sci.,* 61, 652–659, 1955.
4. **Stob, M.,** Estrogens, in *Toxicants Occurring Naturally in Foods,* National Research Council, National Academy of Sciences, Washington, D.C., 1966, 18–23.

k. Fats (Toxic)

1. **Mead, J. F. and Alfin-Slater, R. B.,** Toxic substances present in food fats, in *Toxicants Occurring Naturally in Foods,* National Research Council, National Academy of Sciences, Washington, D.C., 1966, 236–241. (28 ref.)

l. Favism-Producing Factors

1. **Du, S. D.,** Favism in West China, *Chin. Med. J.,* 70, 17–26, 1952.
2. **Liener, I. E.,** Favism, in *Toxicants Occurring Naturally in Foods,* National Research Council, National Academy of Sciences, Washington, D.C., 1966, 47–50.
3. **Mager, J., Razin, A., and Hershko, A.,** Favism, in *Toxic Constituents of Plant Foodstuffs,* Liener, I. E., Ed., Academic Press, New York, 1969, 293–318.
4. **Vella, T.,** Favism in Asia, *Med. J. Aust.,* 2, 196–197, 1959.

m. Goitrogens

1. **Bell, J. M. and Williams, K.,** Growth depressing factors in rapeseed oilmeal, *Can. J. Agric. Sci.,* 33, 201–209, 1953.
2. **Gaitán, E.,** Water-borne goitrogens and their role in the etiology of endemic goiter, *World Rev. Nutr. Diet.,* 17, 53–90, 1973.
3. **Greer, M. A.,** Nutrition and goiter, *Physiol. Rev.,* 30, 513–548, 1950.
4. **Greer, M. A.,** Goitrogenic substances in food, *Am. J. Clin. Nutr.,* 5, 440–444, 1957.
5. **Greer, M. A.,** The significance of naturally occurring antithyroid compounds in the production of goiter in man, *Borden's Rev. Nutr. Res.,* 21, 61–73, 1960. (71 ref.)
6. **Greer, M. A.,** The natural occurrence of goitrogenic agents, *Recent Prog. Hormone Res.,* 18, 187–219, 1962.
7. **Greer, M. A., Kendall, J. W., and Smith, M.,** Antithyroid compounds, in *The Thyroid Gland,* Pitt-Rivers, R. and Trotter, W. R., Eds., Butterworth, London, 1964, 357–389.
8. **Van Etten, C. H.,** Goitrogens, in *Toxic Constituents of Plant Foodstuffs,* Liener, I. E., Ed., Academic Press, New York, 1969, 103–142.
9. **Virtanen, A. I.,** On the chemistry of the *Brassica*-factor, its effect on the function of the thyroid gland and its transfer to milk, *Experientia,* 17, 241–251, 1961.
10. **Wills, J. H.,** Goitrogens, in *Toxicants Occurring Naturally in Foods,* National Research Council, National Academy of Sciences, Washington, D.C., 1966, 3–17. (83 ref.)

n. Gossypol

1. **Adams, R., Geissman, T. A., and Edwards, J. D.,** Gossypol, a pigment of cottonseed, *Chem. Rev.,* 60, 555–574, 1960.
2. **Berardi, L. C. and Goldblatt, L. A.,** Gossypol, in *Toxic Constituents of Plant Foodstuffs,* Liener, I. E., Ed., Academic Press, New York, 1969, 211–266.
3. **Boatner, C. H.,** Pigments of cottonseed, in *Cottonseed and Cottonseed Products. Their Chemistry and Chemical Technology,* Bailey, A. E., Ed., John Wiley & Sons, New York, 1948, 213–363.
4. **Eagle, E.,** Gossypol, in *Toxicants Occurring Naturally in Foods,* National Research Council, National Academy of Sciences, Washington, D.C., 1966, 242–249. (74 ref.)
5. **Markman, A. L. and Rzhekhin, V. P.,** Gossypol and Its Derivatives, Clearinghouse for Federal Scientific and Technical Information, Springfield, Va., 1968, 189 pp.
6. **Pons, W. A., Jr., Hoffpauir, C. L., and Hopper, T. H.,** Gossypol in cottonseed: Influence of variety of cottonseed and environment, *J. Agric. Food Chem.,* 1, 1115–1118, 1953.
7. Proc. Conf. on Inactivation of Gossypol with Mineral Salts, New Orleans, 1966, National Cottonseed Products Association, Memphis, Tenn., 1966.
8. Proc. Conf. on Chemical Structure and Reactions of Gossypol and Nongossypol Pigments of Cottonseed, National Cottonseed Products Association, Memphis, Tenn., 1959.

o. Hemagglutinins

1. **Bird, G. W. G.,** Haemagglutinins in seeds, *Br. Med. Bull.,* 15, 165–168, 1959.
2. **Boyd, W. C.,** The lectins: Their present status, *Vox Sang.,* 8, 1–32, 1963. (128 ref.)
3. **Jaffé, W. G.,** Hemagglutinins, in *Toxic Constituents of Plant Foodstuffs,* Liener, I. E., Ed., Academic Press, New York, 1969, 69–101.
4. **Liener, I. E.,** Seed hemagglutinins, *Econ. Bot.,* 18, 27–33, 1964. (74 ref.)
5. **Liener, I. E.,** Hemagglutinins in foods, in *Toxicants Occurring Naturally in Foods,* National Research Council, National Academy of Sciences, Washington, D.C., 1966, 51–57. (57 ref.)
6. **Liener, I. E.,** Phytohemogglutinins: Their nutritional significance, *J. Agric. Food Chem.,* 22, 17–22, 1974.
7. **Tobiska, J.,** *Die Phytohämagglutinine,* Akademie Verlag, Berlin, 1964.

p. Hepatotoxins

1. **Schoental, R.,** Liver disease and "natural hepatotoxins", *Bull. W. H. O.,* 29, 823–833, 1963.

q. Hydrocarbons

1. **Howard, J. W. and Fazio, T.,** A review of polycyclic aromatic hydrocarbons in foods, *J. Agric. Food Chem.,* 17, 527–531, 1969.

r. Lathyrogens

1. **Dastur, D. K. and Iyer, C. G. S.,** Lathyrism versus odoratism, *Nutr. Rev.,* 17, 33–36, 1959.
2. **Gardner, A. F.,** Experimental lathyrism. Review of the literature, *Am. J. Clin. Nutr.,* 7, 213–223, 1959.
3. **German, W. J.,** Lathyrism. A review of recent developments, *J. Neurosurg.,* 17, 657–663, 1960.
4. **Liener, I. E.,** Lathyrogens, in *Toxicants Occurring Naturally in Foods,* National Research Council, National Academy of Sciences, Washington, D.C., 1966, 40–46. (33 ref.)

5. **Sarma, P. S. and Padmanaban, G.,** Lathyrogens, in *Toxic Constituents of Plant Foodstuffs,* Liener, I. E., Ed., Academic Press, New York, 1969, 267–291.
6. **Selye, H.,** Lathyrism, *Rev. Can. Biol.,* 16, 1–82, 1957.

s. Neurotoxins

1. **Simpson, L. L., Ed.,** *Neuropoisons. Their Pathophysiological Actions.* Vol. 1, *Poisons of Animal Origin,* Plenum Press, New York, 1971, 361 pp. (15 papers)
2. **Tu, A. T.,** Sea snake venoms and neurotoxins, *J. Agric. Food Chem.,* 22, 36–43, 1974.
3. **Watt, D. D., Babin, D. R., and Mlejnek, R. V.,** The protein neurotoxins in scorpion and elapid snake venoms, *J. Agric. Food Chem.,* 22, 43–51, 1974.

t. Nitrates and Nitrites

1. **Fassett, D. W.,** Nitrates and nitrites, in *Toxicants Occurring Naturally in Foods,* National Research Council, National Academy of Sciences, Washington, D.C., 1966, 250–256. (26 ref.)
2. Institute of Nutrition, University of North Carolina, Effects of Nitrates and Nitrites in Food as Related to Human Health, Chapel Hill, N.C., 1974.
3. **Phillips, W. E.,** Naturally occurring nitrate and nitrite in foods in relation to infant methaemoglobinaemia, *Food Cosmet. Toxicol.,* 9, 219–228, 1971. (42 ref.)
4. **Wolff, I. A. and Wasserman, A. E.,** Nitrates, nitrites, and nitrosamines, *Science,* 177, 15–19, 1972.
5. **Wright, M. J. and Davidson, K. L.,** Nitrate accumulation in crops and nitrate poisoning in animals, *Adv. Agron.,* 16, 197–247, 1964.

u. Oxalates

1. **Fassett, D. W.,** Oxalates, in *Toxicants Occurring Naturally in Foods,* National Research Council, National Academy of Sciences, Washington, D.C., 1966, 257–266.

v. Phenolic Substances

1. **Singleton, V. L. and Kratzer, F. H.,** Toxicity and related physiological activity of phenolic substances of plant origin, *J. Agric. Food Chem.,* 17, 497–512, 1969.

w. Pressor Amines

1. **Strong, F. M.,** Pressor amines in foods, in *Toxicants Occurring Naturally in Foods,* National Research Council, National Academy of Sciences, Washington, D.C., 1966, 94–97. (17 ref.)

x. Protease Inhibitors

1. **Birk, Y.,** Chemistry and nutritional significance of proteinase inhibitors from plant sources, in *Chemistry, Pharmacology, and Clinical Applications of Proteinase Inhibitors,* Back, A. N. and Mammen, E. F., Eds., *Ann. N.Y. Acad. Sci.,* 146, 388–399, 1968.
2. **Borchers, R.,** Raw soybean growth inhibitors, *Fed. Proc.,* 24, 1494–1497, 1965.
3. **Dechary, J. M.,** Seed proteases and protease inhibitors, *Econ. Bot.,* 24, 113–122, 1970. (81 ref.)
4. **Elias, P. S.,** Toxic agents and protein availability, in *Proteins as Human Food,* Lawrie, R. A., Ed., Avi Publishing, Westport, Conn., 1970, 65–88.
5. **Laskowski, M. and Laskowski, M., Jr.,** Naturally occurring trypsin inhibitors, *Adv. Protein Chem.,* 9, 203–242, 1954.
6. **Liener, I. E.,** Toxic factors in edible legumes and their elimination, *Am. J. Clin. Nutr.,* 11, 287–298, 1962.
7. **Liener, I. E.,** The effect of heat processing and storage on the nutritive value of proteins of importance in the cereal industry, *Cereal Chem.,* 8, 162–185, 1950.
8. **Liener, I. E.,** Effect of heat on plant proteins, in *Processed Plant Protein Foodstuffs,* Altschul, A. M., Ed., Academic Press, New York, 1958, 79–129.
9. **Liener, I. E.,** Toxic substances associated with seed proteins, in *World Protein Resources,* in *Advances in Chemistry,* Ser. No. 57, American Chemical Society, Washington, D.C., 1966, 178–194.
10. **Liener, I. E. and Kakade, M. L.,** Protease inhibitors, in *Toxic Constituents of Plant Foodstuffs,* Liener, I. E., Ed., Academic Press, New York, 1969, 7–68.
11. **Ofelt, C. W., Smith, A. K., and Mills, J. M.,** Proteases of the soybean, *Cereal Chem.,* 32, 53–63, 1955.
12. **Polanowski, A.,** Trypsin inhibitor from rye seeds, *Acta Biochim. Polon.,* 14, 389–395, 1967.
13. **Pusztai, A.,** Trypsin inhibitors of plant origin, their chemistry and potential role in animal nutrition, *Nutr. Abstr. Rev.,* 37, 1–9, 1967.
14. **Rackis, J. J.,** Isolation and characterization of soybean trypsin inhibitors, Proc. Seed Protein Conf., New Orleans, La., U.S. Dep. Agric., 1963, 98–108.

15. **Rackis, J. J.**, Physiological properties of soybean trypsin inhibitors and their relationship to pancreatic hypertrophy and growth inhibition of rats, *Fed. Proc.*, 24, 1488–1493, 1965.
16. **Rackis, J. J.**, Soybean trypsin inhibitors: Their inactivation during meal processing, *J. Food Technol.*, 20, 102–104, 1966.
17. **Ramirez, J. S. and Mitchell, H. L.**, The trypsin inhibitor of alfalfa, *J. Agric. Food Chem.*, 8, 393–395, 1960.
18. **Steiner, R. F. and Frattalli, V.**, Purification and properties of soybean protein inhibitors of proteolytic enzymes, *J. Agric. Food Chem.*, 17, 513–518, 1969.
19. **Vogel, R., Trautschold, I., and Werle, E.**, *Natürliche Proteinasen-Inhibitoren*, Thieme, Stuttgart, 1966.
20. **Ubatuba, F. B.**, Occurence of a trypsin inhibiting factor in the seeds of Canavalia ensiformis, *Rev. Brasil Biol.*, 21, 168–171, 1955.

y. Proteins and Peptides (Toxic)

1. Symposium on Toxic Proteins, Peptides, and Related Substances, *J. Agric. Food Chem.*, 22, 1–53, 1974.
2. **Van Hayningen, W. E.**, Toxic proteins, in *The Proteins. Chemistry, Biological Activity and Methods*, Vol. 2, part A, Neurath, H. and Bailey, K. Eds., Academic Press, New York, 1954, 345–387. (263 ref.)

z. Saponins

1. **Birk, Y.**, Saponins, in *Toxic Constituents of Plant Foodstuffs*, Liener, I. E., Ed., Academic Press, New York, 1969, 169–210.
2. **Boiteau, P., Pasich, B., and Rakoto Ratsimamanga, A.**, *Les Triterpenoides en Physiologie Vegetale et Animale*, Gauthier-Villard, Ed., Centre National de la Recherche Scientifique, Paris, 1964.
3. **Cole, H. H., Huffman, C. F., Kleiber, M., Olson, T. M., and Shalk, A. F.**, A review of bloat in ruminants, *J. Animal Sci.*, 4, 183–236, 1945.
4. **Coulson, C. B.**, Saponins. I. Triterpenoid saponins from lucerne and other species, *J. Sci. Food Agric.*, 9, 281–288, 1958.
5. **Coulson, C. B. and Davies, T.**, Saponins. II. Fractionation and pharmacological properties of lucerne saponins. The possible relation of these to bloat, *J. Sci. Food Agric.*, 13, 53–57, 1962.
6. **George, A. J.**, Legal status and toxicity of saponins, *Food Cosmet. Toxicol.*, 3, 85–91, 1965.
7. **Gestetner, B., Birk, Y., and Tencer, Y.**, Soybean saponins. Fate of ingested soybean saponins and the physiological aspect of their hemolytic activity, *J. Agric. Food Chem.*, 16, 1031–1035, 1968.
8. **Korbert, R.**, Saponins, their properties, composition and uses, *Pharm. J.*, 86, 244–245, 293–294, 1911.
9. **Tschesche, R. and Wulff, G.**, Konstitution und Eigenschaften der Saponine, *Planta Med.*, 12, 272–292, 1964.

za. Stimulants and Depressants

1. **Blood, F. R. and Randolph, G. G.**, Some naturally occurring stimulants and depressants, in *Toxicants Occurring Naturally in Foods*, National Research Council, National Academy of Sciences, Washington, D.C., 1966, 62–71. (45 ref.)

zb. Thioglucosides

1. **Van Etten, C. H., Daxenbichler, M. E., and Wolff, I. A.**, Natural glucosinolates (Thioglucosides) in foods and feeds, *J. Agric. Food Chem.*, 17, 483–491, 1969.

zc. Unusual Foods

1. **Van Veen, A. G.**, Toxic properties of some unusual foods, in *Toxicants Occurring Naturally in Foods*, National Research Council, National Academy of Sciences, Washington, D.C., 1966, 174–182. (21 ref.)

6. Radioactivity

1. **Bird, P. M.**, Radionuclides in food, *Can. Med. Assoc. J.*, 94, 590–597, 1966.
2. **Comar, C. L.**, Movement of fallout radionuclides through the biosphere and man, *Annu. Rev. Nucl. Sci.*, 15, 175–206, 1965.
3. **Comar, C. L.**, Natural radioactivity in the biosphere and foodstuffs, in *Toxicants Occurring Naturally in Foods*, National Research Council, National Academy of Sciences, Washington, D.C., 1966, 117–125. (5 ref.)
4. **Comar, C. L. and Thompson, J. C., Jr.**, Radioactivity in foods, in *Modern Nutrition in Health and Disease. Dietotherapy*, 5th ed., Goodhart, R. S. and Shils, M. E., Eds., Lea & Febiger, Philadelphia, 1973, 442–454. (22 ref.)
5. **Dunning, G. M.**, Foods and fallout, *Borden's Rev. Nutr. Res.*, 23, 1–15, 1962.
6. **Fowler, E. B., Ed.**, *Radioactive fallout, soils, plants, foods, man*, Elsevier, Amsterdam, 1965.
7. **Garner, R. J.**, *Transfer of Radioactive Materials from the Terrestrial Environment to Animals and Man*, Chemical Rubber Co., Cleveland, Ohio, 1972, 57 pp.
8. National Research Council, *Radionuclides in Foods*, National Academy of Sciences, Washington, D.C., 1973, 97 pp.
9. Nutritional hazards caused by radioactive contamination of foods, *Proc. Nutr. Soc.*, 18, 34–53, 1959. (4 papers)

10. **Russell, R. S., Ed.,** *Radioactivity and Human Diet,* Pergamon Press, Oxford, 1966, 552 pp.
11. Symposium on radioactive fallout in relation to foods, *J. Agric. Food Chem.*, 9, 90–126, 1961. (8 papers)

C. Food Additives

1. General

1. American Chemical Society, *Chemicals in Modern Food and Fiber Production,* National Academy of Sciences, Washington, D.C., 1962, 104 pp.
2. **Barnes, J. M.,** The potential toxicity of chemicals used in food technology, *Proc. Nutr. Soc.,* 15, 148–155, 1956.
3. **Benarde, M. A.,** *Chemicals We Eat,* American Heritage Press, New York, 1971.
4. **Berryman, G. H.,** This chemical world, in *Proc. SOS/70. 3rd International Congress Food Science and Technology,* Institute of Food Technologists, Chicago, 1971, 748–752.
5. **Bing, F. C.,** Chemical and other additives to foods, in *Modern Nutrition in Health and Disease,* 4th ed., Wohl, M. G. and Goodhart, R. S., Eds., Lea & Febiger, Philadelphia, 1968, 496–505.
6. **Cannon, P. R.,** Chemicals in food production, in *Chemical and Biological Hazards in Food,* Ayres, J. C., Kraft, A. A., Snyder, H. E., and Walker, H. W., Eds., Iowa State University Press, Ames, 1962, 13–23.
7. **Frey, C. N.,** Chemicals in foods, in *Handbook of Food and Agriculture,* Blanck, F. C., Ed., Reinhold, New York, 1955, 789–851.
8. **Furia, T. E., Ed.,** *CRC Handbook of Food Additives,* 2nd ed., CRC Press, Cleveland, 1973, 1,000 pp. (15 papers)
9. **Gillet, J. M.,** *Food Additives Manual,* Manufacturing Chemists' Association, Washington, D.C., 1962. (2 vols.)
10. **Goodwin, R. W. L., Ed.,** *A Symposium on Chemical Additives in Food,* Churchill, London, 1967, 128 pp. (7 papers)
11. **Hueper, W. C.,** The potential role of non-nutritive food additives and contaminants as environmental carcinogens, *Acta Un. Int. Cancer,* 13, 221–252, 1957.
12. **Kuehnau, J.,** The problem of foreign substances in foodstuffs, *Triangle,* 4, 313–321, 1960.
13. **Lehman, A. J., Ed.,** *Appraisal of the Safety of Chemicals in Foods, Drugs and Cosmetics,* Association of Food and Drug Officials, Topeka, Kansas, 1959, 107 pp.
14. Manufacturing Chemists' Association, *Food Additives. What They Are, How They Are Used,* Washington, D.C., 1971, 63 pp.
15. **Menon, K. K. G.,** Problem of food additives and contaminants, in Proc. 1st Asian Congress of Nutrition, Tulpule, P. G. and Jaya Rao, K. S., Eds., Nutrition Society of India, Hyderabad, 1972, 496–504. (26 ref.)
16. **Mrak, E. M.,** Foods and chemicals, *Borden's Rev. Nutr. Res.,* 23, 21–29, 1962.
17. National Research Council. Food Protection Committee, *Chemicals Used in Food Processing,* National Academy of Sciences, Washington, D.C., 1965, 295 pp.
18. National Research Council. Food Protection Committee, *Food Chemicals Code,* National Academy of Sciences, Washington, D.C., 1966, 846 pp.
19. **Oser, B. L.,** The proper use of food additives, in *The Safety of Foods,* Graham, H. D., Ed., Avi Publishing, Westport, Conn., 1968, 278–284.
20. **Oser, B. L.,** Additives and pesticides in foods, in *Modern Nutrition in Health and Disease,* 5th ed., Goodhart, R. S. and Shils, M. E., Eds., Lea & Febiger, Philadelphia, 1973, 434–441.
21. **Philp, J. McL.,** *Food Additives. I. Modern Trends in Toxicology,* Butterworths, London, 1968.
22. **Sapeika, N.,** Food additives, in *Food, Nutrition and Health,* Rechcigl, M., Jr., Ed., *World Rev. Nutr. Diet.,* Vol. 16 S. Karger, Basel, 1973, 334–362.
23. **Stoner, H. B.,** The addition of chemicals to food, *Nutr. Diet.,* 2, 1–20, 1960.
24. Survey of analytical methods available for the estimation of some food additives in food, *Pure Appl. Chem.,* 26, 77–120, 1971. (418 ref.)
25. Symposium on additives and residues in human foods, *Am. J. Clin. Nutr.,* 9, 259–303, 1961.
26. Symposium on food additives and contaminants, *Proc. Nutr. Soc.,* 31, 1–44, 1972. (6 papers)

2. Coloring Agents

1. **Golberg, L.,** The toxicology of artificial colouring materials, *J. Soc. Cosmet. Chem.,* 18, 421–432, 1967.
2. Hazelton Laboratories, Inc., Use of certified FD & C colors in food, *Food Technol.,* 22, 946–949, 1968.
3. **Jablonski, C. F.,** Coloring matters in foods, in *Chemistry and Technology of Food and Food Products,* Vol. 1, Jacobs, M. B., Ed., Interscience, New York, 1951, 349–381.
4. **Little, A. C. and Mackinney, G.,** The color of foods, *World Rev. Nutr. Diet,* 14, 59–84, 1972.
5. **Mackinney, G. and Chichester, C. O.,** The color problem in food, *Adv. Food Res.,* 7, 301–351, 1954.
6. **Mackinney, G. and Little, A. C.,** *Color of Foods,* Avi Publishing, Westport, Conn., 1962, 321 pp.
7. **Mackinney, G. and Little, A. C.,** The coloring matters of food, *World Rev. Nutr. Diet.,* 14, 85–99, 1972.
8. **Noonan, J.,** Color additives in food, in *Handbook of Food Additives,* Furia, T. E., Ed., Chemical Rubber Co., Cleveland, 1968, 25–49.
9. **Schramm, A. T.,** Toxicological assessment of food colors, in *Proc. SOS/70. 3rd International Congress Food Science and Technology,* Institute of Food Technologists, Chicago, 1971, 784–792.
10. Specifications for the Identity and Purity and Toxicological Evaluation of Food Colors, FAO Nutr. Meetings Rep. Ser. No. 38B, WHO/Food Add./66.25, FAO/WHO, 1966.

11. **Williams, R. T.,** Biochemical and pharmacological aspects of colors and flavors, *Chem. Ind.* (London), 473–476, 1959.
12. **Winston, R. P.,** Food colors, in *A Symposium on Chemical Additives in Food,* Goodwin, R. W. L., Ed., Churchill, London, 1967, 115–123.

3. Flavoring Agents

1. **Eiserle, R. J. and Downey, W. J.,** A review of the literature concerned with flavor research as it applies to the problems of the flavor industry, *CRC Crit. Rev. Food Technol.,* 2(2), 159–170, 1971. (18 ref.)
2. FAO/WHO Expert Committee on Food Additives, Toxicological Evaluation of Some Flavouring and Non-nutritive Sweetening Agents, FAO Nutr. Meetings Rep. Ser. No. 44A, WHO/Food/Add., 68, 33, 1967.
3. **Furia, T. E. and Bellanca, N., Eds.,** *Fenaroli's Handbook of Flavor Ingredients,* Chemical Rubber Co., Cleveland, Ohio, 1971, 803 pp.
4. **Hall, R. L.,** Toxicants occurring naturally in spices and flavors, *Toxicants Occurring Naturally in Foods,* National Research Council, National Academy of Sciences, Washington, D.C., 1966, 164–173. (54 ref.)
5. **Hall, R. L.,** Some specific problems related to the safety-in-use of flavors, in *Proc. SOS/70. 3rd International Congress Food Science and Technology,* Institute of Food Technologists, Chicago, 1971, 779–783.
6. **Hall, R. L. and Oser, B. L.,** Recent progress in the consideration of flavoring ingredients under the Food Additives Amendment III (generally recognized as safe) substances, *Food Technol.,* 19, 151–197, 1965.
7. **Jenner, P. M., Hogan, E. C., Taylor, J. M., Cookland, E. L., and Fitzhugh, O. G.,** Food flavoring and compounds of related structure. 1. Acute oral toxicity, *Food Cosmet. Toxicol.,* 2, 327–343, 1964.
8. **Kilbuck, J. H.,** Seasoning for the food manufacturer, in *Food Processing Operations,* Vol. 2, Avi Publishing, Westport, Conn., 1963, 174–191.
9. **Lessel, B.,** Carcinogenic and teratogenic aspects of saccharin, in *Proc. SOS/70. 3rd International Congress Food Science and Technology,* Institute of Food Technologists, Chicago, 1971, 764–770.
10. **Merory, J.,** *Food Flavorings. Composition, Manufacture, and Use,* 2nd ed., Avi Publishing, Westport, Conn., 1968, 478 pp.
11. National Research Council. Food Protection Committee, *The Safety of Artificial Sweeteners for Use in Foods,* National Academy of Sciences, Washington, D.C., 1955.
12. **Oser, B. L.,** Comparison of the safety-in-use of natural and synthetic flavors, in *Proc. SOS/70. 3rd International Congress Food Science and Technology,* Institute of Food Technologists, Chicago, 1971, 771–778.
13. **Salant, A.,** Nonnutritive sweeteners, *Handbook of Food Additives,* Furia, T. E., Ed., Chemical Rubber Co., Cleveland, 1968, 501–563.
14. **Sjöström, L. B.,** Flavor potentiators, in *Handbook of Food Additives,* Furia, T. E., Ed., Chemical Rubber Co., Cleveland, 1968, 493–500.
15. **Sjöström, L. B., and Kensler, C. J.,** Flavors and colors, in *Chemical and Biological Hazards in Food,* Ayres, J. C., Kraft, A. A., Snyder, H. E., and Walker, H. W., Eds., Iowa State University Press, Ames, 1962, 50–59.
16. **Swaine, R. L.,** Natural and synthetic flavorings, in *Handbook of Food Additives,* Furia, T. E., Ed., Chemical Rubber Co., Cleveland, 1968, 461–491.
17. **Verbeek, E. M.,** Food flavors, in *A Symposium on Chemical Additives in Food,* Goodwin, R. W. L., Ed., Churchill, London, 1967, 95–113.

4. Antioxidants

1. **Johnson, F. C.,** A critical review of the safety of phenolic antioxidants in foods, *CRC Crit. Rev. Food Technol.,* 2(2), 267–304, 1971. (114 ref.)
2. **Daniel, J. W.,** Toxicological assessment of foodstuff antioxidants, in *Proc. SOS/70. 3rd International Congress Food Science and Technology,* Institute of Food Technologists, Chicago, 1971, 793–798.
3. **Lehman, A. J., Fitzhugh, O. G., Nelson, A. A., and Woodward, G.,** The pharmacological evaluation of antioxidants, *Adv. Food Res.,* 3, 197–208, 1951.
4. **Mattil, F. C.,** Antioxidants, in *Chemical and Biological Hazards in Food,* Ayres, J. C., Kraft, A. A., Snyder, H. E., and Walker, H. W., Eds., Iowa State University Press, Ames, 1962, 60–69.
5. **Ostendorf, J. P.,** Antioxidants, in *A symposium on Chemical Additives in Food,* Goodwin, R. W. L., Ed., Churchill, London, 1967, 63–81.
6. **Pascal, G.,** Physiological and metabolic effects antioxidant food additives, *World Rev. Nutr. Diet.,* 19, 237–299, 1974.
7. **Stuckey, B. N.,** Antioxidants as food stabilizers, in *Handbook of Food Additives,* Furia, T. E., Ed., Chemical Rubber Co., Cleveland, 1968, 209–243.
8. **Ward, R. J.,** Biochemical and pharmacological aspects of antioxidants, *Chem. Ind.,* (Lond.), 498–501, 1959.

5. Antimicrobial Agents

1. **Allen, R. J. and Brook, M.,** Effects of sulphur dioxide in animals and man, in *Proc. SOS/70. 3rd International Congress Food Science Technology,* Institute of Food Technologists, Chicago, 1971, 799–805.
2. **Chichester, D. F. and Tanner, F. W., Jr.,** Antimicrobial food additives, in *Handbook of Food Additives,* Furia, T. E., Ed., Chemical Rubber Co., Cleveland, 1968, 137–207.
3. FAO/WHO Expert Committee on Food Additives, Evaluation of the Toxicity of a Number of Antimicrobials and Antioxidants, WHO Tech. Rep. Ser. No. 228, WHO, Geneva, 1962.
4. FAO/WHO Expert Committee on Food Additives, Specifications for Identity and Purity and Toxicological Evaluation of Some Antimicrobials and Antioxidants, FAO Nutr. Meetings Rep. Ser. No. 38A, 1965.
5. **Gibbs, B. M.,** Food preservatives, in *A Symposium on Chemical Additives in Food,* Goodwin, R. W. L., Ed., Churchill, London, 1967, 21–41.
6. **Ingram, M.,** Technical aspects of the commercial use of antimicrobial chemicals as food preservatives, *Chem. Ind.* (Lond.), 552–557, 1959.
7. **Ingram, M., Ottaway, F. J. H., and Coppock, J. B. M.,** The preservative action of acid substances in food, *Chem. Ind.* (Lond.), 1154–1163, 1956.
8. **Molin, N., Ed.,** *The Action, Use and Natural Occurrence of Microbial Inhibitors in Food,* Almqvist & Wiksell, Stockholm, 1964, 402 pp.
9. **Tarr, H. L.,** Microbial inhibitors, in *Chemical and Biological Hazards in Food,* Ayres, J. C., Kraft, A. A., Snyder, H. E., and Walker, H. W., Eds., Iowa State University Press, Ames, 1962, 202–223.
10. **Vaughn, R. H. and Stewart, G. F.,** Antibiotics as food preservatives, *JAMA,* 174, 1308–1310, 1960.
11. **Weiser, H. H., Kunkle, L. E., and Deatherage, F. E.,** The use of antibiotics in meat processing, *Appl. Micribiol.,* 2, 88–94, 1954.

6. Surface Active Substances

1. **Aylward, F.,** Surface-active substances in food processing. Some biochemical and pharmacological aspects, *Chem. Ind.* (Lond.), 524–531, 1959.
2. FAO/WHO Expert Committee on Food Additives, Specifications for the Identity and Purity of Food Additives and Their Toxicological Evaluation: Emulsifiers, Stabilizers, Bleaching and Maturing Agents, Tech. Rep. Ser. No. 2811, WHO, Geneva, 1964.
3. **Griffin, W. C. and Lynch, M. J.,** Surface active agents, in *Handbook of Food Additives,* Furia, T. E., Ed., Chemical Rubber Co., Cleveland, 1968, 413–445.
4. National Research Council. Food Protection Committee, *The Safety of Mono- and Diglycerides for Use as Intentional Additives in Foods,* National Academy of Sciences, Washington, D.C., 1952.
5. National Research Council. Food Protection Committee, *Safety of Polyoxyethylene Stearates for Use as Intentional Additives in Food,* National Academy of Sciences, Washington, D.C., 1953.
6. Report on the Review of the Emulsifiers and Stabilizers in Food Regulations, 1962, HMSO, London, 1970.

7. Nutritional Supplements

1. **Borenstein, B.,** Vitamins and amino acids, in *Handbook of Food Additives,* Furia, T. E., Ed., Chemical Rubber Co., Cleveland, 1968, 107–136.
2. **Scrimshaw, N. S.,** Specific nutrients, in *Chemical and Biological Hazards in Foods,* Ayres, J. C., Kraft, A. A., Snyder, H. E., and Walker, H. W., Eds., Iowa State University Press, Ames, 1962, 27–49.

8. Miscellaneous

1. **Ellinger, R. H.,** *Phosphates as Food Ingredients,* CRC Press, Cleveland, Ohio, 1972, 190 pp.

D. Food Contaminants

1. General

1. **Harris, K. L.,** Annotated bibliography of methods on the examination of foods and filth, *J. Assoc. Off. Agric. Chem.,* 24, 201–210, 1946.
2. **Chichester, C. O.,** Food pollution, in *Pollution and Environmental Health,* Faith, W. L., Ed., American Institute of Chemical Engineers, New York, 1961, Chem. Eng. Prog. Symp. Ser. No. 35, Vol. 57, 27–30.
3. **Friedman, L.,** Adventitious toxic factors in protein concentrates, *Food Technol.,* 18, 1553–1563, 1964.
4. **Kuehnau, J.,** The problem of foreign substances in foodstuffs, *Triangle,* 4, 313–321, 1960.
5. **Somers, E. and Smith, D. M.,** Source and occurrence of environmental contaminants, *Food Cosmet. Toxicol.,* 9, 185–193, 1971. (46 ref.)
6. Symposium: Clean food, *Proc. Nutr. Soc.,* 16, 135–163, 1957.
7. **Waldbott, G. L.,** *Health Effects of Environmental Pollutants,* C. V. Mosby, St. Louis, 1973, 448 pp.
8. **Wogan, G. N.,** Physiologically significant food contaminants, *Fed. Proc.,* 25, 124–129, 1966. (55 ref.)

2. Chemical Contaminants

a. General

1. **Bicknell, F.**, *Chemicals in Food and in Farm Produce: Their Harmful Effect*, Faber & Faber, London, 1960, 192 pp.
2. **Browning, E.**, *Toxicity of Industrial Metals*, Butterworths, London, 1961.
3. Chemical and Radionuclide Food Contamination, MSS Information Corp., New York, 1973, 188 pp.
4. **Golberg, L.**, Trace chemical contaminants in food. Potential for harm, *Food Cosmet. Toxicol.*, 9, 65–80, 1971. (99 ref.)
5. **Hemphill, D. D.**, *Trace Substances in Environmental Health*, University of Missouri Press, Columbia, 1969.
6. **Lee, D. H. K., Ed.**, *Metallic Contaminants and Human Health*, Academic Press, New York, 1972, 241 pp.

b. Specific Chemicals

(1) Arsenic

1. **Schroeder, H. A. and Balassa, J. J.**, Abnormal trace metals in man: Arsenic, *J. Chron. Dis.*, 19, 85–106, 1966.

(2) Beryllium

1. **Tepper, L. B.**, Beryllium, *CRC Crit. Rev. Toxicol.*, 1(3), 235–259, 1972.
2. **Tepper, L. B., Hardy, H. L., and Chamberlin, R. I.**, *Toxicity of Beryllium Compounds*, Elsevier, Amsterdam, 1961.

(3) Cadmium

1. **Friberg, L. T., Piscator, M., Nordberg, G., and Kjellstrom, T.**, *Cadmium in the Environment*, 2nd ed., CRC Press, Cleveland, Ohio, 1973, 150 pp.
2. **Schroeder, H. A. and Balassa, J. J.**, Abnormal trace metals in man: Cadmium, *J. Chron. Dis.*, 14, 236–258, 1961.

(4) Chromium

1. National Research Council, *Chromium*, National Academy of Sciences, Washington, D.C., 1974, 156 pp.

(5) Lead

1. **Clarkson, T. W.**, Epidemiological and experimental aspects of lead and mercury contamination of food, *Food Cosmet. Toxicol.*, 9, 229–243, 1971. (58 ref.)
2. Environmental Health Aspects of Lead, Proc. of a Symposium, Amsterdam, Oct. 1972, Commission of the European Communities, Luxembourg, 1973, 1168 pp.
3. Lead Poisoning in Man and the Environment, MSS Information Corp, New York, 1973, 226 pp.
4. National Research Council, *Lead. Airborne Lead in Perspective*, National Academy of Sciences, Washington, D.C., 1972, 330 pp.
5. **Schroeder, H. A. and Balassa, J. J.**, Abnormal trace metals in man: Lead, *J. Chron. Dis.*, 14, 408–425, 1961.
6. **Zimdahl, R. L. and Arvik, J. H.**, Lead in soils and plants: A literature review, *CRC Crit. Rev. Environ. Contr.*, 3(2), 213–224, 1973. (129 ref.)

(6) Manganese

1. National Research Council, *Manganese*, National Academy of Sciences, Washington, D.C., 1974, 192 pp.

(7) Mercury

1. **D'Itry, F. M.**, *The Environmental Mercury Problem*, CRC Press, Cleveland, Ohio, 1972, 136 pp.
2. **Eyl, T. B.**, Organic-mercury food poisoning, *N. Engl. J. Med.*, 284, 706–709, 1971.
3. **Friberg, L. T. and Vostal, J. J.**, *Mercury in the Environment*, CRC Press, Cleveland, Ohio, 1972, 232 pp.
4. **Hartung, R. and Dinmam, B. D., Eds.**, *Environmental Mercury Contamination*, Ann Arbor Science Publishers, Ann Arbor, Mich., 1972, 349 pp.
5. **Katz, A.**, Mercury pollution: The making of an environmental crisis, *CRC Crit. Rev. Environ. Contr.*, 2(4), 517–534, 1972. (54 ref.)
6. **Krehl, W. A.**, Mercury, the slippery metal, *Nutr. Today*, 7(6), 4–15, 1972.
7. Mercury Contamination in Man and His Environment, International Atomic Energy Agency, Vienna, 1972, 182 pp.
8. Mercury Poisoning, MSS Information Corp., New York, 1973. (2 vols.)
9. **Newberne, P. M.**, Mercury in fish – A literature review, *CRC Crit. Rev. Food Technol.*, 4(3), 311–335, 1973.
10. **Smart, N. A.**, Mercury residues in animals, *Residue Rev.*, 23, 1–36, 1968.

(8) Nickel

1. **Schroeder, H. A., Balassa, J. J., and Tipton, I. H.**, Abnormal trace metals in man: Nickel, *J. Chron. Dis.*, 15, 51–65, 1962.

(9) Tin

1. **Schroeder, H. A., Balassa, J. J., and Tipton, I. H.**, Abnormal trace metals in man: Tin, *J. Chron. Dis.*, 17, 483–502, 1964.

(10) Titanium

1. **Schroeder, H. A., Balassa, J. J., and Tipton, I. H.,** Abnormal trace metals in man: Titanium, *J. Chron. Dis.*, 16, 55–69, 1963.

(11) Vanadium

1. National Research Council, *Vanadium*, National Academy of Sciences, Washington, D.C., 1974, 118 pp.
2. **Schroeder, H. A., Balassa, J. J., and Tipton, I. H.,** Abnormal trace metals in man: Vanadium, *J. Chron. Dis.*, 16, 1047–1071, 1963.

3. Carcinogenic Contaminants

1. **Dixon, J. R., Lowe, D. B., Richards, D. E., Cralley, L. J., and Stockinger, H. E.,** The role of trace metals in chemical carcinogenesis – asbestos cancers, *Cancer Res.*, 30, 1068–1074, 1970.
2. *Environment and Cancer. Proc. Symposium, University of Texas, Houston, 1971*, Williams & Wilkins, Baltimore, 1972, 476 pp.
3. **Harris, R. J. C.,** Cancer and the environment, *Int. J. Environ. Stud.*, 1, 59–65, 1971.
4. **Hueper, W. C.,** Carcinogenesis in human environment, *Arch. Pathol.*, 71, 355–380, 1961.
5. **Hueper, W. C.,** Environmental and occupational cancer hazards, *Clin. Pharmacol. Ther.*, 3, 776–813, 1962.
6. **Kraybill, H. F.,** Symposium on chemical carcinogenesis. II. Carcinogenesis associated with foods, food additives, food degradation products, and related dietary factors, *Clin. Pharmacol. Ther.*, 4, 73–87, 1963.
7. **Kraybill, H. F. and Shimkin, M. B.,** Carcinogenesis related to foods contaminated by processing and fungal metabolites, *Adv. Cancer Res.*, 8, 191–248, 1964. (111 ref.)
8. **Roe, F. J.,** Carcinogens and food: A broad assessment, *Proc. R. Soc. Med.*, 66, 23–26, 1973.
9. **Scott, T. S.,** *Carcinogenic and Chronic Toxic Hazards of Aromatic Amines*, Elsevier, Amsterdam, 1962.
10. **Shimkin, M. B.,** Environmental carcinogenesis, *Arch. Environ. Health*, 16, 513–521, 1968.

4. Parasites

1. **Brandley, P. J., Migaki, G., and Taylor, K. E.,** Trichinosis, in *Meat Hygiene*, 3rd ed., Lea & Febiger, Philadelphia, 1966, 251–277.
2. **Healy, G. R. and Gleason, N. N.,** Parasitic infections, in *Food-borne Infections and Intoxications*, Riemann, H., Ed., Academic Press, New York, 1969, 175–222.
3. **Jacobs, L.,** Parasites in food, in *Chemical and Biological Hazards in Food*, Ayres, J. C., Kraft, A. A., Snyder, H. E., and Walker, H. W., Eds., Iowa State University Press, Ames, 1962, 248–266.
4. **Morgan, P. M.,** Meat animal parasites and the importance of an effective standardized meat inspection system, in *The Safety of Foods*, Graham, H. D., Ed., Avi Publishing, Westport, Conn., 1968, 102–108.

5. Microbial Toxins

a. General

1. **Aji, S. J. and Motie, T. C., Eds.,** *Microbial Toxins. A Comprehensive Treatise*, Academic Press, New York, 1970–1972. (8 vols.)
2. **Bryan, F. L.,** Infections due to miscellaneous microorganisms, in *Food-borne Infections and Intoxications*, Riemann, H., Ed., Academic Press, New York, 1969, 224–287.
3. **Bryan, F. L.,** Microbiological food hazards today – Based on epidemiological information, *Food Technol.*, 28(9), 52–66, 84, 1974.
4. **Mateles, R. I. and Wogan, G. N., Eds.,** *Biochemistry of Some Foodborne Microbial Toxins*, M.I.T. Press, Cambridge, Mass., 1967, 171 pp.
5. Microbial toxins, in *The Safety of Foods*, Graham, H. D., Ed., Avi Publishing, Westport, Conn., 1968, 111–193. (5 papers)
6. **Mossel, D. A. A.,** Significance of microorganisms in food, in *Chemical and Biological Hazards in Food*, Ayres, J. C., Kraft, A. A., Snyder, H. E., and Walker, H. E., Eds., Iowa State University Press, Ames, 1962, 157–202.
7. **Rašková, H. and Mašek, K.,** Microbial toxins, in *Pharmacology and Toxicology of Naturally Occurring Toxins*, Rašková, H., Ed., *International Encyclopedia of Pharmacology and Therapeutics*, Pergamon Press, Oxford, 1971, 3–212.
8. **Thatcher, F. S. and Clark, D. S., Eds.,** *Microorganisms in Foods. Their Significance and Methods of Enumeration*, University of Toronto Press, Toronto, 1968, 234 pp.

b. Bacteria

1. **Baird-Parker, A. C.,** Bacterial food poisoning toxins, *J. Appl. Bacteriol.*, 34, 181–197, 1971.
2. **Foster, E. M. and Bergdoll, M. S.,** Staphylococcal food poisoning, in *The Safety of Foods*, Graham, H. D., Ed., Avi Publishing, Westport, Conn., 1968, 159–167.
3. **Halbinger, R. E.,** The incidence of *Clostridia*, in *The Safety of Foods*, Graham, H. D., Ed., Avi Publishing, Westport, Conn., 1968, 111–120.

4. **Mossel, D. A. A.,** Bacterial toxins of uncertain oral pathogenicity, in *The Safety of Foods*, Graham, M. D., Ed., Avi Publishing, Westport, Conn., 1968, 168–182.
5. **Pivnick, H. and Thatcher, F. S.,** Microbial problems in food safety with particular reference to *Clostridium botulinum*, in *The Safety of Foods*, Graham, H. D., Ed., Avi Publishing, Westport, Conn., 1968, 121–140.

c. Viruses

1. **Cliver, D. O.,** Viral infections, in *Food-borne Infections and Intoxications*, Riemann, H., Ed., Academic Press, New York, 1969, 73–113.
2. **Cliver, D. O.,** Transmission of viruses through foods, *CRC Crit. Rev. Environ. Contr.*, 1(4), 551–579, 1971.

d. Fungi

(1) General

1. **Bamburg, J. R., Strong, F. M., and Smalley, E. B.,** Toxins from moldy cereals, *J. Agric. Food Chem.*, 17, 443–450, 1969.
2. **Banes, D.,** Mycotoxins as a food problem, *Cereal Sci. Today*, 11(1), 4–6,30, 1966.
3. **Biley, V. I.,** Ed., Mycotoxicoses of Man and Agricultural Animals, U.S. Dep. Commerce, Washington, D.C., 1961.
4. **Borker, E., Insalata, N. F., Levi, C. P., and Witzeman, J. S.,** Mycotoxins in feeds and foods, *Adv. Appl. Microbiol.*, 8, 315–351, 1966 (199 ref.)
5. **Brook, P. J. and White, E. P.,** Fungus toxins affecting mammals, *Annu. Rev. Phytopathol.*, 4, 171–194, 1966.
6. **Buck, R. W.,** Mushroom toxins. A brief review of the literature, *N. Engl. J. Med.*, 265, 681–686, 1961.
7. **Christensen, C. M.,** Mycotoxins, *CRC Crit. Rev. Environ. Contr.*, 2(1), 57–80, 1971. (119 ref.)
8. **Ciegler, A. and Lillehoj, E. B.,** Mycotoxins, *Adv. Appl. Microbiol.*, 10, 155–219, 1968.
9. **Coveney, R. D., Peck, H. M., and Townsend, R. J.,** Recent advances in mycotoxicoses, *Soc. Chem. Ind.* Lond., Monogr., 23, 31–43, 1966.
10. **Forgacs, J. and Carll, W. T.,** Mycotoxicoses, *Adv. Vet. Sci.*, 7, 273–382, 1962.
11. Fungal toxins, in *Biochemistry of Some Foodborne Microbial Toxins*, Mateles, R. I. and Wogan, G. N., Eds., M.I.T. Press, Cambridge, Mass., 1967, 69–168. (6 papers)
12. **Hesseltine, C. W.,** Mycotoxins, *Mycopathol. Mycol. Appl.*, 39, 371-383, 1969.
13. **Kensler, C. J.,** Symposium on food toxins of fungal origin: Present status and future outlook, Institute of Food Technologists, Kansas City, Mo., May 16–20, 1965.
14. Mycotoxicoses – Food-borne fungal diseases, *Nutr. Rev.*, 20, 337–342, 1962.
15. **Purchase, I. F. H., Ed.,** *Symposium on Mycotoxins in Human Health*, Macmillan, London, 1971, 306 pp.
16. **Reiss, J.,** Mycotoxine. I. Mycotoxine von *Aspergillus*-Arten, *Z. Allg. Mikrobiol.*, 8, 303–330, 1968.
17. **Scott, P. M. and Sommers, E.,** Biologically active compounds from field fungi, *J. Agric. Food Chem.*, 17, 430–436, 1969.
18. **Uraguchi, K.,** Pharmacology of mycotoxins, in *Pharmacology and Toxicology of Naturally Occurring Toxins*, Rašková, H., Ed., *International Encyclopedia of Pharmacology and Therapeutics*, Pergamon Press, Oxford, 1971, 143–299.
19. **Wilson, B. J.,** Fungal toxins, in *Toxicants Occurring Naturally in Foods*, National Research Council, National Academy of Sciences, Washington, D.C., 1966, 126–146. (130 ref.)
20. **Wilson, B. J.,** Mycotoxins, *Proc. Western Hemisphere Nutr. Conf.*, American Medical Association, Chicago, Ill., 1966, 145–147.
21. **Wogan, G. N., Ed.,** *Mycotoxins in Foodstuffs*, M.I.T. Press, Cambridge, Mass., 1965, 291 pp. (23 papers)
22. **Wogan, G. N.,** Mycotoxin contamination of foodstuffs, in *World Protein Resources*, in *Advances in Chemistry*, Ser. No. 57, American Chemical Society, Washington, D. C., 1966, 195–215. (88 ref.)
23. **Wogan, G. N.,** Alimentary mycotoxicoses, in *Food-borne Infections and Intoxications*, Riemann, H., Ed., Academic Press, New York, 1969, 395–451.
24. **Wright, D. E.,** Toxins produced by fungi, *Annu. Rev. Microbiol.*, 22, 269–282, 1968. (112 ref.)

(2) Aflatoxin

1. **Barnes, J. M.,** Toxic fungi with special reference to aflatoxin, *Trop. Sci.*, 9, 64–74, 1967.
2. **Goldblatt, L. A.,** Aflatoxin, *Assoc. Food Drug Offic. Q. Bull. U.S.*, 29, 58–69, 1965.
3. **Goldblatt, L. A.,** Some approaches to the elimination of aflatoxin from protein concentrates, in *World Protein Resources*, in *Advances in Chemistry*, Ser. No. 57, American Chemical Society, Washington, D.C., 1966, 216–227. (24 ref.)
4. **Goldblatt, L. A.,** Aflatoxin and its control, *Econ. Bot.*, 22, 51–62, 1968.
5. **Goldblatt, L. A., Ed.,** *Aflatoxin. Scientific Background, Control, and Implications*, Academic Press, New York, 1969, 470 pp. (15 papers)
6. **Hesseltine, C. W., Shotwell, O. L., Ellis, J. J., and Stubblefield, R. D.,** Aflatoxin formation by *Aspergillus flavus*, *Bacteriol. Rev.*, 30, 795–805, 1966.
7. **Hesseltine, C. W.,** Aflatoxins and other mycotoxins, *Health Lab. Sci.*, 4, 222–228, 1967.

8. **Legator, M. S.,** Biological effects of aflatoxin in cell culture, *Bacteriol. Rev.*, 30, 471–477, 1966.
9. **Marth, E. H.,** Aflatoxins and other mycotoxins in agricultural products, *J. Milk Food Technol.*, 30, 192–198, 1967.
10. **Mateles, R. I. and Wogan, G. N.,** Aflatoxins, in *Advances in Microbial Physiology*, Vol. 1, Rose, A. H. and Wilkinson, J. F., Eds., Academic Press, New York, 1967, 25–37.
11. **Meyer, H.,** *Bibliographie der Aflatoxine*, Institut für Bakteriologie und Histologie der Bundesanstalt für Fleischforschung, Kulmback, 1970.
12. **Peers, F. G.,** Aflatoxin – A summary of recent work, *Trop. Sci.*, 9, 186–203, 1967.
13. **Schoental, R.,** Aflatoxins, *Annu. Rev. Pharmacol.*, 7, 343–356, 1967.
14. **Skau, D. B.,** Bibliography of Aflatoxin From 1960, USDA, ARS, Southern Utilization Research and Development Division, New Orleans, La., 1965.
15. **Wogan, G. N.,** Chemical nature and biological effects of the aflatoxins, *Bacteriol. Rev.*, 30, 460–470, 1966.
16. **Wogan, G. N.,** Aflatoxin risks and control measures, *Fed. Proc.*, 27, 932–938, 1968.

(3) Ergot

1. **Barger, G.,** *Ergot and Ergotism*, Gurney & Jackson, London, 1931.
2. **Goodman, L. S. and Gilman, A. Z.,** Ergot and ergot alkaloids, in *The Pharmacological Basis of Therapeutics*, 2nd ed., Macmillan, New York, 1955, 886–897.

e. Algae

1. **Schantz, E. J.,** Biochemical studies on certain algal toxins, in *Biochemistry of Some Foodborne Microbial Toxins*, Mateles, R. I. and Wogan, G. N., Eds., M.I.T. Press, Cambridge, Mass., 1967, 51–65.

6. Agricultural Chemical Residues

a. General

1. International Dairy Federation, *Chemical Residues in Milk*, Bruxelles, 1968, 131 pp.
2. **Jukes, T. H.,** Chemical residues in foods. Fact and fancy in nutrition and food science, *J. Am. Diet. Assoc.*, 59, 203–211, 1971. (33 ref.)
3. **McKinley, W. P., et al.,** Unintentional food additives, in *The Safety of Foods*, Graham, H. D., Ed., Avi Publishing, Westport, Conn., 1968, 291–360.
4. Symposium on additives and residues in human foods, *Am. J. Clin. Nutr.*, 9, 259–303, 1961. (5 papers with discussion)

b. Pesticide Residues

1. **Barnes, J. M.,** Carcinogenic hazards from pesticide residues, *Residue Rev.*, 13, 69–82, 1966.
2. **Biros, F. J., Ed.,** *Pesticides Identification at the Residue Level*, in *Advances in Chemistry*, Ser. No. 104, American Chemical Society, Washington, D.C., 1971, 182 pp. (11 papers)
3. **Boyd, E. M.,** *Protein Deficiency and Pesticide Toxicity*, Charles C Thomas, Springfield, Ill., 1972, 468 pp.
4. **Chung, R. A.,** Metabolism of pesticide residues, in *The Safety of Foods*, Graham, H. D., Ed., Avi Publishing, Westport, Conn., 1968, 319–323.
5. **Cook, J. W.,** Persistence of residues in food, in *Research in Pesticides*, Chichester, C. O., Ed., Academic Press, New York, 1965, 205–212.
6. **Crosby, D. G.,** The intentional removal of pesticide residues, in *Research in Pesticides*, Chichester, C. O., Ed., Academic Press, New York, 1965, 213–223.
7. **Cummings, J. G.,** Pesticides in the total diet, *Residue Rev.*, 16, 30–45, 1966.
8. **Durham, W. F.,** Pesticide residues in foods in relation to human health, *Residue Rev.*, 4, 33–81, 1963.
9. **Durham, W. F.,** Body burden of pesticides in man, *Ann. N. Y. Acad. Sci.*, 160(1), 183–195, 1969.
10. **Edwards, C. A.,** Insecticide residues in soils, *Residue Rev.*, 13, 83–132, 1966.
11. **Edwards, C. A.,** Persistent pesticides in the environment, *CRC Crit. Rev. Environ. Contr.*, 1(1), 5–67, 1970. (297 ref.)
12. **Edwards, C. A.,** *Persistent Pesticides in the Environment*, CRC Press, Cleveland, Ohio, 1973, 150 pp.
13. **Edwards, C. A., Ed.,** *Environmental Pollution by Pesticides*, Plenum Press, New York, 1973, 440 pp.
14. **Egler, F. E.,** Pesticides in our ecosystem, *Am. Sci.*, 52, 110–136, 1964.
15. **England, C. W.,** *Bibliography, Abstracts of Research Literature: Pesticide Residues in Milk and Other Agricultural Products*, Dairy Industry Committee, Washington, D.C., 1961, 92 pp.
16. Food and Agriculture Organization, Evaluation of the Toxicity of Pesticide Residues in Food, FAO, Rome, 1964, 172 pp.
17. Food and Agriculture Organization, Pesticide Residues in Food, FAO Agric. Stud. No. 84, FAO, Rome, 1971, 44 pp.
18. **Gunther, F. A.,** Pesticides, in *Chemical and Biological Hazards in Food*, Ayres, J. C., Kraft, A. A., Snyder, H. E., and Walker, H. W., Eds., Iowa State University Press, Ames, 1962, 77–88.
19. **Gunther, F. A. and Blinn, R. C.,** Persisting insecticide residues in plant materials, *Annu. Rev. Entomol.*, 1, 167–180, 1956.

20. **Henderson, J. L.,** Insecticide residues in milk and dairy products, *Residue Rev.*, 8, 74–115, 1965.
21. **Henish, E.,** Hygienic evaluation of plant protection measures condusive to the contamination of foodstuffs with pesticides, *Nahrung*, 6(1), 1–16, 1962. (21 ref.)
22. **Hodges, R. E.,** The toxicity of pesticides and their residues in food, *Nutr. Rev.*, 23, 225–230, 1965. (12 ref.)
23. **Kraybill, H. F., Ed.,** Biological effects of pesticides in mammalian systems, *Annu. N.Y. Acad. Sci.*, 160(1), 1–422, 1969.
24. **Lindgren, D. L., Sinclair, W. B., and Vincent, L. E.,** Residues in raw and processed foods resulting from post-harvest insecticidal treatment, *Residue Rev.*, 21, 1–128, 1968.
25. **McKinley, W. P.,** Sources of pesticide residues, in *The Safety of Foods*, Graham, H. D., Ed., Avi Publishing, Westport, Conn., 1968, 291–308.
26. **McKinley, W. P. and Charbonneau, S.,** Permitted pesticides: Validation of their safety, in *Proc. SOS/70. 3rd International Congress Food Science and Technology*, Institute of Food Technologists, Chicago, 1971, 753–763.
27. **Thomson, J. and Abbott, D. C.,** Pesticide Residues: Alternatives and Analysis, Royal Institute of Chemistry, London, 1967, 33 pp.
28. U.S. Department of Health, Education, and Welfare, Report of the Secretary's Commission on Pesticides and Their Relationship to Environmental Health, U.S. DHEW, Washington, D.C., 1969.
29. **Van Middelem, C. H.,** Incidental pesticide residues in foods, in *The Safety of Foods*, Graham, H. D., Ed., Avi Publishing, Westport, Conn., 1968, 324–357.
30. **West, I. and Milby, T. H.,** Public health problems arising from the use of pesticides, *Residue Rev.*, 13, 141–159, 1965.
31. **White-Stevens, R.,** *Pesticides in Environment*, Dekker, New York, 1971, 629 pp. (7 papers)
32. World Health Organization, Expert Committee on Insecticides, Specifications for Pesticides Used in Public Health: Insecticides, Rodenticides, Mollescicides, Repellents, Methods, 3rd ed., WHO, Geneva, 1967, 300 pp.
33. World Health Organization, Pesticide Residues, WHO/FAO Tech. Rep. 391, 43, 1968.

c. Antibiotics and Other Drug Residues

1. **Albright, J. L., Tuckey, S. L., and Woods, G. T.,** Antibiotics in milk, A review, *J. Dairy Sci.*, 34, 779–807, 1961.
2. **Bird, H. R.,** Additives and residues in foods of animal origin, *Am. J. Clin. Nutr.*, 9, 260–268, 1961.
3. **Elias, P. S.,** Medical significance of antibiotic usage in food production, in *Proc. SOS/70. 3rd International Congress Food Science and Technology*, Institute of Food Technologists, Chicago, 1971, 729–742.
4. **Kastelic, J.,** Animal growth promoters, in *Chemical and Biological Hazards in Food*, Ayres, J. C., Kraft, A. A., Snyder, H. E., Walker, H. W., Eds., Avi Publishing, Westport, Conn., Iowa State University Press, Ames, 1962, 127–141.
5. **Kiser, J. S., Gale, G. O., and Kemp, G. A.,** Antibiotics as feedstuff additives: Their risk-benefit equation for man, *CRC Crit. Rev. Toxicol.*, 1(1), 55–92, 1971. (83 ref.)
6. **Marth, E. H.,** Antibiotics in foods – naturally occurring, developed and added, *Residue Rev.*, 12, 65–161, 1966.
7. **Myers, R. P.,** Antibiotic residues in milk, *Residue Rev.*, 7, 9–36, 1964.
8. National Research Council, *The Use of Drugs in Animal Feeds*, Proceedings of a Symposium, National Academy of Sciences, Washington, D.C., 1969, 407 pp.
9. **Peoples, S. A. and Pier, A. C.,** Drug levels in milk, *Med. Vet. Pract.*, 41, 26–33, 1960.
10. **Siddique, I. H.,** Antibiotic residues in foods, in *The Safety of Foods*, Graham, H. D., Ed., Avi Publishing, Westport, Conn., 1968, 358–360.
11. **Somogyi, J. C. and François, A. C., Eds.,** *Antibiotics in Agriculture, Bibl. Nutr. Dieta* No. 10, S. Karger, Basel, 1968, 208 pp.
12. Symposium on feed additives and significance of their residues in animal tissues, *J. Agric. Food Chem.*, 11, 362–395, 1963. (13 papers)

d. Plant Growth Regulators

1. **Vlitos, A. J.,** Plant growth regulators, in *Chemical and Biological Hazards in Food*, Ayres, J. C., Kraft, A. A., Snyder, H. E., and Walker, H. W., Eds., Iowa State University Press, Ames, 1962, 89–126.

7. Toxic Factors in Processed Foods

a. General

1. **Chatt, E. M.,** Adventitious metals in processed food, *World Rev. Nutr. Diet.*, 4, 141–156, 1964. (40 ref.)
2. **Friedman, L. and Shibko, S. I.,** Adventitious toxic factors in processed foods, in *Toxic Constituents in Plant Foodstuffs*, Liener, I. E., Ed., Academic Press, New York, 1969, 349–408.
3. **Mateles, R. I. and Wogan, G. N.,** *Biochemistry of Some Foodborne Industrial Toxins*, M.I.T. Press, Cambridge, Mass., 1967.
4. **Riemann, H.,** Food processing and preservation effects, in *Food-borne Infections and Intoxications*, Riemann, H., Ed., Academic Press, New York, 1969, 490–541.

b. Processing Degradation Products

(1) General

1. **Holmes, A. W.,** Chemical and physical changes during food processing, in *Proc. SOS/70. 3rd International Congress Food Science and Technology,* Institute of Food Technologists, Chicago, 1971, 604–612.

(2) Lipids

1. **Halaby, G. A. and Fagerson, I. S.,** Polycyclic aromatic hydrocarbons in heat-treated foods – Pyrolysis of some lipids, beta-carotene, and cholesterol, in *Proc. SOS/70. 3rd International Congress Food Science and Technology,* Institute of Food Technologists, Chicago, 1971, 820–829.
2. **Kaunitz, H.,** Biological effects of autoxidized lipids, in *Symposium on Foods: Lipids and Their Oxidation.* Schultz, H. W., Day, E. A., and Sinnhuber, R. O., Eds., Avi Publishing, Westport, Conn., 1962, 269–293.
3. **Kummerow, F. A.,** Toxicity of heated fats, in *Symposium on Foods: Lipids and Their Oxidation,* Schultz, H. W., Day, E. A., and Sinnhuber, R. O., Eds., Avi Publishing, Westport, Conn., 1962, 294–320.
4. **Lea, C. H.,** The influence of industrial and household handling on fats seen from the chemical point of view, *Nutr. Diet.,* 1, 88–105, 1965.
5. **Lovern, J. A.,** Lipids of fish and changes occuring in them during processing and storage, *Proc. Conf. Fish Nutr.,* Washington, D.C., 1961, 1, 1.8.1–1.8.45. (113 ref.)
6. **Matsuo, N.,** Nutritional effects of oxidized and thermally polymerized fish oils, in *Symposium of Foods: Lipids and Their Oxidation,* Schultz, H. W., Day, E. A., and Sinnhuber, R. O., Eds., Avi Publishing, Westport, Conn., 1962, 321–359.
7. Symposium. Nutritional and toxicity problems associated with fats, *Proc. Nutr. Soc.,* 25, 1–60, 1966. (9 papers)

(3) Proteins

1. **Bender, A. E.,** Processing damage to protein foods, *PAG Bull.,* 2(1), 10–19, 1972. (77 ref.)
2. Effect of processing and storage on the value of protein products, in *Evaluation of Novel Protein Products,* Bender, A. E., Kihlberg, P., Löfqvist, B., and Munk, L., Eds., Pergamon Press, Oxford, 1970, 313–356. (5 papers)
3. **Mauron, J.,** Influence of industrial and household handling on food protein quality, in *Protein and Amino Acid Functions,* Bigwood, E. J., Ed., *International Encyclopedia of Food and Nutrition,* Vol. 11, Pergamon Press, Oxford, 1972, 417–473.

(4) Carbohydrates

1. **Adrian, J.,** Nutritional and physiological consequences of the Milliard reaction, *World Rev. Nutr. Diet.,* 19, 71–122, 1974.

(5) Changes in Specific Foods

1. **Gould, W. A.,** Effect of processing factors on the quality of fruits and vegetables, in *Food Quality,* Irving, G. W., Jr., and Hoover, S. R., Eds., American Association for the Advancement of Science, Washington, D.C., 1965, 57–69.
2. **Koonz, C. H.,** Effect of processing factors on quality of poultry, in *Food Quality,* Irving, G. W., Jr. and Hoover, S. R., Eds., American Association for the Advancement of Science, Washington, D.C., 1965, 223–235.
3. **Lawrie, R. A.,** Chemical changes in meat due to processing. A review, *J. Sci. Food Agric.,* 19, 233–240, 1968.
4. **Pomeranz, Y.,** Biochemical and functional changes in stored cereal grains, *CRC Crit. Rev. Food Technol.,* 2(1), 45–80, 1971. (63 ref.)
5. **Pratt, D. B.,** The effect of processing on quality of flour, in *Food Quality,* Irving, G. W., Jr. and Hoover, S. R., Eds., American Association for the Advancement of Science, Washington, D.C., 1965, 121–132.
6. **Wilson, G. D.,** Effect of processing factors on meat quality, in *Food Quality,* Irving, G. W., Jr. and Hoover, S. R., Eds., American Association for the Advancement of Science, Washington, D.C., 1965, 287–292.

c. Radionuclides

1. Food and Agriculture Organization, Report of the Meeting on the Wholesomeness of Irradiated Foods, with Exclusive Reference to the Evaluation of Nutritional Adequacy and Safety for Consumption, Brussels, Oct. 23–30, 1961, FAO, Rome, 1962, 205 pp.

Exclusive Reference to the Evaluation of Nutritional Adequacy and Safety for Consumption, Brussels, Oct. 23–30,

2. **Goldblith, S. A.,** The wholesomeness of irradiated foods: Past history, present status, international aspects and future outlook, *Food Technol.,* 20, 191–196, 1966.
3. **Kraybill, H. F. and Whitehair, L. A.,** Toxicological safety of irradiated foods, *Annu. Rev. Pharmacol.,* 7, 357–380, 1967.
4. **Lang, K. and Bässler, K. H.,** Biological effects of irradiated fats, in Food Irradiation, Proceedings Ser., IAEA Publ. STI/PUB/127, Vienna, 1966, 147–158.
5. **Lehman, A. J. and Laug, E. P.,** Evaluating the safety of radiation sterilized foods, *Nucleonics,* 12(1), 52–54, 1954.
6. **Ley, F. J.,** Safety of irradiated food – Biological aspects, in *Proc. SOS/70. 3rd International Congress Food Science and Technology,* Institute of Food Technologists, Chicago, 1971, 806–815.

7. **McArdle, F. J. and Desrosier, N. W.,** Influence of ionizing radiations upon some protein components of selected foods, *Food Technol.*, 9, 527–532, 1955.
8. **Raica, N., McDowell, M. E., Darby, W. J., Howie, D. L., and Sherman, J. L., Jr.,** Wholesomeness of irradiated food, Proc. International Conference on Radiation Research, U.S. Army Natick Lab., Natick, Mass., Jan. 14–16, 1963, 168–184.
9. **Raica, N., Jr. and Howie, D. L.,** Review of U.S. Army wholesomeness of irradiated food program 1955–1966, Food Irradiation, Proceedings Ser., IAEA Publ. STI/PUB/127, Vienna, 1966, 119–135.
10. **Read, M. S.,** Current aspects of the wholesomeness of irradiated foods, *J. Agric. Food Chem.*, 8, 342–349, 1960. (63 ref.)
11. **Reber, E. F., Raheja, K., and Davis, D.,** Wholesomeness of irradiated foods. An annotated bibliography, *Fed. Proc.*, 25, 1529–1579, 1966.
12. **Schubert, J.,** Mutagenicity and cytogenicity of irradiated food and food components, *Bull. W.H.O.*, 41, 873–904, 1969.
13. **Teply, L. J. and Kline, B. E.,** Wholesomeness and possible carcinogenecity of irradiated foods, *Fed. Proc.*, 15, 927–929, 1956.
14. **Whitehair, L. A. and Hilmas, D. E.,** Radiological problems in the National Irradiation Program, in *The Safety of Foods*, Graham, H. D., Ed., Avi Publishing, Westport, Conn., 1968, 82–101.

d. Food Packaging Materials

1. **Ingle, G. W.,** Regulating additives from food contact materials, in *The Safety of Foods*, Graham, H. D., Ed., Avi Publishing, Westport, Conn., 1968, 309–318.
2. **Ingle, G. W.,** Problems of investigating the safety-in-use of food packaging, in *Proc. SOS/70. 3rd International Congress Food Science and Technology*, Institute of Food Technologists, Chicago, 1971, 816–819.
3. **McCollister, D. D.,** Toxicological research approach to the development of a plastic package for food, *Food Cosmet. Toxicol.*, 2, 23–29, 1964.
4. National Research Council. Food Protection Committee, Food Packaging Materials – Their Composition and Uses, National Academy of Sciences, Washington, D.C., 1958.

XI. MANUFACTURED AND BIOLOGICAL INPUTS FOR AGRICULTURE

A. General

1. **Mellor, J. W.,** Supply of new forms of inputs, in *The Economics of Agricultural Development,* Cornell University Press, Ithaca, N.Y., 1966, 288–309.
2. President's Science Advisory Committee, Manufactured and biological inputs, in The World Food Problem. Report of the Panel on the World Food Supply, Vol. 2, U.S. Gov. Print. Off., Washington, D.C., 1967, 373–402.

B. Fertilizer

1. **Baum, E. L.,** *Economic and Technological Analysis of Fertilizer Innovations and Resources Use,* Iowa State University Press, Ames, 1957, 393 pp.
2. British Sulphur Corporation, *World Fertilizer Atlas,* 3rd ed., London, 1969.
3. **Callister, G. J.,** The world fertilizer supply and the FAO world program, *Fert. Rev.,* 26(1), 18–22, 1951.
4. **de Guia, E. O.,** A Comparative Study of Fertilizer Distribution Systems, OECD, Paris, 1972, 137 pp.
5. **Delwiche, C. C.,** Nitrogen and future food requirements, in *Research for the World Food Crisis,* Aldrich, D. G., Jr., Ed., American Association for the Advancement of Science, Washington, D.C., 1970, 191–210.
6. **Desai, G. and Mellor, J. W.,** Changing basis of demand for fertilizer, *Econ. Politic. Weekly,* 4(39), 175–187, 1969.
7. **Ewell, R.,** Famine and fertilizer, *Chem. Eng. News,* 14, 106–117, 1964.
8. **Ewell, R.,** Fertilizer and famine, *Co-Op Grain Q.,* 23, 8–16, Fall 1965.
9. **Ewell, R.,** Wanted 30 million tons of fertilizer to avert widespread famine in 1980's, *Agric. Chem.,* 19, 48–49, 1964.
10. Food and Agriculture Organization, Efficient Use of Fertilizers, FAO Agric. Stud. No. 43, FAO, Rome, 1958.
11. Food and Agriculture Organization, The FAO Fertilizer Programme under the Freedom from Hunger Campaign: Its Purposes and Accomplishments, United Nations, Center for Industrial Development, New York, 1965.
12. Food and Agriculture Organization, Fertilizers. An Annual Review of World Production, Consumption, and Trade, FAO, Rome, 1951–.
13. Food and Agriculture Organization, Statistics of Crop Responses to Fertilizers, FAO, Rome, 1966, 112 pp.
14. **Heady, E. O., Baum, E. L., and Blackmore, J., Eds.,** *Methodological Procedures in the Economic Analysis of Fertilizer Use Data,* Iowa State University, Ames, 1956.
15. **Jacob, A. and von Vexküll, H.,** Fertilizer use, *Verlagsgesellschaft für Ackerbau,* Hannover, Germany, 1960.
16. **Jacob, K. D., Ed.,** *Fertilizer Technology and Resources in the United States,* Academic Press, New York, 1953, 454 pp. (14 papers)
17. **Kurian, A. P.,** Impact of the fertilizer levy, *Econ. Politic. Weekly,* 4(23), 943–947, 1969.
18. **Lamer, M.,** *The World Fertilizer Economy,* Stanford University Press, Stanford, Calif., 1957, 715 pp.
19. **Markham, J. W.,** *The Fertilizer Industry: Study of an Imperfect Market,* Vanderbilt University Press, Nashville, 1958.
20. **McCune, D. L., Hignett, T. P., and Douglas, J. R., Jr.,** Estimated World Fertilizer Production Capacity as Related to Fertilizer Needs, Tennessee Valley Authority, Muscle Shoals, Ala., February 1966.
21. **McVickar, M. H.,** *Using Commercial Fertilizers and Crop Production,* Interstate, Danville, Ill., 1961, 286 pp.
22. **Mehring, A. L.,** Fertilizers, in *Handbook of Food and Agriculture,* Blanck, F. D., Ed., Reinhold, New York, 1955, 89–143. (82 ref.)
23. **Nelson, L. B.,** Fertilizer requirements for increased food needs, in The World Food Problem, Report of the Panel on the World Food Supply, Vol. 3, President's Science Advisory Committee, U.S. Gov. Print. Off., Washington, D.C., 1967, 95–118.
24. Organization for Economic Cooperation and Development, Supply and Demand Prospects for Fertilizers in Developing Countries, OECD, Paris, 1968, 206 pp.
25. **Parker, F. W.,** Fertilizer and economic development, in *Fertilizer Technology and Usage,* McVickar, M. H., Bridgar, G. L., and Nelson, L. B., Eds., Soil Science Society of America, Madison, Wis., 1963, 1–21.
26. **Patel, B. B.,** Fertilizer consumption in developing countries: Relevance of the price factor, *Econ. Politic. Weekly,* 3(26–28), 1059–1062, 1968.
27. **Pieters, A. J.,** *Green Manuring,* John Wiley & Sons, New York, 1927.
28. **Sahota, G. S.,** *Fertilizers in Economic Development: An Econometric Analysis,* Praeger, New York, 1968, 240 pp.
29. **Smith, A. M.,** *Manures and Fertilizers,* Nelson, London, 1952, 215 pp.
30. **Sweeney, G. C., Jr.,** World fertilizer production, *Chem. Eng. News,* 41(48), 91–94, 1963.
31. **Tisdale, S. L. and Nelson, W. L.,** *Soil Fertility and Fertilizers,* Macmillan, New York, 1966.
32. United Nations, Directory of Fertilizer Facilities, U.N., Industrial Development Organization, New York, 1970.
33. United Nations, Fertilizer Industry, U.N., Industrial Development Organization, New York, 1969, 76 pp.
34. United Nations, Fertilizer Manual, U.N., Industrial Development Organization, New York, 1967.
35. United Nations, Fertilizer Production and Technology and Use, Interregional Seminar on the Production of Fertilizers, Kiev, 24 August–11 September, 1965, U.N., New York, 1968, 404 pp.
36. Willman–Lord Inc., *World Fertilizer Forecast 1965–1980,* Jacksonville, Fla., 1967, 128 pp.

C. Seed

1. **Barton, L. V.,** *Bibliography of Seeds,* Columbia University Press, New York, 1967, 858 pp.
2. **Barton, L. V.,** *Seed Preservation and Longevity,* Interscience Publishers, New York, 1961, 215 pp.
3. **Chalam, C. V., Singh, A., and Douglas, J. E.,** Seed Testing Manual, Indian Council for Agricultural Research and U.S. Agency for International Development, New Delhi, India, 1967.
4. **Delouche, J. C. and Potts, H. C.,** Seed Program Development, Mississippi State University Seed Technology Laboratory, State College, Miss., 1971, 120 pp.
5. European Productivity Agency, Development of Seed Production and the Seed Trade in Europe, OEEC, Paris, 1955, 138 pp.
6. Food and Agriculture Organization, Agricultural and Horticultural Seeds – Their Productivity, Control and Distribution, FAO Agric. Ser. No. 55, FAO, Rome, 1961, 531 pp.
7. Food and Agriculture Organization, Control of Production and Distribution of Seeds, FAO, Rome, 1969, 19 pp.
8. **Gregg, B. R., Law, A. G., Virdi, S. S., and Balis, J. S.,** Seed Processing, National Seeds Corporation and U.S. Agency for International Development, New Delhi, India, 1970, 396 pp.
9. **Hawthorn, L. R. and Pollard, L. H.,** *Vegetable and Flower Seed Production,* Blakiston, Garden City, N.Y., 1954, 626 pp.
10. **Hill, D. D.,** Seed and the world we live in, *Foreign Agric.,* 25(1), 4–6, 1961.
11. **Holland, R. F.,** Inputs for seed, in The World Food Problem. A Report of the Panel on the World Food Supply, President's Science Advisory Committee, U.S. Gov. Print. Off., Washington, D.C., 1967, 118–130.
12. **Law, A. G., Gregg, B. R., Young, P. B., and Chetty, P. R.,** Seed Marketing, National Seeds Corporation and U.S. Agency for International Development, New Delhi, India, 1971, 183 pp.
13. **Martin, A. C. and Barkley, W. D.,** Seed Identification Manual, University of California Press, Berkeley, 1961, 221 pp.
14. **Mayer, A. M. and Poljakoff-Mayber, A.,** *The Germination of Seeds,* Macmillan, New York, 1963.
15. Organization for European Economic Cooperation, Seed Production, Testing and Distribution in European Countries, OEEC, Paris, 1954.
16. Organization for European Economic Cooperation, Development of Seed Production and the Seed Trade in Europe, OEEC, Paris, 1955.
17. **Owen, E. B.,** The Storage of Seeds for Maintenance of Viability, Commonwealth Agric. Bureaux, Fornham Royal, Bucks, England, Bull. 43.
18. Seed Technology Laboratory, Seed Processing and Handling, Handbook No. 1, Mississippi State University, State College, Miss., 1968, 300 pp.
19. U.S. Department of Agriculture, Seeds. The Yearbook of Agriculture 1961, U.S. Gov. Print. Off., Washington, D.C., 1961, 591 pp.
20. U.S. Department of Agriculture, Identification of Crop and Weed Seeds, Agric. Handb. No. 219, U.S. Dep. Agric., Washington, D.C., 1963, 171 pp.
21. **Wheeler, W. A. and Hill, D. D.,** *Grassland Seeds,* Van Nostrand, Princeton, N.J., 1957, 734 pp.

D. Pesticides

1. General

1. American Chemical Society, *Pesticides in Tropical Agriculture, Advances in Chemistry,* Ser. No. 13, American Chemical Society, Washington, D.C., 1955, 102 pp.
2. American Chemical Society, *Natural Pest Control Agents, Advances in Chemistry,* Ser. No. 53, American Chemical Society, Washington, D.C., 1965.
3. Association of American Pesticide Control Officials, Pesticide Chemicals Official Compendium, AAPCO, 1958.
4. **Chichester, C. O., Ed.,** *Research in Pesticides, Proc. Conf. on Research Needs and Approaches to the Use of Agricultural Chemicals from a Public Health Viewpoint, University of California, Davis, Oct. 1–3, 1964,* Academic Press, New York, 1965, 380 pp.
5. Cooperative Programme of Agro-Allied Industries, *Pesticides in the Modern World. A Symposium,* FAO, 1972, Newgate Press, London, 1972, 59 pp.
6. **Ennis, W. B., Jr., Jansen, L. L., Ellis, I. T., and Newsom, L. D.,** Inputs for pesticides, in The World Food Problem, Report of the Panel on the World Food Supply, Vol. 3, President's Science Advisory Committee, U.S. Gov. Print. Off., Washington, D.C., 1967, 130–175.
7. **Eto, M.,** *Organophosphate Pesticides: Chemistry and Biochemistry,* CRC Press, Cleveland, Ohio, 1973, 250 pp.
8. **Frear, D. E. H.,** *Chemistry of the Pesticides,* 3rd ed., Van Nostrand, New York, 1955, 469 pp.
9. **Frear, D. E. H.,** *Pesticide Handbook – Entoma,* College Science Publishers, State College, Pa., 1970, 284 pp.
10. **Hartley, G. S. and West, T. F.,** *Chemicals for Pest Control,* Pergamon Press, Oxford, 1969, 316 pp.
11. **Hurtig, H.,** Benefits from pesticide use, *World Rev. Pest Contr.,* 4(1), 8–13, 1965.
12. **Martin, H., Ed.,** Pesticide Manual, Basic Information on the Chemicals Used as Active Components of Pesticides, British Crop Protection Council, Worcester, 1971, 495 pp.

13. **Mel'nikov, N. N.,** *Chemistry of Pesticides,* Gunther, F. A. and Gunther, J. D., Eds., Springer-Verlag, New York, 1971, 480 pp.
14. **Muirhead-Thomson, R. C.,** *Pesticides and Freshwater Fauna,* Academic Press, New York, 1971.
15. National Research Council, *Effect of Pesticides on Fruit and Vegetable Physiology,* in *Principles of Plant and Animal Pest Control,* Vol. 6, National Academy of Sciences, Washington, D.C., 1968, 90 pp.
16. **Roan, C. C.,** Insecticides, fungicides and herbicides, in *Handbook of Food and Agriculture,* Blanck, F. C., Ed., Reinhold, New York, 1955, 145–191.
17. **van der Bosch, R.,** Pesticides: Prescribing for the ecosystem, *Environment,* 12(3), 21–25, 1970.
18. **Zweig, G.,** *Analytical Methods for Pesticides, Plant Growth Regulators, and Food Additives,* Academic Press, New York, 1964–1972. (6 vols.)

2. Herbicides

1. **Andus, L. J., Ed.,** *Physiology and Biochemistry of Herbicides,* Academic Press, New York, 1964, 555 pp.
2. **Crafts, A. S.,** Toxicity of certain herbicides in soils, *Hilgardia,* 16, 459–483, 1945.
3. **Crafts, A. S.,** Selectivity of herbicides, *Plant Physiol.,* 21, 345–361, 1946.
4. **Crafts, A. S.,** *The Chemistry and Mode of Action of Herbicides,* Interscience, New York, 1961, 269 pp.
5. **Foy, C. L.,** Foliar penetration – review of herbicide penetration through plant surfaces, *J. Agric. Food Chem.,* 12, 473–476, 1964.
6. Herbicides and Their Use in Forestry, Special Symposium, Oregon State University, School of Forestry, Corvallis, 1961, 122 pp.
7. **Hilton, J. L., Jansen, L. L., and Hull, H. M.,** Mechanisms of herbicide action, *Annu. Rev. Plant Physiol.,* 14, 353–384, 1963.
8. **Lewis, C. J.,** Contribution of herbicides to food production in the next three decades, *Chem. Ind.,* June 29, 1968, 860–863.
9. **Loeffler, J. E. and van Overbeek, J.,** Metabolism of herbicides, in *Pesticides in the Environment,* Dekker, New York, 1971, 237–270.
10. **McDermott, R. E. and Byrnes, W. R.,** *Herbicides and Their Use in Forestry, Proc. Forestry Symposium,* Pennsylvania State University, University Park, 1960, 128 pp.
11. Weed Society of America, *Herbicide Handbook of the Weed Society of America,* Humphrey Press, Geneva, N.Y., 1967, 293 pp.
12. **Woodford, E. K., Holly, K., and McCready, C. C.,** Herbicides, *Annu. Rev. Plant Physiol.,* 9, 311–358, 1959.
13. **Woodford, E. K. and Sagar, G. R.,** *Herbicides and the Soil,* Blackwell, Oxford, 1960, 88 pp.
14. **Zweig, G., Ed.,** *Herbicides,* in *Analytical Methods for Pesticides, Plant Growth Regulators, and Food Additives,* Vol. 4, Academic Press, New York, 1964, 269 pp.

3. Fungicides

1. **Bent, K. J.,** Fungicides in perspective, *Endeavour* (Engl. ed.,) 28, 129–134, 1969.
2. **Domsch, K. H.,** Soil fungicides, *Annu. Rev. Phytopathol.,* 2, 293–320, 1964.
3. **Georgopoulos, S. G.,** The problem of fungicide resistance, *Bioscience,* 19, 971–973, 1969.
4. **Gottlieb, D. and Shaw, P.,** Mechanism of action of antifungal antibiotics, *Annu. Rev. Phytopathol.,* 8, 371–402, 1970.
5. **Horsfall, J. G.,** *Principles of Fungicidal Action,* Chronica Botanica, Waltham, Mass., 1956, 279 pp.
6. **Horsfall, J. G. and Lukens, R. J.,** The strategy of finding fugicides, *Residue Rev.,* 25, 81–91, 1969.
7. **Owens, R. G.,** Chemistry and physiology of fungicidal action, *Annu. Rev. Phytopathol.,* 1, 77–100, 1963.
8. **Sijpesteijn, A. K. and van der Kerk, G. J. M.,** Fate of fungicides in plants, *Annu. Rev. Phytopathol.,* 3, 127–152, 1965.
9. Society of Chemical Industry, *Fungicides in Agriculture and Horticulture,* Society of Chemical Industry, London, 1961, 145 pp.
10. **Torgeson, D. C., Ed.,** *Fungicides: An Advanced Treatise,* Academic Press, New York, 1967, 1969. (2 vols., 29 papers)
11. **van der Kerk, G. J. M.,** The development of synthetic fungicides. Trends and prospects, *Neth. J. Plant Pathol.,* 75, Suppl. 1, 5–20, 1969.

4. Nematocides

1. **Chitwood, B. G.,** Nematocidal action of halogenated hydrocarbons, *Advances in Chemistry,* Ser. No. 7, American Chemical Society, Washington, D.C., 1952, 91–99.
2. **Moje, W.,** The chemistry and nematocidal action of organic halides, *Adv. Pest Contr. Res.,* 3, 181–217, 1960.

5. Insecticides

1. **Braun, J. L., Jr.,** Apparatus for application of insecticides, *Annu. Rev. Entomol.,* 1, 241–260, 1957.
2. **Brooks, G. T.,** *Chlorinated Insecticides,* CRC Press, Cleveland, Ohio, 1973. (2 vols.)

3. **Brown, A. W. A.**, The spread of insecticide resistance in pest species, *Adv. Pest Contr. Res.*, 2, 351–414, 1958.
4. **Brown, A. W. A.**, Insecticide Resistance in Arthropods, WHO Monogr. Ser. 38, 1958, 240 pp.
5. **Brown, A. W.**, Mechanism of resistance against insecticides, *Annu. Rev. Entomol.*, 5, 301–306, 1960.
6. **Casida, J. E.**, The metabolism of insecticides by insects, in *Biochemistry of Insects. Proc. 4th International Congress Biochemistry*, Pergamon Press, New York, 1959, 216–236.
7. **Dahm, P. E.**, The mode of action of insecticides exclusive of organic phosphorous compounds, *Annu. Rev. Entomol.*, 2, 247–260, 1957.
8. **Gunther, F. A. and Jeppson, L. R.**, *Modern Insecticides and World Food Production*, Chapman & Hall, London, 1960, 284 pp.
9. **Hanna, A. D.**, A review of the use of insecticides on the main agricultural crops by the world. I(1). Cash crops, *World Rev. Pest Contr.*, 7, 97–114, 1968.
10. **Hanna, A. D.**, A review of the use of insecticides on the main agricultural crops of the world. I(2). Deciduous fruit trees, *World Rev. Pest Contr.*, 7, 184–203, 1968.
11. **Hanna, A. D.**, A review of the use of insecticides on the main agricultural crops of the world. I(3). Cotton, *World Rev. Pest Contr.*, 8, 23–44, 1969.
12. **Jacobson, M. and Crosby, D. G., Eds.**, *Naturally Occurring Insecticides*, Dekker, New York, 1971.
13. **Kearns, C. W.**, The mode of action of insecticides, *Annu. Rev. Entomol.*, 1, 123–148, 1956.
14. **Kenaga, E. E. and Allison, W. E.**, Commercial and experimental organic insecticides (1969 revision), *Bull. Entomol. Soc. Am.*, 15, 85–148, 1969.
15. **March, R. B.**, The chemistry and action of acaricides, *Annu. Rev. Entomol.*, 3, 355–376, 1958.
16. **Metcalf, R. L.**, *Organic Insecticides, Their Chemistry and Mode of Action*, Interscience, New York, 1955, 392 pp.
17. **Metcalf, R. L.**, Physiological basis for insect resistance to insecticides, *Physiol. Rev.*, 35, 197–232, 1955.
18. **Metcalf, R. L.**, Selective toxicity of insecticides, *World Rev. Pest Contr.*, 3, 28–43, 1964.
19. **Metcalf, R. L.**, Mode of action of insecticide synergists, *Annu. Rev. Entomol.*, 12, 229–256, 1967.
20. **O'Brien, R. D.**, *Insecticides. Action and Metabolism*, Academic Press, New York, 1967, 332 pp.
21. **O'Brien, R. D. and Yamamoto, I., Eds.**, *Biochemical Toxicology of Insecticides*, Academic Press, New York, 1970, 218 pp.
22. **Oppenoorth, F. J.**, Biochemical genetics of insecticide resistance, *Annu Rev. Entomol.*, 10, 185–206, 1965.
23. **Perry, A. S.**, Metabolism of insecticides by various insect species, *J. Agric. Food Chem.*, 8, 266–272, 1960.
24. **Perry, A. S.**, The physiology of insecticide resistance by insects, in *The Physiology of Insects*, Vol. 3, Rockstein, M., Ed., Academic Press, New York, 1964, 286–378.
25. **Roan, C. C. and Hopkins, T. L.**, Mode of action of insecticides, *Annu. Rev. Entomol.*, 6, 333–346, 1961.
26. **Shepard, H. H.**, *The Chemistry and Action of Insecticides*, McGraw-Hill, New York, 1951, 504 pp.
27. **Spencer, E. Y. and O'Brien, R. D.**, Chemistry and mode of action of organophosphorus insecticides, *Annu. Rev. Entomol.*, 2, 261–278, 1957.
28. **Winteringham, F. P. W. and Lewis, S. E.**, On the mode of action of insecticides, *Annu. Rev. Entomol.*, 4, 303–318, 1959.
29. **Winteringham, P.**, Mechanism of selective insecticidal action, *Annu. Rev. Entomol.*, 14, 409–442, 1969.
30. **Zweig, G.**, Insecticides, in *Analytical Methods for Pesticides, Plant Growth Regulators, and Food Additives*, Vol. 2, Academic Press, New York, 1964, 619 pp.

E. Feed Additives and Supplements

1. Antibiotics

1. **Elliot, R. F.**, Uses of antibiotics in animal production, in *Antimicrobial Agents and Chemotherapy – 1966*, Hobby, G. L., Ed., American Society for Microbiology, Ann Arbor, MI, 1967, 739–746.
2. **Jukes, T. H.**, Antibiotics in Nutrition, Antibiotics Monogr. No. 4, Medical Encyclopedia, New York, 1955.
3. **Jukes, T. H.**, Antibiotics in animal feeds and animal production, *BioScience*, 22, 526–534, 1972. (71 ref.)
4. **Jukes, T. H. and Williams, W. L.**, Nutritional effects of antibiotics, *Pharmacol. Rev.*, 5, 381–420, 1953.
5. **Somogyi, J. C. and Francois, A. C., Eds.**, *Antibiotics in Agriculture, Bibl. Nutr. Dieta*, No. 10, S. Karger, Basel, 1968, 208 pp.

2. Hormones

1. **Haard, N. F.**, Potential applications of hormones to plant and animal tissues as food, *CRC Crit. Rev. Food Technol.*, 2(3), 305–353, 1971. (343 ref.)

3. Drugs

1. National Research Council, The Use of Drugs in Animal Feeds, Publ. 1679, National Academy of Science, Washington, D.C., 1969, 407 pp.
2. **Welch, H. and Marti-Ibañez, R., Eds.**, *Symposium on Medicated Feeds*, Medical Encyclopedia, New York, 1956, 214 pp.

4. Nitrogen Supplements

1. **Armstrong, D. G.,** Uses of non-protein nitrogen in animal feeding, *Chem. Ind.,* July 6, 1968, 894–898.
2. **Briggs, M. H., Ed.,** *Urea as a Protein Supplement,* Pergamon Press, Oxford, 1967, 466 pp. (23 papers)
3. **Loosli, J. K. and McDonald, I. W.,** Nonprotein Nitrogen in the Nutrition of Ruminants, FAO Agric. Stud. No. 75, FAO, Rome, 1968, 96 pp.
4. National Research Council, *Alternative Sources of Protein for Animal Production,* National Academy of Sciences, Washington, D.C., 1973, 184 pp.

5. Minerals

1. **Russell, F. C. and Duncan, D. L.,** Minerals in Pasture: Deficiencies and Excesses in Relation to Animal Health, Tech. Comm. No. 15, Commonwealth Agric. Bureaux, Farnham Royal, Sloughs, Bucks, England, 1956, 170 pp.
2. **Underwood, E. J.,** *The Mineral Nutrition of Livestock,* Commonwealth Agric. Bureaux, Farnham Royal, Sloughs, Bucks, England, 1966, 256 pp.

6. Trace Elements

1. **Lamb, C. A., Bentley, O. G., and Beattie, J. M., Eds.,** *Trace Elements,* Academic Press, New York, 1957.
2. **Sauchelli, V.,** *Trace Elements in Agriculture,* Van Nostrand Reinhold, New York, 1969.
3. **Underwood, E. J.,** *Trace Elements in Human and Animal Nutrition,* Academic Press, New York, 1971.

7. Vitamins

1. **Bauernfeind, J. C.,** Vitamins and carotenoids in modern feeds and animal applications, *World Rev. Anim. Prod.,* 5(21), 20–50, 1969.

F. Machinery

1. **Acock, A. M.,** Progress and Economic Problems in Farm Mechanization, FAO, Rome, 1950.
2. **Bainer, R.,** Progress in mechanization, in *Research for the World Food Crisis,* Aldrich, D. G., Jr., Ed., American Association for the Advancement of Science, Washington, D.C., 1970, 181–190.
3. **Bates, W. N.,** *Mechanization of Tropical Crops,* Temple Press, London, 1957.
4. **Behr, A. B. P. G.,** Progress through power farming. 1. Increasing production in the under-developed countries, *World Crops,* 13(3), 87–91, 1961; 2. Achievement of rational mechanization, *World Crops,* 13(4), 134–138, 1961.
5. **Chambers, R. M.,** Mechanization of rice production, *Trop. Agric.,* 33, 182–189, 1956.
6. **Davies, C.,** Considerations and Procedures for Successful Introduction of Farm Mechanization, FAO Agric. Dev. Paper No. 44, FAO, Rome, 1954, 36 pp.
7. Food and Agriculture Organization, Progress and Economic Problems in Farm Mechanization, FAO, Washington, D.C., 1950, 88 pp.
8. Food and Agriculture Organization, Agriculture and Industrialization, FFHC Basic Stud. No. 17, FAO, Rome, 1962, 129 pp.
9. **Giles, G. W.,** Agricultural power and equipment, in The World Food Problem. Report of the Panel on the World Food Supply, Vol. 3, President's Science Advisory Committee, U.S. Gov. Print. Off., Washington, D.C., 1967, 175–216.
10. **Gordon, A. S.,** Essential Consideration in Mechanization of Farming, FAO Agric. Dev. Paper No. 5, FAO, Washington, D.C., 1950, 11 pp.
11. **Hall, C. W.,** *Processing Equipment for Agricultural Products,* Edward Bros. Inc., Ann Arbor, MI, 1963, 265 pp.
12. **Hopfen, H. J.,** Farm Implements for Arid and Tropical Regions, FAO Agric. Dev. Paper No. 67, FAO, Rome, 1960, 158 pp.
13. **Hopfen, H. J.,** International Directory of Agricultural Engineering Institutions, 2nd ed., FAO, Rome, 1962, 393 pp.
14. **Johnson, L.,** Mechanization to improve agriculture of developing countries, *J. Indian Soc. Agric. Eng.,* 2(2), Dec. 1965.
15. **Lönnemark, H.,** Multifarm Use of Agricultural Machinery, FAO Agric. Dev. Paper No. 85, FAO, Rome, 1967.
16. **Mayne, J. E.,** The small farm and mechanization, *World Crops,* 2(11), 465–468, 1950.
17. **Mayne, J. E.,** Progress in the mechanization of farming in the colonial territories, *Trop. Agric.,* 31, 178–183, 1954; 32(2), 95–99, 1955; 33(4), 272–277, 1956.
18. **McColly, H. F. and Martin, J. W.,** *Introduction to Agricultural Engineering,* McGraw-Hill, New York, 1955, 553 pp.
19. **Nowacki, T.,** Agricultural Mechanization: Basic Concepts Connected with the Economics of Agricultural Mechanization, United Nations, New York, 1968, 32 pp.
20. **Stenstrom, L.,** Agricultural engineering. Development of agricultural equipment, in World Food Congress. Summary of Proceedings, FAO, Washington, D.C., June 1963, 177–186.
21. **Stokes, A. R.,** Mechanization and the peasant farmer, *World Crops,* 15, 444–450, 1963.
22. **Stout, B. A.,** Equipment for Rice Production, FAO Agric. Dev. Paper No. 84, FAO, Rome, 1966, 169 pp.

23. U.S. Department of Agriculture, Power to Produce. The Yearbook of Agriculture 1960, U.S. Gov. Print. Off., Washington, D.C., 1960, 480 pp.
24. **Wijewardene, R.,** The machine in tropical agriculture, *World Crops*, 11, 309, 310, 313, 353–356, 407, 408, 437, 438, 444, 1959.
25. **Winkelman, D.,** The Traditional Farmer: Maximation and Mechanization, OECD, Paris, 1972, 94 pp.
26. **Yamanaka, I. and Kaburaki, H.,** Agricultural Machinery and Implements, International Trade Service Bureau, 1961, 115 pp.

XII. FOOD AID

A. Bibliographies

1. **Brown, J. C., Grossman, S. P., and Stubner L., Jr.,** *Foreign Aid: A Selective Working Bibliography,* Pennsylvania State University Libraries, University Park, 1971, 59 pp.
2. **Hald, M.,** *A Selected Bibliography on Economic Development and Foreign Aid,* The Rand Corporation, Santa Monica, 1958.
3. **Henderson, E.,** Food Aid: A Selective Annotated Bibliography on Food Utilization for Economic Development, FAO, Rome, 1964, 203 pp.

B. General References

1. **Allen, G. R. and Smethurst, R. G.,** The Impact of Food Aid on Donor and Other Food-Exporting Countries, World Food Program Stud. No. 2, FAO, Rome, 1965, 52 pp.
2. **Berg, A. D.,** Food for peace – the U.S. effort in meeting world nutrition needs, *J. Am. Diet. Assoc.,* 48, 512–513, 1966.
3. **Beringer, C.,** Real effects of foreign surplus disposal in underdeveloped economies: A comment, *Q. J. Econ.,* 77, 317–323, 1963.
4. **Chakravarty, S. and Rosenstein-Rodan, P. N.,** The Linking of Food Aid with Other Aid, World Food Program Stud. No. 3, FAO, Rome, 1965, 39 pp.
5. **Cochrane, W. W., Mackie, A. B., and Chappell, G. C.,** Potential uses of farm products as aid to developing countries, *J. Farm Econ.,* 45, 961–973, 1963.
6. **Dandekar, V. M.,** The Demand for Food, and Conditions Governing Food Aid During Development, World Food Program Stud. No. 1, FAO, Rome, 1965, 69 pp.
7. **Dessau, J.,** The Role of Multilateral Food Aid Programs, World Food Program Stud. No. 5, FAO, Rome, 1965, 38 pp.
8. **Ezekiel, M.,** Impact and application of foreign surplus disposal on developed economies and foreign competitors: The international perspective, *J. Farm Econ.,* 42, 1063–1077, 1960.
9. **FitzGerald, D. A.,** Operational and Administrative Problems of Food Aid, World Food Programs Stud. No. 4, FAO, Rome, 1965, 63 pp.
10. Food and Agriculture Organization, Development Through Food – A Strategy for Surplus Utilization, FFHC Basic Stud. No. 2, FAO, Rome, 1962, 162 pp.
11. Food and Agriculture Organization, Disposal of Agricultural Surpluses: Principles Recommended by the FAO, 2nd ed., FAO, Rome, 1963, 23 pp.
12. Food and Agriculture Organization, Food Aid and Other Forms of Utilization of Agricultural Surpluses, A Review of Programs, Principles and Consultations, FAO Commodity Policy Stud. No. 15, FAO, Rome, 1964, 55 pp.
13. Food and Agriculture Organization, Food Aid to Education and Training, World Food Program in Action, FAO, Rome, 1970, 44 pp.
14. Food and Agriculture Organization, Functions of a World Food Reserve: Scope and Limitations, FAO Commodity Policy Stud. No. 10, FAO, Rome, 1961, 77 pp.
15. Food and Agriculture Organization, Report on the World Food Program by the Executive Director, FAO, Rome, 1965, 100 pp.
16. Food and Agriculture Organization, Uses of Agricultural Surpluses to Finance Economic Development in Underdeveloped Countries, FAO Commodity Policy Stud. No. 6, FAO, Rome, 65 pp.
17. Food and Agriculture Organization, The Power of Food. A Progress Report on the World Food Program, FAO, Rome, 1969, 32 pp.
18. Food and Agriculture Organization, World Food Program. A Story of Multilateral Aid, FAO, Rome, 1967, 38 pp.
19. **Ginor, F.,** Uses of Agricultural Surpluses: Analysis and Assessment of the Economic Effect of the Public Law 480, Title I Program in Israel, Bank of Israel Research Dep., Jerusalem, 1963.
20. **Kahn, A. E.,** Agricultural aid and economic development: The case of Israel, *Q. J. Econ.,* 76, 568–591, 1962.
21. **Khatkhate, D. R.,** Some notes on the real effects of foreign surplus disposal in underdeveloped economies, *Q. J. Econ.,* 76, 186–196, 1962.
22. **Toma, P. A.,** *The Politics of Food for Peace,* University of Arizona Press, Tucson, 1967.
23. **Witt, L. W.,** Development through food grants and concessional sales, in *Agriculture in Economic Development,* McGraw-Hill, New York, 1964.
24. **Witt, L. W. and Eicher, C.,** The Effect of United States Agricultural Surplus Disposal Programs on Recipient Countries, Michigan State University, Res. Bull. No. 2, 1964.

XIII. FOOD MARKETING AND DISTRIBUTION

A. Bibliographies

1. Food and Agriculture Organization, Bibliography of Food and Agricultural Marketing, FAO, Rome, 1950.
2. Food and Agriculture Organization, Bibliography of Food and Agricultural Marketing: Publications in West European Languages Relating to Countries Other than U.S.A. and Canada, with Summary Comments, FAO, Rome, 1965, 625 pp.
3. Foreign Economic Development Service, Food Marketing in Developing Countries, An Annotated Bibliography of Work Done under Auspices of the U.S. Agency for International Development, A.I.D. Bibliography Ser.: Agric. No. 6, Dec. 1971, 45 pp.
4. **Torrealba, P.,** Food Marketing in Economic Development. An Annotated Bibliography, Dept. of Agric. Econ., Michigan State University, East Lansing, 1971, 60 pp.

B. General References

1. **Abbott, J. C.,** Marketing Problems and Improvement Programs, FAO, Rome, 1958.
2. **Abbott, J. C.,** The role of marketing in the development of backward agricultural economies, *J. Farm Econ.*, 44, 349–362, 1962.
3. **Abbott, J. C.,** Marketing: Its Role in Increasing Productivity, FFHC Basic Stud. No. 4, FAO, Rome, 1967, 116 pp.
4. **Abbott, J. C.,** The development of marketing institutions, in *Agricultural Development and Economic Growth*, Southworth, H. M. and Johnston, B. F., Eds., Cornell University Press, Ithaça, N.Y., 1967, pp. 364–398.
5. **Abbott, J. C. and Creupelandt, H. C.,** Agricultural Marketing Boards: Their Establishment and Operation, FAO Marketing Guides No. 5, FAO, Rome, 1966, 236 pp.
6. **Anschel, K. R., Brannon, R. H., and Smith, E. D., Eds.,** *Agricultural Cooperatives and Markets in Developing Countries*, Praeger, New York, 1969, 373 pp.
7. **Bonnen, J. T., Eicher, C. K., and Schmid, A.,** Marketing in economic development, in *Agricultural Market Analysis*, Michigan State University, East Lansing, 1964, 35–65.
8. **Bowring, J. R., et al.,** *Marketing Policies for Agriculture*, Prentice-Hall, Englewood Cliffs, N.J., 1960, 276 pp.
9. **Brunk, M. E. and Darrah, L. B.,** *Marketing Agricultural Products*, Ronald Press, New York, 1955, 419 pp.
10. **Buzzell, R. D. and Nourse, R. E. M.,** *Products Innovation in Food Processing 1954–1964*, Harvard University Press, Boston, Mass., 1967, 220 pp.
11. **Chaturvedi, J. M.,** Theory of Marketing in Underdeveloped Countries, Allahabad, 1959.
12. **Collins, N. R. and Holton, R. H.,** Programming changes in marketing in planned economic development, *Kyklos*, 16, 123–137, 1963.
13. **Digby, M. and Gretton, R. H.,** Co-operative Marketing for Agricultural Producers, FAO Agric. Dev. Papers No. 53, FAO, Rome, 1968, 80 pp.
14. **Drucker, P. F.,** Marketing and economic development, *J. Market.*, 22, 252–259, 1958.
15. Food and Agriculture Organization, Marketing. A Dynamic Force in Agricultural Development, World Food Problems No. 10, FAO, Rome, 1970, 40 pp.
16. **Hirsch, L. V.,** *Marketing in an Underdeveloped Economy*, Prentice-Hall, Englewood Cliffs, N.J., 1961, 392 pp.
17. **Holton, R. H.,** Marketing structure and economic development, *Q. J. Econ.*, 67, 344–361, 1953.
18. **Jones, W. O.,** Agricultural marketing and economic development, in *Some Issues Emerging from Recent Breakthroughs in Food Production*, Turk, K. L., Ed., Cornell University, Ithaca, New York, 1971, 257–282.
19. **Kohls, R. L.,** *Marketing of Agricultural Products*, 2nd ed., Macmillan, New York, 1961, 424 pp.
20. **Kracmar, J. Z.,** *Marketing Research in Developing Countries. A Handbook*, Praeger, New York, 1971, 322 pp.
21. **Kriesberg, M.,** Marketing foods in developing nations – Second phase of the war on hunger, *J. Market.*, 32 (October), 55–60, 1968.
22. **Kriesberg, M.,** Agricultural Marketing in Developing Nations: Challenge for Innovator and Investor, U.S. Dep. Agric., Foreign Economic Development Service, Washington, D.C., 1968, 18 pp.
23. **Kriesberg, M., Ed.,** The Marketing Challenge, FEDR No. 7, U.S. Dep. Agric., Foreign Economic Development Service, Washington, D.C., 1970, 96 pp.
24. **Kriesberg, M.,** Food Distribution Research in Developing Countries, U.S. Dep. Agric., Foreign Economic Development Service, Washington, D.C., 1970, 22 pp.
25. **Lund, R. G.,** Strategic approach to commodity market development, *FAO Mon. Bull. Agric. Econ. Stat.*, 22(78), 1–8, 1973.
26. **McNair, M. P., Applebarn, W., and Salmon, W. J.,** *Cases in Food Distribution*, R. D. Irwin Publishing, Homewood, Ill., 1964, 715 pp.
27. **Mehren, G. L.,** Market organization and economic development, *J. Farm Econ.*, 51(5) 1307–1315, 1959.
28. **Mehren, G. L.,** Technology of marketing, in *Some Issues Emerging from Recent Breakthroughs in Food Production*, Turk, K. L., Ed., Cornell University, Ithaca, N.Y., 1971, 333–351.
29. **Mellor, J. W.,** Marketing and marketing institutions, in *The Economics of Agricultural Development*, Cornell University, Ithaca, N.Y., 1966, 328–344.

30. **Mellor, J. W.,** Growth of the Market and the Pace of Agriculture Development in Low Income Nations, Dep. of Agricultural Economics, Cornell University, Ithaca, N.Y., 1967, 10 pp.
31. **Moore, J. R. and Walsh, R. G.,** *Market Structure of the Agricultural Industries,* Iowa State University Press, Ames, 1966. 412 pp.
32. **Moyer, R.,** Marketing in Economic Development, Occasional Paper, Institute of International Business Management Studies, Michigan State University, East Lansing, 1965, 63 pp.
33. **Moyer, R. and Hollander, S. C.,** *Markets and Marketing in Developing Economies,* R. D. Irwin Publishing, Homewood, Ill., 1968, 264 pp.
34. **Mueller, W. F.,** Some market considerations in economic development, *J. Farm Econ.,* 41, 414–425, 1959.
35. Organization for Economic Cooperation and Development, Food Marketing and Economic Growth, OECD, Paris, 1970, 130 pp.
36. President's Science Advisory Committee, Marketing processing and distribution of food products, in The World Food Problem, Report of the Panel on the World Food Supply, Vol. 2, U.S. Gov. Print. Off., Washington, D.C., 1967, 537–567.
37. **Pritchard, N. T.,** Framework for analysis of agricultural marketing systems in developing countries, *Agric. Econ. Res.,* 21(3), 78–85, 1969.
38. **Shepard, G. S.,** *Marketing Farm Products,* 4th ed., Iowa State University Press, Ames, 1962, 521 pp.
39. **Solomon, M. R.,** The structure of the market in underdeveloped economics, *Q. J. Econ.,* 62, 519–541, 1948.
40. U.S. Department of Agriculture, Marketing. The Yearbook of Agriculture 1954, U.S. Dep. Agric., Washington, D.C., 1954, 506 pp.
41. **Waugh, F. V., Ed.,** *Readings on Agricultural Marketing,* Iowa State College Press, Ames, 1954, 456 pp.

C. The Role of Transportation

1. **Bredo, W.,** Transportation Problems of Expanding Western Agriculture, Agricultural Marketing Service, Washington, D.C., 1954, 227 pp.
2. **Fromm, G., Ed.,** Transport Investment and Economic Development, Brookings Institute, Washington, D.C., 1965.
3. **Owen, W.,** Transportation and economic development, *Am. Econ. Rev.,* 49(2), 179–187, 1959.
4. **Owen, W.,** Transportation and technology, *Am. Econ. Rev.,* 52(2), 405–413, 1962.
5. President's Science Advisory Committee, Transportation, in The World Food Problem. Report on the Panel on the World Food Supply, Vol. 2, U.S. Gov. Print. Off., Washington, D.C., 1967, 569–592.
6. **Prest, A. R.,** *Transport Economics in Developing Countries,* Praeger, New York, 1969.
7. **Sisler, D. G.,** The role of transportation in the process of economic development, in *Some Issues Emerging from Recent Breakthroughs in Food Production,* Turk, K. L., Ed., N.Y. State College of Agriculture, Cornell University, Ithaca, N.Y., 1971, 333–351.

XIV. FOOD HABITS

A. Bibliographies

1. **Gottlieb, D. and Rossi, P. H.,** A Bibliography and Bibliographic Review of Food and Food Habit Research, Quartermaster Food and Container Institute for the Armed Forces, Chicago, 1961, 112 pp.
2. **Wilson, C. S.,** Food Habits: A selected annotated bibliography, *J. Nutr. Educ.*, 6(1), Suppl. 1, 39–72, 1973.

B. Food Habits and Their Origin

1. General

1. **Cuthbertson, D. P.,** Food selection, in *Nutrition, A Comprehensive Treatise*, Vol. 2, Beaton, G. H. and McHenry, E. W., Eds., Academic Press, New York, 1964, 435–488.
2. **Eppright, E. S.,** Factors affecting food acceptance, *J. Am. Diet. Assoc.*, 23, 579–587, 1947. (33 ref.)
3. **Gift, H. H., Washbow, M. B., and Harrison, G. G.,** *Nutrition, Behavior and Change*, Prentice-Hall, Englewood Cliffs, N.J., 1972, 392 pp.
4. **Graubard, M.,** *Man's Food. Its Rhyme or Reason*, Macmillan, New York, 1943.
5. **Grivetti, L. E. and Pangborn, R. M.,** Food habit research: A review of approaches and methods, *J. Nutr. Educ.*, 5(3), 204–208, 1973. (101 ref.)
6. **Lowenberg, M. E.,** The development of food patterns, *J. Am. Diet. Assoc.*, 65(3), 263–268, 1974.
7. **Lowenberg, M. E., Todhunter, E. N., Wilson, E. D., and Feeney, M. C.,** *Food and Man*, John Wiley & Sons, New York, 1968, 341 pp.
8. **Renner, H. D.,** *The Origin of Food Habits*, Faber & Faber, London, 1944.

2. Role of Taste and Smell

1. **Amerine, M. A., Pangborn, R. M., and Roessler, E. B.,** *Principles of Sensory Evaluation of Food*, Academic Press, New York, 1965, 602 pp.
2. **Henkin, R. J.,** The role of taste in disease and nutrition, *Borden's Rev. Nutr. Res.*, 28, 71–87, 1967.
3. **Kare, M. L. and Maller, O.,** *The Chemical Senses and Nutrition*, Johns Hopkins Press, Baltimore, 1967, 495 pp.
4. **Kouwenhoven, T.,** Olfactory and gustatory problems. An introduction to the technological, nutritional and physiological aspects of the organoleptic assessment of food characteristics, *World Rev. Nutr. Diet.*, 12, 318–376, 1970.

3. Psychological Aspects

1. **Babcock, C. G.,** Food and its emotional significance, *J. Am. Diet Assoc.*, 24, 390–393, 1948.
2. **Bilz, R.,** Ungewöhnliche Ernährungsformen and ihre psychologischen Hintergründe, *Wiss. Veröff. dtsch. Ges. Ernähr.*, 5, 170–197, 1960.
3. **Brožek, J.,** Research on diet behavior, *J. Am. Diet Assoc.*, 57, 321–325, 1970. (31 ref.)
4. **Moore, H. B.,** Psychologic factors and dietary fancies, *J. Am. Diet. Assoc.*, 28, 789–794, 1952.
5. **Morgane, P. J. and Jacobs, H. L.,** Hunger and satiety, *World Rev. Nutr. Diet.*, 10, 101–213, 1969.
6. Symposium. The psychology of eating, *Proc. Nutr. Soc.*, 12, 143–165, 1953. (6 papers)
7. **Townsend, C. W.,** Food prejudices, *Sci. Mon.*, 27, 65–68, 1928.

4. Social and Cultural Aspects

1. **Cassel, J.,** Social and cultural implications of food and food habits, *Am. J. Public Health*, 47, 732–740, 1957.
2. **de Garine, I.,** The social and cultural background of food habits in developing countries, *FAO Nutr. Newsletter*, 8(1), 9–22, 1970.
3. **Fathauer, G. H.,** Food habits. An anthropologist's view, *J. Am. Diet. Assoc.*, 37, 335–338, 1960.
4. **Fliegel, F. C.,** Food habits and national backgrounds, Penn. State Univ. Agric. Exp. Sta. Bull. 684, University Park, 1961.
5. **Lee, D.,** Cultural factors in dietary choice, *Am. J. Clin. Nutr.*, 5, 166–170, 1957.
6. **Lowenberg, M. E.,** Socio-cultural basis of food habits, *Food Technol.*, 24(7), 27–32, 1970.
7. **Mead, M.,** Dietary patterns and food habits, *J. Am. Diet. Assoc.*, 19, 1–5, 1943.
8. **Mead, M.,** Factor of food habits, *Ann. Am. Acad. Pol. Soc. Sci.*, 225, 135–141, 1943.
9. **Mead, M.,** *Food Habits Research, Problems of the 1960's*, National Research Council, National Academy of Sciences, Washington, D.C., 1964, 39 pp.
10. **Read, M.,** *Culture, Health and Disease*, Tavistock, London, 1966.
11. **Wagner, M.,** The irony of affluence, *J. Am. Diet. Assoc.*, 57, 311–315, 1970.
12. **Wenkam, N. S.,** Cultural determinants of nutritional behavior, *Nutr. Program News*, July – Aug., 1969, 4 pp.

5. Geographical Aspects

1. **Bennett, M. K.,** International aspects in food consumption, *Geogr. Rev.*, 31, 365, 1941.

2. **Bennett, M. K.,** *The World's Food. A study of the Interrelations of World Populations, National Diets, and Food Potentials,* Harper & Row, New York, 1954.
3. **Kariel, H. G.,** A proposed classification of diet, *Ann. Assoc. Am. Geogr.,* 56, 68, 1966.
4. **Sorre, M.,** The geography of diet, in *Readings in Cultural Geography,* Wagner, P. L. and Mikesell, M. W., Eds., University of Chicago Press, Chicago, 1962, 445–456.

6. Historical Aspects

1. **Barker, T. C., McKenzie, J. C., and Yudkin, J., Eds.,** *Our Changing Fare: Two Hundred Years of British Food Habits,* MacGibbon & Kee Soc., London, 1966.
2. **Brothwell, D. and Brothwell, P.,** *Food in Antiquity. Á Survey of the Diet of Early People,* Thames & H. Publishing, London, 1969.
3. **Burnett, J.,** *Plenty and Want: A Sociological History of Diet in England, From 1815 to Present Day,* Nelson, London, 1966.
4. **Curtis-Bennett, N.,** *The Food of the People, Being the History of Industrial Feeding,* Faber & Faber, London, 1949,
5. **Drummond, J. C. and Wilbraham, A.,** *The Englishman's Food: A History of Five Centuries of English Diet,* J. Cape London, 1939, 574 pp.
6. **Jensen, L. B.,** *Man's Foods. Nutrition and Environments in Food Gathering Times and Food Producing Times,* Garrard Publishing, Champaign, Ill., 1953, 278 pp.
7. **Prakash, O.,** *Food and Drinks in Ancient India,* Munshi Ram Manohar Lal, Delhi, 1961.
8. **Simmons, F. J.,** *Eat Not This Flesh. Food Avoidances in the Old World,* University of Wisconsin Press, Madison 1961, 241 pp.
9. **Tannahill, R.,** *Food in History,* Stein & Day, New York, 1973, 448 pp.

C. Misinformation, Food Cultism and Quackery

1. American Institute of Nutrition, Symposium: What can we do about food faddism? *Fed. Proc.,* 13, 780–796, 1954 (4 papers)
2. American Medical Association. Council on Foods and Nutrition, Proc. Resources Conference on Food Faddism and Cultism, October 1971, American Medical Association, Chicago, Ill., 1972.
3. **Blix, G., Ed.,** *Food Cultism and Nutrition Quackery, 8th Symposium of the Swedish Nutrition Foundation* Almqvist & Wilksells, Uppsala, 1970, 103 pp.
4. **Bruch, H.,** The allure of food cults and nutrition, *J. Am. Diet. Assoc.,* 57, 316–320, 1970.
5. Conference on quackery, *Science,* 134, 1957–1958, 1961.
6. **Deutsch, R. M.,** *Nuts Among the Berries. An Expose of America's Food Fads,* Ballantine Books, New York, 1961
7. **Deutsch, R. M.,** Nutritional nonsense and food fanatics, Proc. 3rd National Congress on Medical Quackery, Chicago Ill., 1966, 15–24.
8. **Engel, R. W.,** Food faddism, *Nutr. Rev.,* 17, 353–355, 1959.
9. **Henderson, L. M.,** Programs to combat nutritional quackery, *J. Am. Diet. Assoc.,* 64, 372–375, 1974.
10. **Huenemann, R. L.,** Combating food misinformation and quackery, *J. Am. Diet. Assoc.,* 32, 623–626, 1956.
11. **Mitchell, H. S.,** Food fads. What protection have we? *J. Home Econ.,* 53, 100–103, 1961.
12. **New, P. K. and Priest, R. P.,** Food and thought. A sociological study of food cultists, *J. Am. Diet. Assoc.,* 51 13–18, 1967.
13. Nutrition Foundation, The Role of Nutrition Education in Combating Food Fads, Nutrition Foundation, New York, 1962.
14. Nutrition Misinformation and Food Faddism, *Nutr. Rev.,* 32 (Special Suppl.) July 1974, 73 pp.
15. **Olson, R. E.,** Food faddism. Why? *Nutr. Rev.,* 16, 97–99, 1958.
16. **Pyke, M.,** Food facts and fallacies, *R. Soc. Health J.,* 87, 110–114 passim, 1967.
17. **Rosenberg, R. S.,** Nutritional claims in food advertising, *J. Am. Diet. Assoc.,* 32, 631–635, 1956.
18. **Stare, F. J.,** Quackery in the field of nutrition, Proc. 1st National Congress on Medical Quackery, Washington, D.C., 1961, 66–71.
19. **Todhunter, E. N.,** Food habits, food faddism and nutrition, in *Food, Nutrition and Health,* Rechcigl, M. J., Ed. *World Rev. Nutr. Diet.,* Vol. 16, S. Karger, Basel, 1973, 286–317.
20. **Young, C. M.,** Food habits and faddism, in *The Science of Nutrition and its Application in Clinical Dentistry,* 2nd ed., Nizel, A. E., Ed., W. B. Saunders, Philadelphia, 1966, 219–231.

D. Malnutrition and Food Habits

1. **Burgess, A. and Dean, R. F. A., Eds.,** *Malnutrition and Food Habits,* Tavistock, London, 1962, 210 pp.
2. **Clark, F. L. G.,** Food habits as a practical nutrition problem, *World Rev. Nutr. Diet.,* 9, 56–84, 1968.
3. **Cooper, M.,** *Pica,* Charles C Thomas, Springfield, Ill., 1957.
4. **Devadas, R. P.,** Social and cultural factors influencing malnutrition, *J. Home Econ.,* 62, 164–171, 1970.
5. **Halsted, J. A.,** Geophagia in man. Its nature and nutritional effects, *Am. J. Clin. Nutr.,* 21, 1384–1393, 1968.

6. **Hundley, J. M.,** Nutrition problems associated with food habits and environment, in Proc. Borden Centennial Symposium on Nutrition. The Nutritional "Ages of Man." Nutrition: Past, Present, and Future, Borden Co. Foundation, New York, 1958, 1–11.
7. **Jenner, A.,** Social, emotional and cultural influences as related to eating patterns and malnutrition, *Can. Nutr. Notes*, 24(4), 37–43, 1968.
8. **Livingston, S. K.,** What influences malnutrition? *J. Nutr. Educ.*, 3, 18–27, 1971.

E. Changing Food Habits

1. **Adams, R. N.,** Changing dietary and health practices, in Science, Technology and Development, Vol. 5, Agriculture, U.S. Gov. Print. Off., Washington, D.C., 1963, 86–96.
2. Committee on Food Habits, The Problem of Changing Food Habits, Bull. National Research Council No. 108, National Academy of Sciences, Washington, D.C., 1943, 177 pp.
3. Committee on Food Habits, Manual for the Study of Food Habits, Bull. National Research Council No. 111, National Academy of Sciences, Washington, D.C., 1945, 142 pp.
4. **Gallop, R. A.,** Redirecting food habits – a food scientist's viewpoint, *J. Can. Diet. Assoc.*, 31, 9–16, 1970.
5. **Mackinnon, C. F.,** Changing food habits: The dietician's dilemma, *J. Am. Diet. Assoc.*, 31(6), 566–569, 1955.
6. **Niehoff, A.,** Changing food habits, *J. Nutr. Educ.*, 1(1), 10–11, 1969.
7. **Parrish, J. B.,** Implications of changing food habits for nutrition educators, *J. Nutr. Educ.*, 2, 140–146, 1971.
8. **Sweeney, M.,** Changing food habits, *J. Home Econ.*, 34, 457–462, 1942.
9. **Teulon, H. P.,** Changes in dietary habits: Their importance in nutrition education programs, *FAO Nutr. Newsletter*, 6(1), 53–55, 1968.
10. **Yudkin, J. and McKenzie, J. C., Eds.,** *Changing Food Habits*, MacGibbon & Kee Soc., London, 1964, 144 pp.

XV. NUTRITION, NATIONAL DEVELOPMENT, AND PLANNING

A. Nutrition and National Development

1. **Barg, B.,** Nutrition and national development, in *Nutrition, National Development and Planning,* Berg, A., Scrimshaw, N. S., and Call, D. L., Eds., M.I.T. Press, Cambridge, Mass., 1973, 49–69.
2. **Berg, A.,** Malnutrition and national development, *Foreign Affairs,* 46, 126–136, 1967; *J. Trop. Pediatr.,* 14, 116–123, 1968.
3. **Berg, A.,** Priority of nutrition in national development, *Indian J. Med. Educ.,* 8(8), 329–337, 1969. (43 ref.); *Indian J. Nutr. Diet.,* 8, 208–218, 1971; *Technol. Rev. Inst. Technol.,* Boston, Mass., 1970, 44–51; *Nutr. Rev.,* 28, 199–204, 1970.
4. **Berg, A.,** Nutrition as a national priority: Lessons from the India experiment, *Am. J. Clin. Nutr.,* 23, 1396–1408, 1970.
5. **Berg, A. and Muscat, R.,** Nutrition and development: The view of the planner, *Am. J. Clin. Nutr.,* 25, 186–209, 1972.
6. **Berg, A. and Muscat, R. J.,** *The Nutrition Factor. Its Role in National Development,* Brookings Institution, Washington, D.C., 1973, 290 pp.
7. **Berg, A., Scrimshaw, N. S., and Call, D. L., Eds.,** *Nutrition, National Development and Planning,* M.I.T. Press, Cambridge, Mass., 1973, 401 pp.
8. **Burk, M. C. and Ezekiel, M.,** Food and nutrition in developing economies, in *Agricultural Development and Economic Growth,* Southworh, H. M. and Johnston, B. F., Eds., Cornell University Press, Ithaca, N.Y., 1967, 327–367. (Includes comment by D. M. Hegsted.)
9. **Correa, H.,** The reassured influence of nutrition on personal and social development, *Life Newsletter,* March 1974, 3–7. (43 ref.)
10. **Correa, H. and Cummings, G.,** Contribution of nutrition to economic growth, *Am. J. Clin. Nutr.,* 23, 560–565, 1970.
11. **de Esquef, L.,** Food and nutrition in the overall development of Latin America, *FAO Nutr. Newsletter,* 9(4), 9–15, 1971.
12. **de Garine, I.,** The socio-cultural aspects of nutrition, *Ecol. Food Nutr.,* 1, 143–163, 1972.
13. **Farris, D. K.,** Adequate Nutrition: An Indispensable Base for Economic Development, UNICEF (India), 1966.
14. **Ginor, F.,** Importance of the nutrition component in economic planning, in *Nutrition, National Development, and Planning,* Berg, A., Scrimshaw, N. S., and Call, D. L., Eds., M.I.T. Press, Cambridge, Mass., 1973, 275–281.
15. **Kallen, D. J.,** Nutrition and society, *JAMA*, 215(1), 94–100, 1971.
16. **Lawas, J. M.,** Nutrition in government planning, *Philipp. J. Nutr.,* 25(2), 75–80, 1972.
17. **Mellor, J. W.,** Nutrition and economic growth, in *Nutrition, National Development and Planning,* Berg, A., Scrimshaw, N. S., and Call, D. L., Eds., M.I.T. Press, Cambridge, Mass., 1973, 70–73.
18. **Oshima, H. T.,** Food consumption, nutrition and economic development in Asian countries, 11th Pacific Science Congress, Tokyo, August 1966, 21–22; *Econ. Dev. Cult. Change,* 15, 385–397, 1967.
19. **Pant, P.,** Food, nutrition, population and economic development, *Proc. 1st Asian Nutrition Congress,* Tulpule, P. G. and Jaya Rao, K. S., Eds., Nutrition Society of India, Hyderabad, 1972, 286–310.
20. **Pascual, C. R.,** Social and cultural factors in malnutrition, in *Proc. 1st Asian Congress of Nutrition,* Tulpule, P. G. and Jaya Rao, K. S., Eds., Nutrition Society of India, Hyderabad, 1972, 560–570. (29 ref.)
21. **Sebrell, W. H., Jr.,** Nutrition planning and national development, in CENTO Conference on Combating Malnutrition in Preschool Children, Istamabad, Pakistan, March 18–22, 1968, CENTO, Ankara, 1968, 13–26.
22. **Trémolieres, J.,** Nutrition and underdevelopment, in *Progress in Human Nutrition,* Vol. 1, Margen, S., Ed., Avi Publishing, Westport, Conn., 1971, 1–28.

B. Economics of Malnutrition

1. General

1. **Belli, P.,** The economic implications of malnutrition: The dismal science revisited, *Econ. Dev. Cult. Change,* 20(1), 1–23, 1971.
2. **Call, D. L.,** Research Findings on the Social Consequences of Malnutrition, Cornell Agric. Econ. Staff Paper No. 8, Cornell University, Ithaca, N.Y., August 1969, 20 pp.
3. **Call, D. L. and Longhurst, R.,** Evaluation of the Economic Consequences of Malnutrition, Cornell Agric. Econ. Staff Paper No. 43, Cornell University, September 1971.
4. **Call, D. L. and Longhurst, R.,** Evaluation of the economic consequences of malnutrition, *Proc. 3rd Western Hemisphere Nutrition Congress,* Miami Beach, Florida, 1971, Futura Publishing, Mount Kisco, N.Y., 1972, 312–317.
5. **Cesario, F. J., Simon, S. R., and Kinne, S. L.,** The Economics of Malnutrition, Final Report to A.I.D., Battelle Memorial Institute, Columbus, Ohio, 1970. (2 vols.)

6. **Cook, R.,** The financial cost of malnutrition in the Commonwealth Caribbean, *J. Trop. Pediatr.*, 14, 60–65, 1968 *Nutr. Rev.*, 27(2), 39–41, 1968.
7. **Cook, T.,** The primary costs of malnutrition and its impact on a society, *Proc. 3rd Western Hemisphere Nutritior Congress*, Miami Beach, Florida, 1971, Futura Publishing, Mount Kisco, N.Y., 1972, 324–327.
8. **Lowenstein, F. W.,** The cost of malnutrition – A brief consideration of two major problems, *Food Nutr. Afr.*, 9 3–5, 1971.
9. **Muscat, R. and Berg, A.,** The economics of malnutrition, *Proc. 1st Asian Congress of Nutrition*, Tulpule, P. G. anc Jaya Rao, K. S., Eds., Nutrition Society of India, Hyderabad, 1972, 603–621.
10. **Pollack, H.,** Nutritional problems as part of the total economy, *Am. J. Clin. Nutr.*, 19, 285–290, 1966.
11. **Popkin, B. M.,** Preliminary Report of Proceedings of Methodologies Conference on Economics of Malnutrition Office of Nutrition, Technical Assistance Bureau, A.I.D., Washington, D.C., 1973, 28 pp.
12. **Selowsky, M. and Taylor, L.,** The economics of malnourished children: An example of disinvestment in humar capital, *Econ. Dev. Cult. Change*, 22(1), 17–30, 1973.
13. **Wilson, D.,** The economic analysis of malnutrition, in *Nutrition, National Development and Planning*, Berg, A. Scrimshaw, N. S., and Call, D. L., Eds., M.I.T. Press, Cambridge, Mass., 1973, 129–144.

2. Economic Benefits from Elimination of Malnutrition

1. **Popkin, B. M.,** Economic Benefits from the Elimination of Malnutrition, U.S. Senate Committee on Nutritional anc Human Needs, July 1969.
2. **Popkin, B. M.,** Economic Benefits from the Elimination of Hunger in America, Institute for Research on Poverty Discussion Paper 102–71, University of Wisconsin, Madison, 1971.
3. **Popkin, B. M.,** Economic benefits from the elimination of hunger in America, *Publ. Policy*, 20, 133–153, 1972.

C. Nutrition, Human Development, and Productivity

1. Nutrition, Growth, and Development

1. **Aldrich, R. A.,** Nutrition and human development, *J. Am. Diet. Assoc.*, 46, 453–456, 1965.
2. **Blix, G., Ed.,** *Nutrition in Preschool and School Age*, 7th Symposium of the Swedish Nutrition Foundation Almqvist & Wiksells, Uppsala, 1969, 154 pp.
3. **Cheek, D. B., Graystone, J. E., and Read, M. S.,** Cellular growth, nutrition and development, *Pediatrics*, 45 315–334, 1970.
4. **Childers, V. E.,** Infant nutrition. Priority for development, *Int. Dev. Rev.*, 11(2), 13–15, 1969.
5. **Ebbs, J. H.,** The nutrition and feeding of infants, in *Nutrition. A Comprehensive Treatise*, Vol. 3, Beaton, G. H. anc McHenry, E. W., Eds., Academic Press, New York, 1966, 1–41.
6. **Foman, S. J.,** *Infant Nutrition*, W. B. Saunders, Philadelphia, 1967.
7. **György, P.,** Nutrition in infancy, *Fed. Proc.*, 18(2), II, 9–16, 1959.
8. **György, P.,** The late effects of early nutrition, *Am. J. Clin. Nutr.*, 8, 344–345, 1961.
9. **Hahn, P. and Koldovsky, O.,** *Utilization of Nutrients During Postnatal Development*, Pergamon Press, Oxford, 1966 177 pp.
10. **Heald, F. P., Ed.,** *Adolescent Nutrition and Growth*, Appleton-Century Crofts, New York, 1969.
11. **Holt, L. E., Jr.,** Nutrition and infant feeding, in *Pediatrics*, 14th ed., Barnett, H. L., Ed., Appleton-Century Crofts New York, 1968, 119–170.
12. **Holt, L. E., Jr. and Snyderman, S. E.,** Nutrition in infancy and adolescence, in *Modern Nutrition in Health anc Disease, Dietotherapy*, 4th ed., Wohl, M. G. and Goodhart, R. S., Eds., Lea & Febiger, Philadelphia, 1968 1107–1130; 5th ed., Goodhart, R. S. and Shils, M. E., Eds., 1973, pp. 659–680.
13. **Jackson, R. L., Hanna, F. M., and Flynn, M. A.,** Nutritional requirements of infants and children, *Pediatr. Clin. N Am.*, 9, 879–910, 1962.
14. **Jelliffe, D. B.,** Child Nutrition in Developing Countries. A Handbook for Field Workers, PHS Publ. No. 1822, U.S Gov. Print. Off., Washington, D.C., 1968, 200 pp.
15. **Jelliffe, D. B.,** Infant Nutrition in the Subtropics and Tropics, 2nd ed., WHO Monogr. Ser. 29, WHO, Geneva, 1968 335 pp.
16. **Jelliffe, D. B.,** Nutrition in early childhood, in *Food, Nutrition and Health. A Multidisciplinary Treatise Addressec to the Major Nutrition Problems from a World Wide Perspective*, Vol. 16, Rechcigl, M., Jr., Ed., *World Rev. Nutr Diet.*, S. Karger, Basel, 1973, 1–21.
17. **McCance, R. A. and Widdowson, E. M.,** Nutrition and growth, *Proc. R. Soc.*, 156B, 326–337, 1962.
18. **Macy, I. G.,** *Nutrition and Chemical Growth in Childhood*, Charles C Thomas, Springfield, Ill., 1942. (3 vols.)
19. **Manocha, S. L.,** *Malnutrition and Retarded Human Development*, Charles C Thomas, Springfield, Ill., 1972, 382 pp
20. **Martin, H. C.,** Nutrition: Its relationship to children's physical, mental, and emotional development, *Am. J. Clin Nutr.*, 26, 766–775, 1973. (78 ref.)
21. **Meyer, L. F. and Nassan, E.,** *Physiology and Pathology of Infant Nutrition*, Charles C Thomas, Springfield, Ill. 1955, 533 pp.
22. Nutrition Society, Symposium. Nutrition and cell development, *Fed. Proc.*, 29, 1489–1521, 1970. (5 papers)

23. **Robertson, E. C.**, The nutrition of children and adolescents, in *Nutrition. A Comprehensive Treatise*, Vol. 3, Beaton, G. H. and McHenry, E. W., Eds., Academic Press, New York, 1966, 43–101. (233 ref.)
24. **Swaminathan, M.**, The nutrition and feeding of infants and preschool children in the developing countries, *World Rev. Nutr. Diet.*, 9, 85–123, 1968.
25. Symposium. Nutrition and early growth, *Proc. 6th International Congress of Nutrition*, Livingstone, Edinburgh, 1964, 59–99.
26. Symposium. Nutrition and the very young, *Proc. Nutr. Soc.*, 17, 50–77, 1958.
27. Symposium. Nutritional problems in infancy and childhood, *Proc. Nutr. Soc.*, 22, 119–158, 1963. (7 papers)
28. **Widdowson, E. M.**, Early nutrition and later development, in *Diet and Bodily Constitution*, Wolstenholme, G. E. and O'Conner, M., Eds., Little, Brown, & Co., Boston, 1964, 3–18.
29. **Winick, M.**, *Nutrition and Development*, John Wiley & Sons, New York, 1972.

2. Nutrition and Mental Development

a. General

1. **Baldzs, R.**, Effects of hormones and nutrition on brain development, *Adv. Exp. Med. Biol.*, 30, 385-415, 1972. (85 ref.)
2. **Barnes, R. H.**, Nutrition and man's intellect and behavior, *Fed. Proc.*, 30(4), 1429–1433, 1971. (13 ref.)
3. **Brožek, J.**, Soviet studies on nutrition and higher nervous activity, *Ann. N.Y. Acad. Sci.*, 93, 667–714, 1962.
4. **Brožek, J.**, Experimental investigations of nutrition and human behavior. A postcript, *Am. Sci.*, 57, 139–163, 1963.
5. **Brožek, J.**, Food as an essential. Experimental studies on behavioral fitness, in *Food and Civilization*, Farber, S. M. and Wilson, R. H. L., Eds., Howe, 1966, 29–60.
6. **Canosa, C. A.**, The influence of nutrition and environment on brain function, *IBRO Bull.*, 7(3), 69–74, 1968.
7. **Coursin, D. B.**, Nutrition and brain functions, in *Modern Nutrition in Health and Disease. Dietotherapy*, 4th ed., Wohl, M. G. and Goodhart, R. S., Eds., Lea & Febiger, Philadelphia, 1968, 1072–1085.
8. **Craviato, J., DeLicardie, E. R., and Birch, H. G.**, Nutrition, growth, and neuro-integrative development. An experimental and ecologic study, *Pediatrics*, 38(2), Suppl. II, 319–372, 1966.
9. Early weaning, diet, and intelligence, *Nutr. Rev.*, 23, 211–213, 1965. (6 ref.)
10. **Eichenwald, H. F. and Fry, P. C.**, Nutrition and learning, *Science*, 163, 644–648, 1969. (33 ref.)
11. **Galdston, I.**, Nutrition from the psychiatric viewpoint, *J. Am. Diet. Assoc.*, 28, 405–409, 1952.
12. **Kallen, D. J., Ed.**, Nutrition, Development and Social Behavior, DHEW Publ. No. (NIH) 73–242, U.S. Gov. Print. Off., Washington, D.C., 1973, 386 pp.
13. **Mönckeberg, F.**, Nutrition and mental development, Proc. Conf. on Nutrition and Human Development, East Lansing, MI, 1969.
14. National Research Council. Subcommittee on Nutrition, Brain Development, and Behavior of the Committee on International Nutrition Programs, *The Relationship of Nutrition to Brain Development and Behavior*, National Academy of Sciences, Washington, D.C., 1973, 15 pp.
15. **Norman, E. C. and Robson, J. R. K.**, Nutrition and mental health, in Mental Health Considerations in Public Health, Goldston, S. E., Ed., Publ. No. 1898, U.S. Public Health Service, Washington, D.C., 1969.
16. *Nutrition and Intellectual Growth in Children*, Association for Childhood Education International, Washington, D.C., 1969.
17. Nutrition Society, Symposium. Relationship of nutrition to central nervous system development and function, *Fed. Proc.*, 26(1), 134–151, 1967.
18. Pan American Health Organization, Nutrition, the Nervous System, and Behavior, Proc. of the Seminar, Mona, Jamaica, Jan. 10–14, 1972, Sci. Publ. No. 251, PAHO, Washington, D.C., 1972.
19. **Rajalakshmi, R. and Ramakrishnan, C. V.**, Nutrition and brain function, *World Rev. Nutr. Diet.*, 15, 35–85, 1972.
20. **Read, M. S.**, Human studies on nutrition and mental development: Problems, pitfalls and progress, *Proc. 8th International Congress of Nutrition*, Prague, September 1969, Mašek, J., Ošancová, K., and Cuthbertson, D. P., Eds., Experta Medica, Amsterdam, 1970, 226–230.
21. **Read, M. S.**, Malnutrition, hunger, and behavior, *J. Am. Diet. Assoc.*, 63, 379–391, 1973.
22. **Somogyi, J. C. and Fidanza, F., Eds.**, *Nutrition and Nervous System*, Bibl. Nutr. Dieta, No. 17, S. Karger, Basel, 1972, 202 pp.
23. Symposium, Diet and the central nervous system, *Proc. Nutr. Soc.*, 27, 83–112, 1968. (6 papers)
24. Symposium. Nutrition and mental development, in *Proc. 7th International Congress of Nutrition*, Vol. 1, Pergamon Press, Oxford, 1967, 336–351. (7 papers)
25. Symposium. Nutrition, growth, and mental development, 8th Annual Meeting Am. Pediatr. Soc., in *Am. J. Dis. Child.*, 120(5), 395–418, 1970.
26. Symposium on nutrition and behavior, *Am. J. Clin. Nutr.*, 5, 103–211, 1957.
27. **Winick, M.**, Nutrition and mental development, *Med. Clin. N. Am.*, 54, 1413–1429, 1970.
28. **Winick, M.**, Nutrition and nerve cell growth, *Fed. Proc.*, 29, 1510–1515, 1970. (22 ref.)
29. **Winick, M. and Coombs, J.**, Nutrition, environment, and behavioral development, *Annu. Rev. Med.*, 23, 149–160, 1972. (90 ref.)

b. Malnutrition and Mental Development

1. **Adams, R. D.,** Nutritional diseases of the nervous system in the alcoholic patient, *Trans. Am. Clin. Climatol. Assoc.*, 1, 59–94, 1959.
2. **Altman, J.,** Nutritional deprivation and neural development, in *Brain Development and Behavior,* Sterman, M. B., McGinty, D. J., and Adinolfi, A. M., Eds., Academic Press, New York, 1971, 359–368.
3. **Barnes, R. H.,** Effects of malnutrition on mental development. Truths and half-truths, *J. Home Econ.*, 61(9), 671, 1969.
4. **Birch, H. G. and Craviato, J.,** Infection, nutrition and environment in mental development, in The Prevention of Mental Retardation through Control of Infectious Diseases. Proc. Conf. on Prevention of Mental Retardation through Control of Infectious Diseases, June 9–11, 1966, Cherry Hill, N.J., Eichenwald, H. F., Ed., PHS Publ 1692, U.S. Gov. Print. Off., Washington, D.C., 1968, 227–248, (43 ref.)
5. **Birch, H. G. and Gussow, J. D.,** *Disadvantaged Children: Health, Nutrition, and School Failure,* Harcourt, Brace & World, New York, 1970, 322 pp.
6. **Brožek, J.,** Experimental studies on the impact of deficient diet on behavior, *Borden's Rev. Nutr. Res.*, 20, 75–88, 1959.
7. **Brožek, J. and Vaes, G.,** Experimental investigations on the effects of dietary deficiencies on animal and human behavior, *Vitam. Horm.*, 19, 43–94, 1961.
8. **Chase, H. P.,** The effect of intrauterine and postnatal undernutrition on normal brain development, *Ann. N.Y. Acad. Sci.*, 205, 231–244, 1973. (43 ref.)
9. **Cobos, F.,** Malnutrition and mental retardation: Conceptual issues, in *Lipids, Malnutrition, and the Developing Brain,* A Ciba Foundation Symposium, Elsevier, Amsterdam, 1972, 227–248.
10. **Coursin, D. B.,** Undernutrition and brain function, *Borden's Rev. Nutr. Res.*, 26, 1–16, 1965. (63 ref.)
11. **Coursin, D. B.,** Effects of undernutrition on central nervous system function, *Nutr. Rev.*, 23(3), 65–68, 1965.
12. **Dobbing, J.,** Undernutrition and the developing brain, in *Developmental Neurobiology,* Himwich, W. A., Ed. Charles C Thomas, Springfield, Ill., 1970, 241–261.
13. **Dobbing, J.,** Undernutrition and the developing brain, in *Handbook of Neurochemistry,* Vol. 6, Lajtha, A., Ed. Academic Press, New York, 1971, 255–266.
14. **Dobbing, J.,** Undernutrition and the developing brain. The relevance of animal models to the human brain problem *Am. J. Dis. Child.*, 120, 411–415, 1970. (20 ref.)
15. **Frisch, R. E.,** Present status of the supposition that malnutrition causes permanent mental retardation, *Am. J. Clin. Nutr.*, 23, 189–195, 1970. (29 ref.)
16. **Frisch, R. E.,** Malnutrition and mental development. A critical review, *Indian J. Nutr. Diet.*, 8, 149–162, 1971. (67 ref.)
17. **Frisch, R. E.,** Does malnutrition cause permanent mental retardation in human beings? *Psychiatr. Neurol. Neurochir.*, 74, 463–479, 1971. (74 ref.)
18. **Herbert, V. and Tisman, G.,** Effects of deficiencies of folic acid and Vitamin B_{12} on central nervous system function and development, in *Biology of Brain Dysfunction,* Gaull, G. E., Ed., Plenum Press, New York, 1973 373–392. (105 ref.)
19. **Kallen, D. J.,** Malnutrition, learning, and behavior, *Ecol. Food Nutr.*, 2, 133–141, 1973.
20. **Klein, R. E. and Gilbert, O.,** Malnutrition and intellectual development, Proc. 11th International Congress of Psychology, Mexico City, Mexico, 1967.
21. *Lipids, Malnutrition and the Developing Brain,* A Ciba Foundation Symposium, Elsevier, Amsterdam, 1972, 326 pp. (17 papers)
22. **Mönckeberg, F.,** Malnutrition and mental behavior, *Nutr. Rev.*, 27, 191–193, 1969.
23. **Mönckeberg, F.,** Malnutrition and mental capacity, *Bol. Oficina Sanit. Panam., Eng. Ed.*, 7(1), 87–93, 1973.
24. **Morgane, P. J. and Jacobs, H. L.,** Hunger and satiety, *World Rev. Nutr. Diet.*, 10, 101–213, 1969.
25. **Osuntokun, B. O.,** The effects of malnutrition on the development of cognitive functions of the nervous system in childhood, *Trop. Geogr. Med.*, 24, 311–326, 1972. (132 ref.)
26. Pan American Health Organization, Nutrition, the Nervous System, and Behavior. Proc. Seminar on Malnutrition in Early Life and Subsequent Mental Development, Mona, Jamaica, Jan. 10–14, 1972, Sci. Publ. No. 251, PAHO Washington, D.C., 1972, 148 pp.
27. **Picasso de Oyagüe, A.,** Scientific Study of Malnutrition as a Limiting Factor in the Development of Education UNESCO Doc. NS/ROU/212Rev., Feb. 15, 1971.
28. **Platt, B. S. and Stewart, R. J. C.,** Reversible and irreversible effects of protein–calorie deficiency on the central nervous system of animal and man, *World Rev. Nutr. Diet.*, 13, 43–85, 1971.
29. **Pollitt, E.,** Ecology, malnutrition, and mental development, *Psychosom. Med.*, 31, 193–200, 1969. (43 ref.)
30. **Pollitt, E.,** Behavioral correlates of severe malnutrition in man, in Nutrition, Growth and Development of North American Indian Children, Moore, W. M., Silverberg, M. M., and Read, M. S., Eds., DHEW Publ. No. (NIH) 72-26 U.S. Gov. Print. Off. Washington, D.C., 1972, 151–166. (22 ref.)
31. **Ricciuti, H. N.,** Malnutrition, learning, and intellectual development: Research and remediation, in *Psychology and the Problems of Society*, Korten, F. F., Cook, W., and Lacey, J. I., Eds., American Psychological Association Washington, D.C., 1970.

32. *The Role of Nutritional Deficiency in Nervous and Mental Disease.* Proc. Assoc. for Research in Nervous and Mental Disease, New York 1941, Williams & Wilkins, Baltimore, 1943, 215 pp.
33. **Scrimshaw, N. S.,** Malnutrition, learning and behavior, *Am. J. Clin. Nutr.*, 20, 493–502, 1967.
34. **Scrimshaw, N. S.,** Early malnutrition and central nervous system function, *Annual Progress Children and Psychiatric Child Development,* 1970, Chess, S. and Thomas, A., Eds., Brunner/Mazel, New York, 1970, 246–260.
35. **Scrimshaw, N. S. and Gordon, J. E., Eds.,** *Malnutrition, Learning and Behavior,* M.I.T. Press, Cambridge, Mass., 1968.
36. **Spillane, J. D.,** *Nutritional Disorders of the Nervous System,* Williams & Wilkins, Baltimore, 1947, 280 pp.
37. **Spillane, J. D. and Wells, C. E. C.,** Nutritional disorders of the nervous system, in *Neurochemistry,* 2nd ed., Elliott, K. A. C., Page, I. H., and Quastel, J. H., Eds., Charles C Thomas, Springfield, Ill., 1962, 930–953. (179 ref.)
38. **Springer, N. S.,** Nutrition and Mental Retardation. An Annotated Bibliography, 1964–1970, Institute for the Study of Mental Retardation, University of Michigan, Ann Arbor, 1970.
39. **Springer, N. S., Ed.,** *Nutrition and the Mentally Retarded,* University of Michigan Press, Ann Arbor, 1972.
40. **Stein, Z. A. and Kassab, H.,** Nutrition, in *Ment. Retard., Annu. Rev.*, 2, 92–116, 1970. (123 ref.)
41. Symposium on malnutrition and behavior, Amsterdam, Jan. 29, 1971, *Psychiatr. Neurol. Neurochir.*, 74(6), 417–470, 1971. (7 papers)
42. **Vaes, G.,** L'influence des malnutrition sur le comportement. Revue 1. Comportement général apprentissage et intelligence, *Encéphale,* 49, 30–73, 1960.
43. **Winick, M.,** Malnutrition and brain development, *J. Pediatr.*, 74, 667–679, 1969. (44 ref.)
44. **Winick, M.,** Neurological correlates of malnutrition in animals and man, in Nutrition, Growth and Development of North American Indian Children, Moore, W. M., Silverberg, M. M. and Read, M. S., Eds., DHEW Publ. No. (NIH) 72–76, U.S. Gov. Print. Off., Washington, D.C., 1972, 139–150. (29 ref.)
45. **Winick, M. and Rosso, P.,** Malnutrition and cellular growth in the brain, in *Proc. 8th International Congress of Nutrition,* Prague, 1969, Excerpta Medica, Amsterdam, 1970, 531–538.
46. **Winick, M. and Rosso, P.,** Effects of malnutrition on brain development, in *Biology of Brain Dysfunction,* Vol. 1, Gaull, G. E., Ed., Plenum Press, New York, 1973, 301–317. (49 ref.)

3. Nutrition and Productivity

1. **Åstrand, P. O.,** Nutrition and physical performance, in *Food, Nutrition and Health. A Multidisciplinary Treatise Addressed to the Major Nutrition Problems from a World Wide Perspective,* Rechcigl, M., Jr., Ed., *World Rev. Nutr. Diet.*, Vol. 16, S. Karger, Basel, 1973, 59–79.
2. **Blix, G., Ed.,** *Nutrition and Physical Activity,* 5th Symp. of the Swedish Nutrition Foundation, Almqvist & Wiksells, Uppsala, 1967, 143 pp.
3. **Consolazio, C. F.,** Nutritional variation in world populations and performance potential, *Ann. N.Y. Acad. Sci.*, 134, 885–896, 1966.
4. Food and Agriculture Organization, Nutrition and Working Efficiency, FFHC Basic Study No. 5, FAO, Rome, 1962, 47 pp.
5. **Forbes, W. H.,** The effects of hard work upon nutritional requirements, *Milbank Memorial Fund Q.*, 23, 89–96, 1945. (26 ref.)
6. **Hutchinson, R. C. and Krehl, W. A.,** The effect of food intake on performance, *Borden's Rev. Nutr. Res.*, 15(3), 33–52, 1959.
7. **Keller, W. D. and Kraut, H. A.,** Work and nutrition, *World Rev. Nutr. Diet.*, 3, 65–81, 1959.
8. **Kraut, H.,** Arbeit und Ernährung, in *Vollwertige Ernährung und Gemeinschaftsverpflegung, Bibl. Nutr. Diet.*, Vol. 2, S. Karger, Basel, 1961, 126–136.
9. **Lewis, S. and Gutin, B.,** Nutrition and endurance, *Am. J. Clin. Nutr.*, 26, 1011–1014, 1973. (14 ref.)
10. **Lowenstein, F. W.,** Nutrition and working efficiency, in Food and Nutrition in Africa, FAO Scientific, Technical and Research Commission Report No. 6, FAO, Rome, 1968.
11. **Mayer, J. and Bullen, B.,** Nutrition and athletic performance, *Physiol. Rev.*, 40, 369–397, 1960.
12. **Mayer, J. and Bullen, B.,** Nutrition and athletics, in *Proc. 6th International Congress on Nutrition,* Edinburgh, 1963, Livingstone, Edinburgh, 1964, 27–39.
13. **McKigney, J. I.,** Nutrition and productivity, *Cajanus,* 2(3), 247–249, 1969.
14. **Rodahl, K.,** Food intake and performance, in Feeding the Military Man, Proc. Symp. U.S. Army Natick Laboratories, Natick, Mass., Oct. 1969, 133.
15. **Simonson, E.,** Influence of nutrition on work performance, in Nutrition in Public Health, Nutr. Symp. Ser. No. 3, National Vitamin Foundation, New York, 1951, 72–117.
16. Symposium. Nutrition and physical performance (6 papers), *Proc. Nutr. Soc.*, 25, 87–119, 1966.
17. **Weisberg, S. M., Reese, K. M., and McDonald, P.,** Nutrition and Productivity. Their Relationship in Developing Countries. Based on Survey as Conducted by the LIFE for the Office of Nutrition, A.I.D., LIFE, Washington, D.C., 1972, 22 pp.
18. **Young, D. R. and Spector, H.,** Physical performance capacity and nutriture. Evaluation of rations by animal experimentation, *Am. J. Clin. Nutr.*, 5, 129–140, 1957.

D. Nutrition in Relation to Other Sectors

1. Agriculture

a. General

1. **Aberg, E.,** Nutrition and the planning of agricultural production, in *Nutrition. A Priority in African Development,* Vahlquist, B., Ed., Almqvist & Wiksell, Stockholm, 1972, 69–82. (7 ref.)
2. **Adolph, W. H.,** Agricultural science and improved human nutrition, *Borden's Rev. Nutr. Res.,* 18, 15–27, 1957.
3. **Brady, N. C.,** Agriculture, in *Proc. Western Hemisphere Nutrition Congress,* American Medical Association, Chicago, 1965, 226–229.
4. **Dema, I. S.,** Nutrition in Relation to Agricultural Production, FAO, Rome, 1965.
5. **Dema, I. S.,** The improvement of nutrition through the development of agriculture in Nigeria, in *Proc. 6th International Congress of Nutrition,* Livingstone, Edinburgh, 1964, 164–175.
6. **King, C. G.,** Agriculture and health, *Food Technol.,* 18, 1162–1163, 1964.
7. **Mahadevan, P.,** Nutrition and agriculture, in *Nutrition, National Development, and Planning,* Berg, A., Scrimshaw, N. S., and Call, D. L., Eds., M.I.T. Press, Cambridge, Mass., 1973, 150–153.
8. **Smith, V. E.,** Agricultural planning and nutrient availability, *Nutr. Rev.,* 28(6), 143–150, 1970.
9. Symposium on agricultural development in relation to national needs, *Proc. 1st Asian Congress of Nutrition,* Tulpule, P. G. and Jaya Rao, K. S., Eds., Nutrition Society of India, Hyderabad, 1972, 17–99. (5 papers)

b. Effects of Green Revolution on Nutrition

1. **Berg, A.,** Nutritional effects of agricultural advances, in *The Nutrition Factor. Its Role in National Development,* The Brookings Institution, Washington, D.C., 1973, 50–73.
2. **Lele, U.,** The Green Revolution: Income distribution and nutrition, in *Proc. 3rd. Western Hemisphere Nutrition Congress,* Futura Publishing, Mount Kisco, N.Y., 1972, 20–25.

2. Food Technology

1. **Altschul, A. M.,** Combating malnutrition: New strategies through food science, 2nd Joint Meeting of the American Institute of Chemical Engineers and Institutio de Ingenieros Quinicos de Puerto Rico, Tampa, Florida, May 21, 1968.
2. **Aylward, F.,** Nutrition in relation to food science and the food industries, in *Nutrition. A Priority in African Development,* Vahlquist, B., Ed., Almqvist & Wiksell, Stockholm, 1972, 83–96. (12 ref.)
3. **Berg, A.,** Industry's struggle with world malnutrition, *Harv. Business Rev.,* January–February 1972, 130–141.
4. **Heinz, H. J., II,** A "five-point program" for industries role in improving global nutrition, *Fed. Proc.,* 20(1), III, Suppl., 411–414, 1961.
5. **Mrak, E. M.,** Food technology, in *Proc. Western Hemisphere Nutrition Congress,* American Medical Association, Chicago, 1965, 229–234.
6. **Mrak, E. M.,** Some of the developments in food production and their impact on nutrition, *J. Nutr.,* 91, 55–61, 1967.
7. **Murphy, W. B.,** The challenge to technology, in *Proc. Western Hemisphere Nutrition Congress,* American Medical Association, Chicago, 1965, 13–18.
8. **Scrimshaw, N. S.,** Applications of nutritional and food science to meeting world food needs, *Proc. Natl. Acad. Sci. U.S.A.,* 56, 352–359, 1966.
9. **Subrahmanyan, V.,** The place of food technology in combating hunger and malnutrition, *Food Technol.,* 16, 23–27, 1962.
10. **Swaminathan, M.,** Role of food technology in solving problems of nutrition in India, *Proc. 6th International Congress of Nutrition,* Livingstone, Edinburgh, 1964, 149–164.
11. Symposium. The impact of food technology on the nutrition of man and animals, *Proc. Nutr. Soc.,* 15, 142–167, 1956. (4 papers)
12. Symposium. The place of food science and technology in the campaign against malnutrition, *Proc. Nutr. Soc.,* 20, 91–137, 1961.
13. **Tape, N. W. and Sabry, Z. I.,** The marriage of food technology and nutrition, *World Rev. Nutr. Diet.,* 10, 1–12, 1969.

3. Health

1. **Bengoa, J. M.,** Nutritional significance of mortality statistics, *Proc. 3rd Western Hemisphere Nutrition Congress,* Futura Publishing, Mount Kisco, N.Y., 1972, 270–279.
2. **Bengoa, J. M.,** Priorities in public health nutrition problems, *Proc. 7th International Congress of Nutrition,* Vol. 4, Pergamon Press, Oxford, 1967, 811–822.
3. **Bengoa, J. M.,** Significance of malnutrition and priorities for its prevention, in *Nutrition, National Development, and Planning,* Berg, A., Scrimshaw, N. S., and Call, D. L., Eds., M.I.T. Press, Cambridge, Mass., 1973, 103–128.
4. **Birch, G. G., Green, L. F., and Plaskett, L. G., Eds.,** *Health and Food,* John Wiley & Sons, New York, 1972, 224 pp.

5. **Browe, J. H.,** Provision of suitable and sufficient nutrition, in *Preventive Medicine,* 2nd ed., Hilleboe, H. E. and Larimore, G. W., Eds., W. B. Saunders, Philadelphia, 1965, 278–293.
6. **Goldsmith, G. A.,** Nutrition in medicine, *Fed. Proc.,* 17, 734–739, 1958.
7. **Goldsmith, G. A.,** Nutrition and world health, *J. Am. Diet. Assoc.,* 65, 513–518, 1973.
8. **Hartroft, W. S. and Porta, E. A.,** Nutritional factors in disease, in *Concepts of Disease. A Textbook of Human Pathology,* Brunson, J. G. and Gall, E. A., Eds., Macmillan, New York, 1971, 254–269.
9. **Jelliffe, D. B. and Jelliffe, E. F. P.,** Nutrition and public health planning, in *Nutrition. A Priority in African Development,* Vahlquist, B., Ed., Almqvist & Wiksell, Stockholm, 1972, 126–133. (7 ref.)
10. **King, C. G.,** Contributions of nutrition research to world health problems, *Proc. 6th International Congress of Nutrition,* Livingstone, Edinburgh, 1964, 456–459.
11. **Lubby, A. L., Ed.,** *Nutrition and Metabolism in Medical Practice,* Futura Publishing, Mount Kisco, N.Y., 1972, 144 pp.
12. **Mayer, J.,** Nutritional aspects of preventive medicine, in *Preventive Medicine,* Clark, D. W. and MacMahon, B., Eds., Little, Brown & Co., Boston, 1967, 187–207.
13. **Merrill, M. H.,** Medicine, in *Proc. Western Hemisphere Nutrition Congress,* American Medical Association, Chicago, 1966, 235–237.
14. **Munro, H. N.,** Impact of nutritional research on human health and survival, *Fed. Proc.,* 30, 1403–1407, 1971.
15. **Patwandhan, V. N.,** Public health, *Proc. Western Hemisphere Nutrition Congress,* American Medical Association, Chicago, 1966, 238–244.
16. **Pokrovsky, A. A.,** The problem of the influence of nutrition on the health of people, *Proc. 7th International Congress of Nutrition,* Vol. 1, Pergamon Press, Oxford, 1967, 12–24.
17. **Pollack, H. and Sheldon, D. R.,** The factor of disease in the world food problems, *JAMA,* 212, 598–603, 1970.
18. **Pyke, M.,** Some general considerations of nutrition and health, in *Proc. 6th International Congress of Nutrition,* Livingstone, Edinburgh, 1964, 54–58.
19. **Rechcigl, M., Jr., Ed.,** *Food, Nutrition and Health. A Multidisciplinary Treatise Addressed to the Major Nutrition Problems from a World Wide Perspective,* Vol. 16, *World. Rev. Nutr. Diet.,* S. Karger, Basel, 1973, 511 pp.
20. **Sebrell, W. H., Jr., Nutrition in preventive medicine, in** *Maxcy-Rosenan Preventive Medicine in Public Health,* Sartwell, P. E., Ed., Appleton-Century-Crofts, New York, 1973, 623–631.
21. **Stare, F. J.,** Nutritional improvement and world health potential, *J. Am. Diet. Assoc.,* 57, 107–110, 1970.
22. Symposium. Nutrition as an instrument of preventive medicine, *Proc. 7th International Congress of Nutrition,* Vol. 1, Pergamon Press, Oxford, 1967, 44–88. (14 papers)
23. Symposium. Nutrition and public health, *Proc. Nutr. Soc.,* 27, 1–43, 1968.(8 papers)
24. Symposium. Nutrition in public health, *Proc. Western Hemisphere Nutrition Congress,* American Medical Association, Chicago, 1966, 151–174. (7 papers)
25. Symposium. Nutrition in the Practice of Medicine: With Comments on Nutrition, Disease and Geography, Nutr. Symp. Ser. No. 4, National Vitamin Foundation, New York, 1952, 163 pp. (7 papers)
26. Symposium. Nutrition Fronts in Public Health, Nutr. Symp. Ser. No. 3, National Vitamin Foundation, New York, 1951, 168 pp. (6 papers)
27. **Trémlières, J.,** Nutrition and public health, *World Rev. Nutr. Diet.,* 18, 275–319, 1973. (32 ref.)
28. **Wadsworth, G. R.,** Nutrition and public health, in *The Theory and Practice of Public Health,* 3rd ed., Hobson, W., Ed., Oxford University Press, New York, 1969, 132–142.
29. **Wohl, M. G. and Goodhardt, R. S., Eds.,** *Modern Nutrition in Health and Disease. Dietotherapy,* 4th ed., Lea & Febiger, Philadelphia, 1968, 1240 pp., 5th ed., Goodhart, R. S. and Shils, M. E., Eds., 1973, 1153 pp.

4. Family Planning

1. **Berg, A. D.,** Toward survival: Nutrition and the population dilemma, *Interplay,* 3(6), 24–27, 1970.
2. **Berg, A. D.,** Improved nutrition and the population dilemma, in *The Nutrition Factor,* Brookings Institution, Washington, D.C., 1973, 31–39.
3. **Chopra, J. G., Comacho, R., Kevany, J., and Thomson, A. M.,** Maternal nutrition and family planning, *Am. J. Clin. Nutr.,* 23, 1043–1058, 1970. (35 ref.)
4. **Correa, H.,** Population, Nutrition, Health and Development, Graduate School of Public and International Affairs, University of Pittsburgh (Mimeo).
5. **Gopalan, C.,** Nutrition and family planning, *Popul. Rev.,* 12, 33–38, 1968.
6. **Gopalan, C. and Nadamuni Naidu, A.,** Nutrition and fertility, *Lancet,* 2, 1077–1079, 1972.
7. **Hodges, R. E.,** Nutrition and the "pill," *J. Am. Diet. Assoc.,* 59, 212–217, 1971.
8. **Oelhaf, R. C.,** Interaction between Nutrition and Family Planning, Prepared for the Office of Nutrition, TAB, A.I.D. Center for Disease Control, U.S. Public Health Service, January 1971, 50 pp.
9. Pan-American Health Organization, Maternal Nutrition and Family Planning in the Americas, Sci. Publ. No. 204, PAHO, Washington, D.C., 1970, 53 pp.
10. **Raza, B.,** Importance of child nutrition programmes to family planning by their contribution to family viability, in CENTO Conference on Combating Malnutrition in Preschool Children, Istamabad, Pakistan, March 18–22, CENTO, Ankara, 1968, 137–145.

11. Symposium on nutrition and family planning, in *Proc. 1st Asian Congress of Nutrition*, Tulpule, P. G. and Jaya Rao, K. S., Eds., Nutrition Society of India, Hyderabad, 1972, 285–348. (5 papers)
12. **Taylor, C. E.**, Nutrition and population, in *Nutrition, National Development, and Planning*, Berg, A. D., Scrimshaw, N. S., and Call, D. L., Eds., M.I.T. Press, Cambridge, Mass., 1973, 74–80.
13. **Visweswara Rao, K. and Gopalan, C.**, Nutrition and family size, *J. Nutr. Diet.* (India), 6, 258–266, 1969.
14. **Williams, C. D.**, Nutrition and population, Newsletter of the Caribbean Food and Nutrition Institute, No. 7, 1969, 3–11.
15. **Wishik, S. M.**, Nutrition, family planning and fertility, *PAG Bull.*, 2(4), 8–16, 1972. (33 ref.)

5. Education

1. **Beeuwkes, A. M.**, Nutrition education, *World Rev. Nutr. Diet.*, 5, 1–20, 1965. (53 ref.)
2. **Clements, F. W.**, Report of an International Seminar on Education in Health and Nutrition, Baguio, Philippines, October 13–November 3, 1955, FAO Nutr. Meet. Rep. Ser. No. 13, FAO, Rome, 1958, 92 pp.
3. Food and Agriculture Organization, Education and Training in Nutrition, FFHC Basic Stud. No. 6, FAO, Rome, 1962, 56 pp.
4. **Fuglesang, A.**, Nutrition implementation programmes as communication systems, in *Nutrition. A Priority in African Development*, Vahlquist, B., Ed., Almqvist & Wiksell, Stockholm, 1972, 112–125. (10 ref.)
5. **Hill, M. M.**, School lunch – A tool for nutrition education, *World Rev. Nutr. Diet.*, 14, 258–268, 1972.
6. **Holmes, A. C.**, Visual Aids in Nutrition Education: A Guide to Their Preparation and Use, FAO, Rome, 1968, 154 pp.
7. **Jacobson, W. J., Boyd, F. L., and Hill, M. H.**, *Promising Practices in Nutrition Education in the Elementary School*, Columbia University Teachers College, New York, 1959, 46 pp.
8. **McKenzie, J. C. and Mumford, P.**, The evaluation of nutrition education programmes. A review of the present situation, *World Rev. Nutr. Diet.*, 5, 21–31,1965. (34 ref.)
9. Nutrition Society, Symposium. Nutrition education, *Fed. Proc.*, 26, 151–171, 1967.
10. **Ritchie, J. A. S.**, Learning Better Nutrition, FAO Nutr. Stud. No. 20, FAO, Rome, 1967, 264 pp.
11. **Ritchie, J. A. S. and Kung-Wagner, B.**, Education in Nutrition: A Selected List of Materials, FAO, Rome, 1957.
12. **Roberts, L. J.**, *Nutrition Work with Children*, University of Chicago Press, Chicago, Ill., 1927.
13. Symposium. Education and the media of mass communications, *Proc. 3rd Western Hemisphere Nutrition Congress*, Futura Publishing, Mount Kisco, N.Y., 1972, 155–186. (5 papers)
14. Symposium. Education in nutrition in the United Kingdom, *Proc. Nutr. Soc.*, 12, 166–199, 1953. (11 papers)
15. Symposium. Nutrition in medical and paramedical education, *Proc. Western Hemisphere Nutrition Congress*, American Medical Association, Chicago, 1966, 205–223. (7 papers)
16. **Whitehead, F. E.**, Nutrition education research, *World Rev. Nutr. Diet.*, 17, 91–149, 1973. (268 ref.)

6. Economics

a. Income and Diet

1. **François, P. J.**, Effects of income projection on the protein structure of the diet, *FAO Nutr. Newsletter*, 7(4), 1–15, 1969.
2. **Périssé, J., Sizaret, F., and François, P. J.**, The effect of income on the structure of the diet, *FAO Nutr. Newsletter*, 7(3), 1–9, 1969.

b. Income and Nutrition

1. **Berg, A.**, Increased income and improved nutrition: A Shibboleth examined, *Int. Dev. Rev.*, 12(3), 3–7, 1970.
2. **Berg, A.**, Economic growth, income, and nutrition, in *The Nutrition Factor*, Brookings Institution, Washington, D.C., 1973, 40–49.
3. **Lele, U.**, The Green Revolution: Income distribution and nutrition, *Proc. 3rd Western Hemisphere Nutrition Congress*, Miami Beach, Florida, 1971, Futura Publishing Co., Mount Kisco, N.Y., 1972, 20–25.

c. Urbanization and Food Demand

1. **de Nigris, M. and Coccia, A.**, The impact of urbanization on food demand, *FAO Mon. Bull. Agric. Econ. Stat.*, 22(9), 1–16, 1973.
2. Food and Agriculture Organization, Urbanization Effects on Food Demand: A Preliminary Report, FAO, Rome, 1969.

E. Food and Nutrition Policy

1. General

1. Agency for International Development, Planning National Nutrition Programs: A Suggested Approach, Vol. 1 – Summary of the Methodology, Office of Nutrition, Bureau for Technical Assistance, A.I.D., Washington, D.C., 1973.

2. **Bengoa, J. M.,** Priorities in public health nutrition problems, *Proc. 7th International Congress of Nutrition,* Hamburg, 1966, Vol. 4, Pergamon Press, Oxford, 1967, 811–822.
3. **Bengoa, J. M.,** Significance of malnutrition and priorities for its prevention, in *Nutrition, National Development, and Planning,* Berg, A., Scrimshaw, N. S., and Call, D. L., Eds., M.I.T. Press, Cambridge, Mass., 1973, 103–128.
4. **Berg, A. D. and Levinson, F. J.,** Combating malnutrition: A working model, *Int. Dev. Rev.,* 10(2), 15–17, 1968.
5. **Berg, A. D. and Levinson, F. J.,** A new need: The nutrition programmer, *Am. J. Clin. Nutr.,* 22, 893–895, 1969.
6. **Berg, A. and Muscat, R.,** An approach to nutrition planning, *Am. J. Clin. Nutr.,* 25, 939–954, 1972. (44 ref.)
7. **Berg, A. and Muscat, R.,** Nutrition program planning: An approach, in *Nutrition, National Development, and Planning,* Berg, A., Scrimshaw, N. S., and Call, D. L., Eds., M.I.T. Press, Cambridge, Mass., 1973, 247–274.
8. **Call, D. L. and Levinson, F. J.,** A systematic approach to nutrition intervention programs, in *Nutrition, National Development, and Planning,* Berg, A., Scrimshaw, N. S., and Call, D. L., Eds., M.I.T. Press, Cambridge, Mass., 1973, 165–197.
9. **Donoso, G. and Greaves, J. P.,** Food and nutrition policy in developing countries, *FAO Nutr. Newsletter,* 9(3), 26–33, 1971.
10. Food and Agriculture Organization, Formulation of food and nutrition policies, *FAO Nutr. Newsletter,* 10(2), 1–13, 1972.
11. **François, P. J.,** Nutrition – A new operational technique in national planning, *Food Nutr. Afr.,* 7, 93–103, 1969.
12. **François, P. J.,** Consumption levels and requirements – General strategies within the framework of a food policy, *FAO Nutr. Newsletter,* 9(4), 1–8, 1971.
13. **Francois, P. J.,** Nutrition – A new operational technique in planning, *FAO Nutr. Newsletter,* 6(1), 46–52, 1968.
14. **Ganzin, M., Périssé, J., and François, P.,** Need for food and nutrition policies, in *Man, Food and Nutrition. Strategies and Technological Measures for Alleviating the World Food Problem,* Rechcigl, M., Jr., Ed., CRC Press, Cleveland, Ohio, 1973, 275–286.
15. **Goldsmith, G. A.,** A national nutrition policy and program, *Am. J. Clin. Nutr.,* 26, 901–906, 1973.
16. **Greaves, J. P.,** The role of the nutritionist in the assessment and attainment of nutritional goals, *Nutr. Rev.,* 23, 193–196, 1965. (3 ref.)
17. **Greaves, J. P.,** Organizational structures for the improvement of nutrition, *FAO Nutr. Newsletter,* 8(4), 6–9, 1970.
18. **Hegsted, D. M.,** The development of a national nutrition policy, *J. Am. Diet. Assoc.,* 62, 394–398, 1973.
19. **Hegsted, D. M.,** Food and nutrition policy – now and in the future, *J. Am. Diet. Assoc.,* 64, 367–371, 1974. (14 ref.)
20. **Hollingsworth, D. F. and Greaves, J. P.,** Nutrition policy with regard to protein, in *Proteins as Human Food,* Lawrie, R. A., Ed., Avi Publishing, Westport, Conn., 1970, 32–45.
21. **Johnston, B. F. and Greaves, J. P.,** Manual on Food and Nutrition Policy, FAO, Rome, 1969, 95 pp.
22. **Joy, L.,** Nutrition intervention programs: Identification and selection, in *Nutrition, National Development, and Planning,* Berg, A., Scrimshaw, N. S., and Call, D. L., Eds., M.I.T. Press, Cambridge, Mass., 1973, 198–206.
23. **Latham, M.,** Planning and Evaluation of Applied Nutrition Programmes, FAO Nutr. Stud. No. 26, FAO, Rome, 1972.
24. **Levinson, F. J. and Call, D. L.,** Nutrition intervention in low income countries. Its economic role and alternative strategies, PAG Document 1.13/1, Protein Advisory Group, United Nations, 1970, 25 pp.
25. **Levinson, F. J. and Call, D. L.,** Nutrition Intervention in Low Income Countries: A Planning Model and Case Study, Cornell International Agricultural Development Mimeogr. No. 34, Cornell University Department of Agricultural Economics and Graduate School of Nutrition, Ithaca, New York, 1971.
26. **Levinson, F. J. and Call, D. L.,** A model for nutrition planning, 17th UN/PAG Meeting, New York City, May 1970.
27. **Mayer, J.,** Toward a national nutrition policy, *Science,* 176, 237–241, 1972.
28. **Mayer, J.,** *U.S. Nutrition Policies in the Seventies,* W. H. Freeman Publishing, San Francisco, 1973, 256 pp.
29. **McRoberts, M. R.,** Food and nutrition policy, Proc. 1st South East Asian Regional Seminar on Nutrition, Djakarta, Indonesia, Oct. 1969.
30. National Research Council, Food and Nutrition Board, Recommendations on Administrative Policies for International Food and Nutrition Programs. A statement of Food and Nutrition Board, Washington, D.C., 1963, 6 pp.
31. **Ohlin, G.,** An overall strategy of assistance in the field of nutrition, Proc. 1971 Dag Hammarskjöld Seminar on Nutrition as a Priority in African Development, Uppsala, Sweden, July 1971.
32. **Oke, O. L.,** A nutrition policy for Nigeria, *World Rev. Nutr. Diet.,* 14, 1–47, 1972.
33. Pan American Health Organization, Elements of Food and Nutrition Policy in Latin America, Sci. Publ. No. 194, PAHO, Washington, D.C., 1970.
34. **Périssé, J.,** The nutritional approach in food policy planning, *FAO Nutr. Newsletter,* 6(1), 30–45, 1968.
35. **Recalde, F.,** Guidelines for the establishment of national food and nutrition policies in Latin America, *FAO Nutr. Newsletter,* 9(4), 16–26, 1971.
36. **Reuda-Williamson, R.,** The applied nutrition program, the basis of the National Nutrition Plan, *PAHO Bull.,* 68(3), 187–200, 1970.
37. **Vamoer, A. P.,** Nutrition, planning and coordination, Proc. Eastern African Conference on Nutrition and Child Feeding, Gov. Print. Off., Washington, D.C., 1969, 274–282.

2. Economics in Nutrition Planning

1. **Berg, A. and Muscat, R.,** Macronutrition, *Proc. 3rd Western Hemisphere Nutrition Congress,* Miami Beach, Florida, 1971, Futura Publishing, Mount Kisco, N.Y., 1972, 318–323.
2. **Joy, L.,** Economic aspects of food and nutritional planning, *Proc. 1st Asian Conference on Nutrition,* Tulpule, P. G. and Jaya Rao, E. S., Eds., Nutrition Society of India, Hyderabad, 1972, 267–274.
3. **McKigney, J. I.,** Food economics in nutrition policy and planning, *FAO Nutr. Newsletter,* 7(4), 16–24, 1969.
4. **Popkin, B.,** The economics of nutrition. Unpublished manuscript, Institute for Research on Poverty, University of Wisconsin, Madison, Wis., May 1967.
5. **Popkin, B. and Lidman, R.,** Economics as an Aid in Nutritional Change, Institute for Research on Poverty, Discussion Paper 113–72, University of Wisconsin, Madison, Wis., 1972.
6. **Popkin, B. and Lidman, R.,** Economics as an aid in nutritional change, *Am. J. Clin. Nutr.,* 25, 331–344, 1972. (43 ref.)

XVI. FAMILY PLANNING AND POPULATION CONTROL

A. Bibliographies

1. *Bibliography of Family Planning and Population*, Simon Population Trust, Cambridge, England, Vol. 1, 1972.
2. **Goode, S. M.,** *Population and the Population Explosion. A Bibliography for 1970–*, Whitston Publishing, Troy, N.Y., 1973–.
3. Population Reference Bureau, A source book on population, *Popul. Bull.*, 25(5), 1–51, 1969.
4. Population growth and its implications: A bibliography, in *Population Ethics*, Quinn, F. X., Ed., Corpus Books, Washington, D.C., 1968, 109–140.
5. Princeton University, Office of Population Research, *Population Index Bibliography. Cumulated 1935–1968*, G. K. Hall & Co., Boston, Mass., 1971. (9 vols.)
6. **Zelinsky, W.,** *A Bibliographic Guide to Population Geography*, University of Chicago Dep. of Geography, Chicago, 1962, 257 pp.

B. General References Relating to Population Growth

1. **Brown, H., Holdren, J. P., Sweezy, A., and West, B., Eds.,** *Population: Perspective*, 1973, Freeman, Cooper & Co., San Francisco, 1974.
2. **Caldwell, J. C. and Okonjo, C., Eds.,** *The Population of Tropical Africa*, Columbia University Press, New York, 1968.
3. **Carr-Saunders, A. M.,** *World Population, Past Growth and Present Trends*, Barnes & Noble, New York, 1965, 336 pp.
4. **Clark, C.,** *Population Growth and Land Use*, Macmillan, London, 1967, 406 pp.
5. **Demko, G. J., Rose, H. M., and Schnell, G. A., Eds.,** *Population Geography: A Reader*, McGraw-Hill, New York, 1970.
6. **Durand, J. D., Ed.,** *World Population*, American Academy of Political and Social Science, Philadelphia, 1967, 254 pp.
7. **Ehrlich, P. R.,** *The Population Bomb*, Ballantine Books, New York, 1968.
8. Foreign Policy Association, The world population crisis: what is it and where to get information about it, *Intercom* 10(4), 18–56, 1968.
9. **Frejka, T.,** *The Future of Population Growth. Alternative Paths to Equilibrium*, John Wiley & Sons, New York, 1973, 268 pp.
10. **Hardin, G., Ed.,** *Population, Evolution, Birth Control. A Collage of Controversial Readings*, W. H. Freeman Publishing, San Francisco, 1938, 341 pp.
11. **Hauser, P. M., Ed.,** *The Population Dilemma*, 2nd ed., Prentice-Hall, Englewood Cliffs, N.J., 1969, 211 pp.
12. **Hauser, P. M.,** World Population Problems, Foreign Policy Association, New York, 1965.
13. **Kammeyer, K. C. W., Ed.,** *Population Studies: Selected Essays and Research*, Rand McNally, Chicago, 1969, 481 pp.
14. **Keyfitz, N. and Flieger, W.,** *Population: Facts and Methods of Demography*, W. H. Freeman Publishing, San Francisco, 1971.
15. **Keyfitz, N. and Flieger, W.,** *World Population. An Analysis of Vital Data*, University of Chicago Press, Chicago, 1968, 672 pp.
16. **Peterson, W.,** *Population*, 2nd ed., MacMillan, New York, 1969, 735 pp.
17. **Stycos, J. M.,** *Population Dilemma in Latin America*, Potomac Books, Washington, D.C., 1966.
18. **Stycos, J. M.,** *Human Fertility in Latin America*, Cornell University Press, Ithaca, N.Y., 1968.
19. **Thompson, W. S. and Lewis, D. J..** *Population Problems*, 5th ed., McGraw-Hill, New York, 1965.
20. United Nations, World Population Situation, Economic and Social Council, Population Commission Report (E/CN. 9/231), Sept. 23, 1969.

C. Determinants of Fertility Rate

1. **Cain, G. G. and Weininger, A.,** Economic determinants of fertility: Results from cross-sectional aggregate data, *Demography*, 10, 205–223, 1973.
2. **Freedman, R.,** The sociology of human fertility: A trend report and bibliography, *Cur. Sociol.*, 10, 11(2), 1961–62, 121 pp.
3. **Friedlander, S. L. and Silver, M. A.,** Quantitative study of determinants of fertility behavior, *Demography*, 4(1), 32–61, 1967.
4. **Greep, R. O.,** *Human Fertility and Population Problems*, Schenkman Publishing, Cambridge, Mass., 1963, 278 pp.
5. **Gregory, P. R., Campbell, J. M., and Cheng, B. S.,** Differences in fertility determinants: developed and developing countries, *J. Dev. Stud.*, 9, 233–341, 1972.
6. **Henschaw, P.,** *Adaptive Human Fertility*, McGraw-Hill, New York, 1955, 322 pp.
7. Interdisciplinary Communications Program, The Policy Relevance of Recent Social Research on Fertility, Occasional Monogr. Ser. No. 2, Smithsonian Institution, Washington, D.C., 1974, 107 pp.

8. **Kirk, D.,** Factors affecting Moslem natality, in *Family Planning and Population Programs: A Review of World Developments,* Berelson, B., Ed., University of Chicago Press, Chicago, 1966, 561–579.
9. **Liu, W. T., Ed.,** *Family and Fertility,* University of Notre Dame Press, Notre Dame, In., 1967, 257 pp.
10. **Lorimer, F.,** *Culture and Human Fertility,* UNESCO, Zurich, 1954, 514 pp.
11. **Nag, M.,** *Factors Affecting Human Fertility in Non-Industrialized Societies. A Cross-Cultural Study,* Human Relations Area Files Press, New Haven, Conn., 1968, 227 pp.
12. Research Triangle Institute, Social and Economic Correlates of Family Fertility. A Survey of the Evidence, Research Triangle, N.C., 1971.
13. **Schultz, T. P.,** Economic Factors Affecting Population Growth: A Preliminary Survey of Economic Analyses of Fertility, University of Minnesota, Minneapolis, 1972.
14. **Schultz, T. P.,** Fertility Determinants: A Theory, Evidence, and an Application to Policy Evaluation, The Rand Corp., Santa Monica, Calif., 1974.
15. **Sweezy, A.,** The economic explanation of fertility changes in the U.S., *Popul. Stud.*, 25(2), 255–267, 1971.
16. **Szabady, E., Ed.,** *Studies on Fertility and Social Mobility,* Akademie Kiado, Budapest, 1964, 331 pp.
17. United Nations, The Determinants and Consequences of Population Trends, Vol. 1, Popul. Stud. No. 50, United Nations, New York, 1973.
18. **Westoff, C. G., Potter, R. G., and Sagi, P. C.,** *The Third Child: A Study in the Prediction of Fertility,* Princeton University Press, Princeton, N.J., 1963, 293 pp.
19. **Whelpton, P. K. and Kiser, C. V., Eds.,** Social and Psychological Factors Affecting Fertility, Milbank Memorial Fund, New York, 1950. (5 vols.)

D. Population Growth and Economic Development

1. **Black, E. R.,** Population increase and economic development, in *Our Crowded Planet,* Osborn, F., Ed., Doubleday, Garden City, N.Y., 1962, 83–91.
2. **Bourgeois-Pichat, J.,** Population Growth and Development, Carnegie Endowment for International Peace, New York, 1966.
3. **Coale, A. J. and Hoover, E. M.,** *Population Growth and Economic Development of Low Income Countries,* Princeton University Press, Princeton, N.J., 1958.
4. **Enke, S.,** Population growth and economic development, The Public Interest, No. 32, National Affairs, Inc., New York, 1973, 87–96.
5. **Gray, H. P.,** *Economic Development and Population Growth. A Conflict,* Heath, Lexington, Mass., 1970, 162 pp.
6. **Heer, D. M.,** Economic development and fertility, *Demography,* 3(2), 423–444, 1966.
7. **Hoselitz, B. F. and Hargreaves, H. W.,** Population growth and economic development, in *Perspectives in Developmental Change,* University of Kentucky Press, Lexington, 1968, 101–129.
8. **Keeley, M. C.,** Economic Analysis of Fertility: The Effect of Development on Fertility, General Electric TEMPO, Washington, D.C., 1974.
9. **Sweezy, A.,** Recent light on the relation between socioeconomic development and fertility decline, Caltech Population Program Occasional Papers No. 1: 1, California Institute of Technology, Pasadena, 1973.
10. **Williams, A. D.,** Feedback from Development to Fertility: Review and Evaluation of the Literature, General Electric TEMPO, Santa Barbara, Calif., 1973.

E. Population Growth and Its Control

1. **Borrie, W. D.,** *The Growth and Control of World Population,* Weidenfeld & Nicolson (Fred B. Rothman), South Hackensack, N.J., 1970, 340 pp.
2. **Kirk, D.,** Prospects for reducing natality in the underdeveloped world, *Annals of the American Academy of Political and Social Science,* January 1967, 51–60.
3. **Notestein, F. W.,** Population growth and its control, in *Overcoming World Hunger,* Hardin, C. M., Ed., Prentice-Hall, Englewood Cliffs, N.J., 1969, 9–39.
4. **Tyler, E. T., Ed.,** *Birth Control. A Continuing Controversy,* Charles C Thomas, Springfield, Ill., 1967, 182 pp.

F. Population Control and Economic Development

1. **Holmes, D. N., Jr.,** The Economic Benefits of Slowing Population Growth, General Electric TEMPO, Santa Barbara, Calif., 1971.
2. **Jones, G. W.,** The Economic Effect of Declining Fertility in Less Developed Countries, Population Council, New York, 1969, 30 pp.
3. **Ohlin, G.,** Population Control and Economic Development, OECD, Paris, 1967, 138 pp.
4. **Ruprecht, T. K. and Wahren, C.,** Population Programmes and Economic and Social Development, OECD, Paris, 1970, 171 pp.
5. World Bank, *Population Policies and Economic Development,* Johns Hopkins Press, Baltimore, 1974.

G. Family Planning

1. Current Programs: World Developments

1. **Behrman, S. J., Corsa, L., Jr., and Freedman, R., Eds.,** *Fertility and Family Planning: A World View,* University of Michigan Press, Ann Arbor, 1969, 503 pp.
2. **Berelson, B., Ed.,** *Family Planning and Population Programs: A Review of World Developments,* University of Chicago Press, Chicago, 1966, 848 pp.
3. **Berelson, B., Ed.,** *National Programmes in Family Planning: Achievements and Problems,* Meenakshi Prakashan, Beguin Bridge, Meerut, 1969, 319 pp.
4. **Berelson, B.,** Beyond family planning, *Science,* 163, 533–543, 1969.
5. **Berelson, B.,** The present state of family planning programs, in *Are Our Descendants Doomed? Technological Change and Population Growth,* Brown, H. and Hutchings, E., Jr., Eds., Viking Press, New York, 1972, 201–242.
6. **Kasdon, D. L.,** *International Family Planning, 1966–1968. A Bibliography,* National Institute of Mental Health, Chevy Chase, MD, 1969, 62 pp.
7. **Kiser, C. V., Ed.,** *Research in Family Planning,* Princeton University Press, Princeton, N.J., 1962, 662 pp.
8. **Nortman, D.,** Population and Family Planning Programs: A Factbook. Reports on Population and Family Planning, No. 2, Population Council, New York, 1969.
9. **O'Brien, J. A.,** *Family Planning in an Exploding Population,* Hawthorne Books, New York, 1968.
10. Organization for Economic Cooperation and Development, An Assessment of Family Planning Programmes. Summary of Proc., OECD, Paris, 1972, 193 pp.
11. Population Association of America, Progress and problems of fertility control around the world, *Demography,* 5(2), 1968 (special issue).
12. **Stycos, J. M.,** Family planning: Reform and resolution, *Fam. Plann. Perspect.,* 3(1), 49–50, 1971.

2. Attitudes Toward Population Control

1. **Blake, J.,** Are babies consumer durables? *Popul. Stud.,* 22(1), 5–25, 1968.
2. **Blake, J.,** Reproductive motivation and population policy, *Bioscience,* 21(5), 215–220, 1971.
3. **Blake, J.,** Income and reproductive motivation, *Popul. Stud.,* 21, 185–206, 1967.
4. **Davis, K.,** Population policy: Will current programs succeed? *Science,* 158, 730–739, 1967.
5. **Day, A. T.,** Population control and personal freedom: Are they compatible? *The Humanist,* Nov.–Dec. 1968, 7–10.
6. **Mauldin, W. P.,** Fertility Studies: Knowledge, Attitude and Practice, Population Council Studies in Family Planning No. 7, June 1965, 1–10.
7. **Newman, L. F.,** Family planning: An anthropological approach, Proc. International Congress on Anthropological and Ethnological Sciences, Tokyo, Sept. 6, 1968.

3. Motivation and Incentives for Population Control

1. **Barnett, L. D.,** Population policy: Payments for fertility limitation in the U.S., *Social Biol.,* 16(4), 239–248, 1969.
2. **Bogue, D. J., Ed.,** Mass Communications and Motivation for Birth Control, Proc. of the Summer Workshop, University of Chicago, Chicago, Ill., 1967, 551 pp.
3. **Kangas, L. W.,** Integrated incentives for fertility control, *Science,* 169, 1278–1283, 1970.
4. **Rich, W.,** Smaller Families Through Social and Economic Progress, Overseas Development Council Monogr. No. 7, Washington, D.C., 1973.

4. Costs of Population Control Measures

1. **Curtin, T. R. C.,** The economics of population growth and control in developing countries, *Rev. Soc. Econ.,* 27(2), 139–153, 1969; 28(1), 101–105, 1970.
2. **Demeny, P.,** The economics of population control, in *Rapid Population Growth Consequences and Policy Implications,* Office of the Foreign Secretary, National Academy of Sciences, Johns Hopkins Press, Baltimore, 1971, 199–221.
3. **Enke, S.,** The economic aspects of slowing population growth, *Econ. J.,* 76(March), 44–56, 1966.
4. **Ross, J.,** Cost of family planning programs, in *Family Planning and Population Programs. A Review of World Developments,* Berelson, B., Ed., University of Chicago Press, Chicago, 1966, 759–778.

5. Population Policy

1. **Back, K. W. and Winsborough, H. H.,** Population policy: Opinions and actions of governments, *Public Opinion Q.,* 32(4), 634–645, 1968/69.
2. **Barnett, H. J.,** Population and world politics, *World Polit.,* 12(4), 640–650, 1960.
3. **Berelson, B., Ed.,** *Population Policy in Developed Countries,* McGraw-Hill, New York, 1974, 793 pp.
4. **Blake, J.,** Population policy for Americans: Is the government being misled? *Science,* 164, 522–529, 1969.
5. **Borrie, W. D.,** *Population Trends and Policies. A Study in Australian and World Demography,* Australasian Publishing, Sydney, 1948, 263 pp.
6. **Burch, G. I.,** Relation of population to national and international policies, *World Affairs,* 3(4), 217–226, 1948.

7. **Davis, K.,** Population policy. Will current programs succeed? *Science*, 153, 730–739, 1967.
8. **Driver, E.,** *World Population Policy: An Annotated Bibliography*, Lexington Books, Lexington, Mass., 1972, 1280 pp.
9. **Eldridge, H. T.,** Population Policies. A Survey of Recent Developments, International Union for the Scientific Study of Population, Washington, D.C., 1954, 154 pp.
10. **Eldridge, H. T.,** Population policies, in *International Encyclopeadia of Social Sciences*, Vol. 12, Sills, D. L., Ed., Macmillan, Free Press, New York, 1968, 381–388.
11. **Gardner, R. N.,** The politics of population, *Dep. of State Bull.*, 48, 906–914, 1963.
12. **Grech, P.,** Selected bibliography on population policy, in *The Role of Law in Population Planning*, Berman, D. and Fox, D. I., Eds., Oceana Publications, Dobbs Ferry, N.Y., 1972, 69–85.
13. **Harkavy, O., Jaffe, F. S., and Wishik, S. M.,** Family planning and public policy: Who is misleading whom? *Science*, 165, 367–373, 1969.
14. **Hauser, P. M.,** *Population and World Politics*, Free Press, Glencoe, Ill., 1958, 297 pp.
15. **Kamerschen, D. R.,** Population policies in underdeveloped countries, *Soc. Econ. Stud.*, 17(1), 70–74, 1968.
16. **Katz, M.,** Legal dimensions of population policy, *Soc. Sci. Q.*, 50(3), 731–741, 1969.
17. **Kirk, D. and Nortman, D.,** Population policies in developing countries, *Econ. Dev. Cult. Change*, 15, 129–142, 1967.
18. **Mayer, J.,** Toward a non-Malthusian population policy, *Columbia Forum*, 12(2), 5–13, 1969.
19. **Meadows, P.,** Toward a socialized population policy, *Psychiatry*, 11(2), 193–202, 1948.
20. **Means, C. C.,** The constitutional aspects of a national population policy, *Vill. L. Rev.*, 15, 854–862, 1970.
21. **Peterson, W.,** *The Politics of Population*, Doubleday, New York, 1964, 350 pp.
22. Population Reference Bureau, Needed – A population policy for the world, *Popul. Bull.*, 21(2), 17–47, 1965.
23. **Smith, T. E., Ed.,** *The Politics of Family Planning in the Third World*, George Allen & Unwin, London, 1973, 352 pp.
24. **Spengler, J. J. and Duncan, O. D., Eds.,** *Population Theory and Policy: Selected Readings*, Free Press, Glencoe, Ill., 1956, 522 pp.
25. **Taylor, C. E.,** Five stages in a practical population policy, *Popul. Rev.*, 13(1), 37–44, 1969.
26. **Zlotnick, J.,** Population pressure and political indecision, *Foreign Affairs*, 39, 683–694, 1961.

H. Contraception

1. Bibliographies

1. **Tietze, C.,** *Bibliography of Contraception 1940–1956*, National Committee on Maternal Health, New York, 1960.
2. **Tietze, C.,** *Selected Bibliography on Contraception 1940–1960*, National Committee on Maternal Health, New York, 1960, 76 pp.
3. **Tietze, C.,** *Bibliography of Fertility Control 1950–1965*, National Committee on Maternal Health, New York, 1965, 198 pp.
4. **Tietze, C.,** *Selected Bibliography of Contraception, Supplement 1960–1963*, National Committee on Maternal Health, 1963, 59 pp.

2. General References

1. **Bender, S. J. and Fellers, S.,** *Contraception. By Choice and by Chance*, W. C. Brown & Co., Dubuque, IA, 1972, 60 pp.
2. **Calderone, M. S., Ed.,** *Manual of Family Planning and Contraceptive Practice*, Williams & Wilkins, Baltimore, 1971.
3. **Diczfalusy, E. and Borell, U., Eds.,** *Control of Human Fertility, Proc. 15th Nobel Symposium*, Almqvist & Wiksell, Stockholm, 1971, 354 pp.
4. **Draper, E.,** *Birth Control in the Modern World. The Role of the Individual in Population Control*, Penguin Books, Baltimore, 1972, 446 pp.
5. **Duncan, G. W., Hilton, E. J., Kraeger, P., and Lumsdaine, A. A., Eds.,** *Fertility Control Methods. Strategies for Institutions*, Academic Press, New York, 1973, 230 pp.
6. **Finch, B. E. and Green, M.,** *Contraception Through the Ages*, Peter Owen, London, 1963.
7. **Guttmacher, A. F.,** *The Complete Book of Birth Control*, Ballantine Books, New York, 1966.
8. **Hafez, E. S. E. and Evans, T. N.,** *Human Reproduction: Conception and Contraception*, Harper & Row, New York, 1973, 778 pp.
9. **Hardin, G.,** *Birth Control*, Pegasus, New York, 1970.
10. **Hardy, N. R. and Kestelman, P.,** Directory of Contraceptives, IPPF, London, 1971.
11. **Himes, N. E.,** *Medical History of Contraception*, Gamut Press, New York, 1963.
12. **Levin, H. L. and Peers, R.,** Directory of Contraceptives, IPPF/Population Council, London, 1967.
13. **Loraine, J. A. and Bell, E. T.,** *Fertility and Contraception in the Human Female*, Livingstone, Edinburgh, 1968, 384 pp.
14. **Peel, J. and Potts, M.,** *Textbook of Contraceptive Practice*, Cambridge University Press, New York, 1969.
15. **Pincus, G.,** *The Control of Fertility*, Academic Press, New York, 1965, 360 pp.

16. **Potts, M. and Wood, C., Eds.,** *New Concepts in Contraception*, University Park Press, Baltimore, MD, 1972, 231 pp.
17. **Roland, M.,** *Response to Contraception, Major Problems in Obstetrics and Gynecology,* Vol. 5, W. B. Saunders, Philadelphia, 1973, 162 pp.
18. **Rudel, H. W., Kincl, F. A., and Henzl, M. R.,** *Birth Control. Contraception and Abortion*, Macmillan, New York, 1973, 372 pp.
19. **Segal, S.,** Contraceptive Technology: Current and Prospective Methods, Reports on Population. Family Planning, Population Council, October 1969.
20. **Smith, A. J., Ed.,** Contraception Today, Family Planning Association, London, 1971, 189 pp.
21. **Tietze, C. et al.,** Birth Control and Abortion, MSS Information Corp., New York, 1972, 289 pp.
22. **Tietze, C.,** The current status of fertility control, *Law Contemp. Probl.,* 25(3), 425–444, 1960.
23. World Health Organization, Developments in Fertility Control, WHO Tech. Rep. Ser. 424, WHO, 1969, 36 pp.
24. World Health Organization, Methods of Fertility Regulations. Advances in Research and Clinical Experience, WHO Tech. Rep. Ser. 473, WHO, 1971, 48 pp.

3. Specific Measures

a. The Rhythm Method

1. **Hartman, C. G.,** *Science and the Safe Period,* Williams & Wilkins, Baltimore, 1962.
2. **Rodriguez, A.,** The rhythm method, in *Family Planning and Population Programs: A Review of World Developments,* Berelson, B., Ed., University of Chicago Press, Chicago, 1966, 397–401.
3. World Health Organization, Biology of Fertility Control by a Periodic Abstinence, WHO Tech. Rep. Ser. 360, WHO, 1967, 20 pp.

b. Chemical Contraceptives

(1) General

1. **Bennett, J. P.,** *Chemical Contraception,* Columbia University Press, New York, 1974, 230 pp.
2. **Hardy, N. R.,** Chemical contraception, in Contraception Today, Smith, A. J., Ed., Family Planning Association, London, 1971, 84–87.
3. **Jackson, H.,** *Antifertility Compounds in the Male and Female,* Charles C Thomas, Springfield, Ill., 1966, 214 pp.
4. **Jackson, H.,** Chemical interference with spermatogenesis and fertility, in *Advances in Reproductive Physiology,* Vol. 4, McBaren, A., Ed., Academic Press, New York, 1969, 65–92.
5. **Jackson, H.,** Chemical methods of male contraception, *Am. Sci.,* 61, 188–193, 1973.
6. Pharmacology Society Symposium, Antifertility drugs, *Fed. Proc.,* 29(3), 1209–1242, 1970.
7. **Lednicer, D., Ed.,** *Contraception: The Chemical Control of Fertility,* Dekker, New York, 1969, 269 pp.

(2) Oral Contraceptives

1. **Barber, H. R. K., Graber, E. A., and O'Rourke, J. J.,** *Are the Pills Safe?* Charles C Thomas, Springfield, Ill., 1969, 103 pp.
2. **Boffa, P. S., et al.,** Pathological Effects of Oral Contraceptives, MSS Information Corp., New York, 1973, 152 pp.
3. **Drill, V. A.,** *Oral Contraceptives,* McGraw-Hill, New York, 1966.
4. **Garcia, C. R., Goldzieher, J. W., and Massey, J. B.,** Oral contraceptives, in *Human Reproduction: Conception and Contraception,* Hafez, E. S. E. and Evans, T. N., Eds., Harper & Row, New York, 1973, 335–359.
5. **Goldzieher, J. W.,** Newer drugs in oral contraception, *Med. Clin.,* 48, 529–545, 1964.
6. **Goldzieher, J. W. and Rice-Wray, E.,** *Oral Contraceptives, Mechanism and Management,* Charles C Thomas, Springfield, Ill., 1966, 146 pp.
7. **Jackson, M. C. N.,** Oral contraception in practice, *J. Reprod. Fertil.,* 6, 153–173, 1963.
8. **Kalman, S. M.,** Effects of oral contraceptives, *Annu. Rev. Pharmacol.,* 9, 363–378, 1969.
9. **Mears, E.,** The mode of action of oral contraceptives, *Eugen. Rev.,* 65, 195–201, 1965.
10. **Mears, E.,** *Handbook of Oral Contraceptives,* Churchill, London, 1965, 107 pp.
11. **Rose, D. P. et al.,** Oral Contraceptives: Psychological and Physiological Effects, MSS Information Corp., New York, 1973, 177 pp.
12. **Satterthwaite, A. P.,** Oral contraceptives, in *Family Planning and Population Programs. A Review of World Developments,* Berelson, B., Ed., University of Chicago Press, Chicago, Ill., 1966, 411–424.

(3) Hormonal Contraceptives

1. **Haller, J.,** Hormonal Contraception, Geron-X, Los Altos, Calif., 1969, 288 pp.
2. **Moghissi, K. S.,** Newer approaches to steroidal contraceptives, in *Human Reproduction: Conception and Contraception,* Hafez, E. S. E. and Evans, T. N., Eds., Harper & Row, New York, 1973, 447–469.
3. **Petrow, V.,** The contraceptive progesterones, *Chem. Rev.,* 70(6), 713–726, 1970.
4. **Salhanick, H. A., Kipnis, D. M., and Van der Wiele, R.,** *Metabolic Effects of Gonadal Hormones and Contraceptive Steroids,* Plenum Press, New York, 1969.

5. **Szontágh, F. E.,** *Mechanism of Action of Oral Progestogens*, Akadimia Kiado, Budapest, 1970, 176 pp.
6. World Health Organization, Hormonal Steroids in Contraception, WHO Tech. Rep. Ser. 386, WHO, 1968, 28 pp.
7. World Health Organization, Clinical Aspects of Oral Gestogens, WHO Tech. Rep. Ser. 326, WHO, 1966, 24 pp.

(4) Prostaglandins

1. **Karim, S. M. M.,** Prostaglandins in fertility control, in *New Concepts in Contraception*, Potts, M. and Wood, C., Eds., University Park Press, Baltimore, MD, 1972, 125–142.
2. **Karim, S. M. M., Ed.,** The Prostaglandins: Progress in Research, Wiley-Interscience, New York, 1972, 327 pp.
3. **Ramwell, P. and Shaw, J. E., Eds.,** Prostaglandins, *Ann. N.Y. Acad. Sci.*, 180, 1–560, 1971.
4. **Speroff, L. and Anderson, G. G.,** Prostaglandins and fertility control, in *Human Reproduction: Conception and Contraception*, Hafez, E. S. E. and Evans, T. N., Eds., Harper & Row, New York, 1973, 361–382.

c. Intrauterine Contraceptives

1. **Davis, H. J.,** *Intrauterine Devices for Contraception – The IUD*, Williams & Wilkins, Baltimore, 1971.
2. Food and Drug Administration, Report on Intrauterine Devices, Advisory Committee on Obstetrics and Gynecology, Washington, D.C., 1968.
3. **Hall, R. E.,** Intrauterine devices: Clinical aspects, in *Manual of Family Planning and Contraceptive Practice*, Calderone, M. S., Ed., Williams & Wilkins, Baltimore, 1970, 331–337.
4. **Segal, S. J., Southam, A. L., and Shafer, K. D., Eds.,** *Proceedings of the Second International Conference on Intra-uterine Contraception*, Excerpta Medica, Amsterdam, 1964, 250 pp.
5. **Sobrero, A. J.,** Intrauterine devices in clinical practice, *Fam. Plann. Perspect.*, 3, 16–24, 1971.
6. **Tietze, C.,** Intrauterine devices: Clinical aspects, in *Human Reproduction: Conception and Contraception*, Hafez, E. S. E. and Evans, T. N., Eds., Harper & Row, New York, 1973, 293–308.
7. **Tietze, C. and Lewit, S., Eds.,** *Intra-uterine Contraceptive Devices, Proc. of the Conference*, Excerpta Medica, Amsterdam, 1962, 154 pp.
8. World Health Organization, Basic and Clinical Aspects of Intra-uterine Devices, WHO Tech. Rep. Ser. 332, WHO, 1966, 25 pp.
9. World Health Organization, Intra-uterine Devices: Physiological and Clinical Aspects, WHO Tech. Rep. Ser. 397, WHO, 1968, 32 pp.

d. Vaginal Contraceptives

1. **Dickinson, R. L.,** The diaphragm, in *Manual of Family Planning and Contraceptive Practice*, Calderone, M. S., Ed., Williams & Wilkins, Baltimore, 1970, 351–367.
2. **Hardy, N. R. and Wood, C.,** Vaginal contraception, in *New Concepts of Contraception*, Potts, M. and Wood, C., Eds., University Park Press, Baltimore, MD, 1972, 103–123.
3. **Langmyhr, G. J.,** Female barrier contraceptives, in *Human Reproduction: Conception and Contraception*, Hafez, E. S. E. and Evans, T. N., Eds., Harper & Row, New York, 1973, 285–292.
4. **Lehfeldt, H.,** Cervical cap, in *Manual of Family Planning and Contraceptive Practice*, Calderone, M. S., Ed., Williams & Wilkins, Baltimore, 1970, 368–375.
5. **Tietze, C.,** The condom as a contraceptive, *Fam. Plann. News*, 8(4), 83–91, 1961.

e. Abortion

1. **af Geijerstam, G. K., Ed.,** An Annotated Bibliography of Induced Abortion, Center for Population Planning, University of Michigan, Ann Arbor, 1969, 359 pp.
2. **Blake, J.,** Abortion and public opinion: the 1960–1970 decade, *Science*, 171, 540–549, 1971.
3. **Callahan, D.,** *Abortion, Law, Choice and Morality*, Macmillan, London, 1970.
4. **Connell, E. B.,** Abortion: Patterns, technics, and results, *Fertil. Steril.*, 24(1), 78–91, 1973.
5. **Dollen, C.,** *Abortion in Context: A Select Bibliography*, Scarecrow Press, Metuchen, N.J., 1970, 150 pp.
6. **Floyd, M. K.,** *Abortion Bibliography for 1970–*, Whitston Publishing Co., Troy, New York, 1972–.
7. **Guttmacher, A. F., Ed.,** *The Case of Legalized Abortion, Now*, Diablo Press, Berkeley, Calif., 1967.
8. **Hardin, G.,** Abortion – compulsory pregnancy, *J. Marriage Fam.*, 30(2), 246–251, 1968.
9. **Lader, L.,** *Abortion*, Beacon Press, Boston, 1966.
10. **Lowe, D.,** *Abortion and the Law*, Pocket Books, New York, 1966.
11. **Nathanson, B. N.,** Drugs for the production of abortion: A review, *Obstet. Gynaecol. Surv.*, 25, 727–731, 1970.
12. **Noonan, J. T., Jr.,** Abortion and the Catholic Church: A summary history, *Nat. Law Forum*, 12, 85–131, 1967.
13. **Potts, M.,** Legal abortion in Eastern Europe, *Eugen. Rev.*, 59(4), 232–250, 1967.
14. **Potts, D. M. and Diggory, P.,** Termination of pregnancy, in *Human Reproduction: Conception and Contraception*, Hafez, E. S. E. and Evans, T. N., Eds., Harper & Row, New York, 1973, 405–425.
15. **Ransil, B. J.,** *Abortion*, Paulist Press Dens Books, New York, 1968.
16. **Rosen, H., Ed.,** *Abortion in America: Medical, Psychiatric, Legal, Anthropological and Religious Considerations*, Beacon Press, Boston, 1967.

17. **Schwartz, R. A.,** Psychiatry and the abortion laws: An overview, *Compr. Psychiatry,* 9(2), 99–117, 1968.
18. **Tietze, C.,** Induced abortion as a method of fertility control, in *Fertility and Family Planning, A World View,* Behrman, S. J., Corsa, L., Jr., and Freedeman, R., Eds., The University of Michigan, Ann Arbor, 1969, 311–337.
19. **Tietze, C. and Lewit, S.,** Abortion, *Sci. Am.,* 220(1), 21–27, 138, 1969.

f. Sterilization

(1) General

1. **Presser, H.,** Voluntary Sterilization. A World View, Population Council, New York, August 1970.
2. **Richart, R. M. and Prager, D. J., Eds.,** *Human Sterilization,* Charles C Thomas, Springfield, Ill., 1972, 402 pp.
3. **Speidel, J. J., Perry, M. I., and Duncan, G. W., Eds.,** *Research Approaches to Sterilization,* Proc. of the 2nd International Conference on Voluntary Sterilization, Geneva, Feb. 25, 1973, Excerpta Medica (in press).
4. **Tietze, C.,** *Surgical Sterilization of Men and Women. A Selected Bibliography,* National Committee on Maternal Health, New York, 1962, 38 pp.
5. **Wood, H. C., Jr.,** *Sex Without Babies,* Whitmore Publishing, Philadelphia, 1967.

(2) Male Sterilization

1. **Blandy, J.,** Male sterilization, in Contraception Today, Smith A. J., Ed., Family Planning Association, London, 1971, 101–107.
2. **Hulka, J. F. and Davis, J. E.,** Vasectomy and reversible vasocclusion, *Fertil. Steril.,* 23, 683–696, 1972.
3. **Hulka, J. F. and Davis, J. E.,** Sterilization of men, in *Human Reproduction: Conception and Contraception,* Hafez, E. S. E. and Evans, T. N., Eds., Harper & Row, New York, 1973, 427–446.
4. **Johnson, D. S.,** Reversible male sterilization: Current status and future directions, *Contraception,* 5, 327–338, 1972.
5. **Kleinman, R., Ed.,** Vasectomy, IPPF, London, 1972.
6. **Lee, H. Y.,** *Sterilization in the Male,* Haimoonsa, Seoul, Korea, 1963.

(3) Female Sterilization

1. **Duncan, G. W., Falb, R. D., and Speidel, J. J.,** *Female Sterilization: Prognosis for Simplified Outpatient Procedures,* Academic Press, New York, 1972, 199 pp.
2. **Evans, T. N.,** Sterilization of women, in *Human Reproduction: Conception and Contraception,* Hafez, E. S. E. and Evans, T. N., Eds., Harper & Row, New York, 1973, 383–404.
3. **Hulka, J. F. and Omrau, K. F.,** New methods of female sterilization, in *New Concepts in Contraception,* Potts, M. and Wood, C., Eds., University Park Press, Baltimore, MD, 1972, 57–67.
4. **Porter, C. W. and Hulka, J. F.,** Female sterilization in current clinical practice, *Fam. Plann. Perspect.,* 6, 30–38, 1974.
5. **Sciarra, J. J., Butter, J. C., Jr., and Speidel, J. J., Eds.,** *Hysteroscopic Sterilization,* Intercontinental Medical Book Corp., New York, 1974, 174 pp.
6. **Steptoe, P. C.,** Female sterilization, in Contraception Today, Smith, A. J., Ed., Family Planning Association, London, 1971, pp. 108–112.

LIST OF ACRONYMS AND ABBREVIATIONS

AAAS	American Association for the Advancement of Science
AEC	Atomic Energy Commission (U.S.)
AFMA	American Feed Manufacturers Association
AID	Agency for International Development (U.S.)
AMA	American Medical Association
ARC	Agricultural Research Council (U.K.)
ARS	Agricultural Research Service (U.S. Dep. Agric.)
ASA	American Society of Agronomy
ASHRAE	American Society of Heating, Refrigerating and Air-Conditioning Engineers
CA	Central America
CAB	Commonwealth Agricultural Bureaux
CENTO	Central Treaty Organization
CFNI	Caribbean Food and Nutrition Institute
CFTRI	Central Food Technological Research Institute (Mysore)
CICRA	Centre International pour la Coordination des Recherches en Agriculture
CIMMYT	Centro Internacional de Mejoramiento de Maiz y Trigo (Mexico)
CSAV	Čzechoslovak Academy of Sciences
CSIRO	Commonwealth Scientific and Industrial Research Organization
CS(S)R	Čzechoslovak (Socialist) Republic
DC	District of Columbia (U.S.)
DHEW	Department of Health, Education, and Welfare (U.S.)
ECAFE	Economic Commission for Asia and the Far East (U.N.)
EEC	European Economic Community
ERS	Economic Research Service (U.S. Dep. Agric.)
FAER	Foreign Agricultural Economic Report (U.S. Dep. Agric.)
FDA	Food and Drug Administration (U.S.)
FEDS	Foreign Economic Development Service (U.S. Dep. Agric.)
FFHC	Freedom from Hunger Campaign (FAO)
FPC	Fish Protein Concentrate
HMSO	Her (His) Majesty's Stationary Office (U.K.)
IA	Iowa (U.S.)
IAEA	International Atomic Energy Agency
IASH	International Association of Scientific Hydrology
IBRO	International Brain Research Organization
ICNND	Interdepartmental Committee on Nutrition for National Defense (U.S.)
ID	Idaho (U.S.)
ILL	Illinois (U.S.)
INCAP	Institute of Nutrition of Central America and Panama
IPPF	International Planned Parenthood Federation (London)
IRRI	International Rice Research Institute
IUD	Inter-uterine Device
IWP	Indicative World Plan (FAO)
J	Journal
JAMA	Journal of the American Medical Association
LDC	Less Developed Country
LIFE	League for International Education
MD	Maryland (U.S.)
M.I.T.	Massachusetts Institute of Technology (U.S.)
MO	Missouri (U.S.)

NAS	National Academy of Sciences (U.S.)
NIH	National Institutes of Health (PHS)
NJ	New Jersey (U.S.)
NRC	National Research Council (NAS)
NY	New York (U.S.)
OECD	Organization for Economic Cooperation and Development
OEEC	Organization for European Economic Cooperation
PAG	Protein Advisory Group (U.N.)
PAHO	Pan American Health Organization
PCM	Protein-Calorie Malnutrition
PHS	Public Health Service (DHEW)
Q	Quarterly
R	Royal
UFAW	Universities Federation for Animal Welfare
U.K.	United Kingdom
U.N.	United Nations
UNDP	U.N. Development Programme
UNESCO	U.N. Educational, Scientific and Cultural Organization
INICEF	U.N. International Children's Fund (formerly U.N. International Children's Emergency Fund)
U.S.	United States
USDA	U.S. Department of Agriculture
U.S.S.R.	Union of Soviet Socialist Republic
UT	Utah (U.S.)
WAAS	World Academy of Arts and Sciences
WHO	World Health Organization
WMO	World Meteorological Organization

AUTHOR INDEX

A

C

D

E

F

G

H

I

J

K

L

M

N

O

P

Q

R

S

T

U

V

W

Y

Z

SUBJECT INDEX

A

B

C

D

E

F

G

H

I

N

O

P

R

S

T

U

V